# Study Guide

for

Sherwood's

# Human Physiology
## From Cells to Systems

Sixth Edition

**John P. Harley**
*Eastern Kentucky University*
and
*University of Kentucky, Chandler Medical Center*

Australia • Brazil • Canada • Mexico • Singapore • Spain • United Kingdom • United States

Printed in the United States of America
1 2 3 4 5 6 7 10 09 08 07 06

Printer: Thomson/West

0-495-01998-4
Cover Image: Nick Clements/Getty Images

For more information about our products, contact us at:
**Thomson Learning Academic Resource Center**
**1-800-423-0563**

For permission to use material from this text or product, submit a request online at
**http://www.thomsonrights.com.**
Any additional questions about permissions can be submitted by email to **thomsonrights@thomson.com.**

**Thomson Higher Education**
**10 Davis Drive**
**Belmont, CA 94002-3098**
**USA**

ABA 0074641

# Contents

# Introduction

The physiology professor has selected an excellent text: *Human Physiology: From Cells to Systems*, Sixth Edition, by Lauralee Sherwood. It is a state-of-the-art book designed to adequately prepare the student in the knowledge of human physiology. Most of the ancillaries, for any text, aid the instructor in the presentation of the material. This *Study Guide* is designed solely for the student.

The goal of this study guide is to help the student gain a better understanding of the material covered in the textbook. Throughout, the concepts, ideas, and phenomena are restated in various forms to help the student understand human physiology. From the student's perspective, this guide is designed to improve test grades and elevate the grade point average. With proper use of this guide, these interrelated goals should become a reality.

The **Chapter Overview** links the chapter to the main fundamental phenomenon of physiology: homeostasis. It mentions broad general statements to relate the chapter to other chapters. Several specifics may be noted or listed to indicate how the physiology covered by the chapter is achieved. In short, the Chapter Overview is a very brief summary. It identifies some of the important concepts that should be understood upon completion of the chapter.

The **Chapter Outline** is a synopsis of the text chapter. The format follows that of the textbook. In each division and subdivision of a chapter, the most important and succinct sentences have been taken directly from the text. The Chapter Outline should be used as a list of important entries that must be thoroughly understood to have a good working knowledge of the material in the chapter. It should be remembered that this is an outline. Additional explanation and examples are found in the textbook.

The **Key Terms** section is a reiteration of the important ideas, concepts, and structures in a one-word format. This should be used for learning by association. A definition for each key term should be prepared and the role of the term, in the physiology depicted in the chapter, should be duly noted.

The questions found in the **Review Exercises** are not intended to indicate the manner in which a student will be tested on particular information. Between the textbook and the *Study Guide* most of the material has been presented in test form. This is another method to check for an accurate understanding of the information presented in the chapter.

The information found under **Points to Ponder** may or may not be answered with the knowledge gained from the current chapter. Each entry is designed to stimulate the use of certain material presented in the chapter. In some instances additional information may be required to answer the questions, whereas, in others it may involve a social or cultural perspective. Each Point is included as a possible application of the new material. These Points to Ponder are just that, something to think about.

In **Clinical Perspectives**, the application of the new material must be employed. Unlike Points to Ponder, these questions can be answered with the material gained from the chapter. Usually, these examples are common applications of the physiology under consideration.

The **Experiment of the Day** section utilizes another method for learning a concept or phenomenon. While most of the science learned today is the result of experimentation, it is not always easy to understand the knowledge gained from such tests. Frequently, a simple activity can aid in the learning process. Although these activities do not qualify as experiments in the purest sense of the word, they encourage the use of curiosity to help understand the physiological material.

The **PhysioEdge Activites and Media Resource Sections** contain the relevant material that is on the PhysioEdge 2 CD-ROM, Book Companion Web Site, and InfoTrac® College Edition. All of these are completely integrated with the textbook.

**PhysioEdge CD-ROM**

PhysioEdge, the CD-ROM packaged with your text, focuses on the concepts students find most difficult to learn. Figures marked with this icon have associated activities on the CD. A diagnostic quiz component allows you to receive immediate feedback on your understanding of the concept and to advance through various levels of difficulty.

**Book Companion Website**

The website for this book contains a wealth of helpful study aids, as well as many ideas for further reading and research. For each chapter of the text, the website contains:

- **Case Histories** introduce the clinical aspects of human physiology.
- Two- and three-dimensional (2-D and 3-D) graphic illustrations and animations of physiological concepts in the **Visual Learning Resource**.
- Review material in the form of **Chapter Outlines**, **Learning Objectives**, and **Chapter Summaries.**
- A **Glossary** and electronic **Flash Cards** help you test your mastery of important terminology.

- For testing your knowledge and preparing for in-class examinations, look for **Fill-in-the-Blank**, **Matching**, **Multiple-Choice**, and **True-False Quizzes** based on each chapter.
- **Web Links** take you to an extensive list of current links to Internet sites with news, research, and images related to individual topics in the chapter.
- **Internet Exercises** are critical-thinking questions that involve research on the Internet with starter URLs (web addresses) provided.
- **InfoTrac® Exercises**, projects that use InfoTrac® College Edition as a research tool.
- Suggested Readings using **InfoTrac College Edition/Research,** your online research library.

The author would personally like to thank Kari Hopperstead, Assistant Editor, Life Sciences, for allowing me to work with her on this project. A special thanks goes to Melanie Bentley, Coordinator/Systems Network Manager, Eastern Kentucky University, for proofing all aspects of this student study guide.

# Homeostasis: The Foundation of Physiology

## Chapter Overview

The cells, the fundamental units of life, are organized into tissues, organs, and systems, which in turn maintain homeostasis. This homeostasis is essential for the survival of the cells. The multi-cellular organism provides this discrete internal environment through its structural organization. The functions of tissues, organs, and systems are specializations of specific cellular capabilities. A cyclic phenomenon exists: The organism, its tissues, organs, and systems, is composed of cells maintained through the homeostasis of an internal environment, which is controlled by body systems. This is the basic foundation of physiology.

## Chapter Outline

LEVELS OF ORGANIZATION IN THE BODY

*The chemical level: Various atoms and molecules make up the body.*

- Atoms form molecules.

*The cellular level: Cells are the basic units of life.*

- The smallest unit capable of performing the basic functions essential for life is the cell.
- Whether they exist as solitary cells or as part of a multi-cellular organism, all cells perform certain basic functions essential for survival of the organism. Organisms are independent living entities. As a result of cell differentiation, your body is made up of many different specialized cell types.

*Basic cell functions.*

- These functions include: (1) obtaining nutrients and oxygen, (2) performing certain chemical reactions that use nutrients and oxygen to provide energy, (3) eliminating from the cell carbon dioxide and other by products or wastes, (4) synthesizing proteins and other components, (5) controlling the exchange of material between the cell and its environment,(6) moving materials within the cell and even the cell itself, (7) being sensitive to changes in the environment, and (8) reproducing
- At the multi-cellular organismal level cells usually become specialized in one or several of the basic functions.

*Specialized cell functions.*

- Each cell also performs a specialized function such as protein synthesis, muscle contraction, and the transmission of information.

*The tissue level: Tissues are groups of cells of similar specialization.*

- Cells of similar origin, structure, and function are referred to as tissues. The four
- primary types are muscle, nervous, epithelial and connective tissue.
- There are three types of muscle tissue: skeletal, cardiac and smooth muscle. Muscle cells are specialized for contraction and force generation.
- Nervous tissue consists of cells specialized for initiation and transmission of electrical impulses, sometimes over long distances.
- These impulses are transmitted to the brain and subsequently to muscles or glands for the appropriate response.
- Epithelial tissue lines or covers the body and its various organs, vessels, and cavities.
- The specialization of epithelial tissue is the exchange of materials.
- Highly specialized epithelial cells form glands, which provide essential secretions.
- Glands may be either exocrine or endocrine.
- Connective tissue, as its name implies, connects, supports, and anchors the various body parts.

*The organ level: An organ is a unit made up of several tissue types.*

- Organs are composed of two or more types of primary tissues organized to perform some specific function within the body.
- Organs are further organized into body systems, each of which is a collection of organs that perform related functions and interact to accomplish a common activity.

*The body system level: A body system is a collection of related organs*

- The total body is composed of various organ systems linked together as an entity that is separate from the external environment.

*The organism level: The body systems are packaged together into a functional whole body.*

- The human body is made up of cells organized into life-sustaining systems.

## Introduction to Physiology

- Physiology is the study of the functions of the body or how the body works.

*Physiology focuses on mechanisms of action.*

- The teleological approach to explaining the events that occur in the body explains body functions in terms of meeting body needs.
- The mechanistic approach states that mechanisms of action can be explained in terms of cause-and-effect sequences of physical and chemical processes.

*Structure and function are inseparable.*

- Physiology is closely related to anatomy.

## Concept of Homeostasis

- Body cells are in contact with a privately maintained internal environment instead of with the external environment that surrounds the body.
- The cells in a multi-cellular organism must contribute to the survival of the organism as a whole and cannot live and function without contributions from the other body cells, because the vast majority of cells are not in direct contact with the external environment.
- The key is the presence of an aqueous internal environment with which the body cells are in direct contact.
- It consists of the extracellular fluid, which is made up of plasma, and interstitial fluid, which surrounds and bathes the cells.
- Various body systems accomplish exchanges between the external environment and the internal environment.
- Body systems perform the necessary functions to maintain homeostasis.
- Homeostasis is essential for the survival of each cell, and each cell, through its specialized activities, contributes as part of a body system to the maintenance of the internal environment shared by all cells.
- To maintain homeostasis, the body must be able to detect deviations in the internal environment, factors that need to be held within narrow limits, and it must be able to control the various body systems responsible for adjusting these factors.
- Control systems that operate to maintain homeostasis can be grouped into two classes: intrinsic and extrinsic controls.
- Intrinsic controls are built in or inherent to an organ.
- Extrinsic controls are regulatory mechanisms initiated outside an organ.
- Extrinsic control of the various organs and systems is accomplished by the nervous and endocrine systems.
- The body's homeostatic control mechanisms primarily operate on the principle of negative feedback.
- Negative feedback occurs when a change in a controlled variable triggers a response that opposes the change, driving the variable in the opposite direction of the initial change.
- Positive feedback occurs less frequently in the body.
- With positive feedback the output is continually enhanced so that the controlled variable continues to be moved in the direction of the initial change.

## Homeostatic Control Systems

*Homeostatic control systems may operate locally or bodywide.*

- Body systems maintain homeostasis, a dynamic steady state in the internal environment.
- The 11 body systems contribute to homeostasis in many different ways.

*Negative feedback opposes an initial change and is widely used to maintain homeostasis.*

- In negative feedback, a change in a homeostatically controlled factor triggers a response that seeks to restore the factor to normal by moving the factor in the opposite direction of its initial change.

*Positive feedback amplifies an initial change.*

- In positive feedback, the output enhances or amplifies a change so that the controlled variable continues to move in the direction of the initial change.
- Positive feedback plays an important role in the birth of a baby.

*Feed-forward mechanisms initiate responses in anticipation of change.*

- The body less frequently employs feed-forward mechanisms, which bring about a response in anticipation of a change in a regulated variable.

*Disruptions in homeostasis can lead to illness and death.*

- When one or more of the body's systems fail to function properly, homeostasis is disrupted, and all cells suffer.
- Various pathophysiological states ensue.

## Key Terms

anatomy
body systems
cell
connective tissue
control system
controlled variable
effector
elastin
endocrine gland
epithelial tissue
exocrine gland
extracellular fluid
extrinsic controls
feedforward
gland
homeostasis
integrator
internal environment
interstitial fluid
intrinsic controls
mechanistic approach
multicellular organism
muscle tissue
negative feedback
organs
pathophysiology
plasma
positive feedback
secretion
sensor
set point
teleological approach
total body
unicellular organism

## Review Exercises

*Answers are in the appendix.*

**True/False**

_____ 1. Tissues perform functions that are specializations of the normal function of individual cells.

_____ 2. Homeostasis maintains body systems to enhance the survival of cells.

_____ 3. Each body system is composed of different organs.

_____ 4. The endocrine system is a major control system.

_____ 5. The nervous system is an example of intrinsic control.

_____ 6. Feed-forward mechanisms are frequently used in the digestive system.

_____ 7. All definitive actions initiated by nervous tissue are either muscle contractions or glandular secretions.

_____ 8. A mechanistic explanation of why a person sweats is to "cool off."

_____ 9. Cells require homeostasis, body systems maintain homeostasis, and cells make up body systems.

_____ 10. With positive feedback, a control system's input and output continue to enhance each other.

_____ 11. To sustain life, the internal environment must be maintained in an absolutely unchanging state.

_____ 12. Glands are formed during embryonic development by pockets of epithelial tissue that dip inward from the surface.

_____ 13. Most human cells can be cultured; that is, when removed from the body, they will continue to thrive and reproduce in laboratory dishes when supplied with appropriate nutrients and other supportive materials.

_____ 14. Endocrine glands have ducts.

_____ 15. Control systems that operate to maintain homeostasis can be grouped into two classes—intracellular and extracellular.

_____ 16. Humans are multi-cellular organisms.

_____ 17. With a mechanistic approach, phenomena that occur in the body are explained in terms of their particular purpose in fulfilling a bodily need, without considering how this outcome is accomplished.

_____ 18. Connective tissue is distinguished by having relatively few cells dispersed within an abundance of extracellular material.

_____ 19. The uterus is in the pelvic cavity.

_____ 20. In the human body, there are five primary tissue types.

_____ 21. Defense against microorganisms is one of the functions of the immune system.

_____ 22. The nervous and endocrine systems play a major role in integration and coordination withinthe human body.

_____ 23. There are four major types of muscle tissue found within the human body.

_____ 24. The integumentary system plays a role in protecting the body.

_____ 25. Connective tissue may also be called supporting tissue.

**Fill in the Blank**

26. ____________ glands have ducts.

27. A(n) __________ is the cavity within the interior of hollow organs and tubes.

28. The ______________ approach explains "how" events in the body occur.

29. ________________ is essential for the survival of cells in the body.

30. ________________ fluid is part of the extracellular fluid.

31. The ____________ system transfers water and electrolytes from the external environment into the internal environment.

32. ____________________ is a common regulatory mechanism for maintaining homeostasis.

33. The smallest unit capable of carrying out the processes associated with life is the ________________

34. The body cells are in direct contact with, and make life-sustaining exchanges, with the ____________________.

35. The internal environment consists of the ____________________, which is made up of ____________________, the fluid portion of the blood, and ____________________, which surrounds and bathes all cells.

36. ____________________ exists when a change in a regulated variable triggers a response that opposes the change.

37. _______________ refers to the abnormal functioning of the body associated with disease.

38. The basic unit of both structure and function is the __________.

39. __________ is a naturally occurring protein that promotes development of nervous tissue and skin in embryos.

40. __________ are epithelial-tissue derivatives that are specialized for secretion.

41. Organs are further organized into ____________________, each of which is a collection of organs that perform related functions and interact to accomplish a common activity that is essential for survival of the whole body.

42. There are 11 major body systems. List them: (1) ______________, (2) _________________ (3) _______________, (4) ________________, (5) _______________, (6) ________________, (7) _________________, (8) _________________, (9) _______________, (10)________________, (11) _________________.

43. __________ is the study of all of the vital functions within the human body.

44. The fluid outside of the body cells is called ___________ fluid.

45. When an initial stimulus is reinforced, this is a good example of __________ feedback.

46. The internal and external forces that create imbalances within the body are called __________.

47. __________ tissue may also be called supporting tissue.

48. The lymphatic system is made up of many small vessels that carry __________.

49. The fluid inside the cell is called __________.

50. The tonsils are part of the __________ system.

**Matching**

*Match the following organ or body part to the correct body system.*

| | |
|---|---|
| _____ 51. Skin | a. Reproductive |
| _____ 52. Joints | b. Circulatory system |
| _____ 53. Trachea | c. Digestive system |
| _____ 54. Adrenals | d. Immune system |
| _____ 55. Special sense organs | e. Integumentary system |
| _____ 56. Uterus | f. Muscular system |
| _____ 57. Tonsils | g. Skeletal system |
| _____ 58. Skeletal muscles | h. Endocrine system |
| _____ 59. Blood | i. Urinary system |
| _____ 60. Urethra | j. Nervous system |
| _____ 61. Liver | k. Respiratory system |

*Match the system to the function.*

| | |
|---|---|
| 62. _____ Circulatory system | a. Transports food, vitamins, minerals, and other nutrients |
| 63. _____ Respiratory system | b. The chemical and mechanical breakdown of food |
| 64. _____ Digestive system | c. Responsible for continuation of the human race |
| 65. _____ Urinary system | d. Excretes nitrogenous waste |
| 66. _____ Reproductive system | e. Exchange of oxygen and carbon dioxide |

**Multiple Choice**

67. These cells, when destroyed by trauma or disease, cannot be replaced.
a. glandular tissue
b. epithelial tissue
c. muscles and nerves
d. connective tissue
e. none of the above

68. Which system is important in maintaining the proper pH of the internal environment?
a. circulatory
b. digestive
c. muscular
d. respiratory
e. immune

69. This tissue lines the stomach.
a. connective
b. muscle
c. epithelial
d. nervous
e. mesodermal

70. Which type of control is exemplified by an air conditioner, its thermostatic device, and their electrical connections?
a. positive feedback
b. negative feedback
c. feedforward
d. biofeedback
e. cybernetic

71. Which muscle type moves the contents of food through the intestines?
a. cardiac
b. skeletal
c. smooth
d. extrinsic
e. both b and c

72. Which approach would give "to keep warm" as a reason for shivering?
a. holistic approach
b. teleological approach
c. mechanistic approach
d. all of the above
e. none of the above

73. Which is not a basic cell function?
a. Maintain a storage reservoir for potassium which will maintain a relatively constant concentration of potassium in the extracellular fluid.
b. Perform various chemical reactions that use nutrients and oxygen to provide energy for the cell.
c. Obtain food and oxygen from the environment surrounding the cell.
d. Be sensitive and responsive to changes in the environment surrounding the cell.
e. Control to a large extent the exchange of materials between the cell and its surrounding environment.

74. Which type of gland loses its cellular connection during fetal development?
a. holocrine
b. endocrine
c. exocrine
d. mesocrine
e. endoderm

75. Which type of tissue utilizes the inherent capability of cells to produce intracellular movement?
a. epithelial
b. muscle
c. nervous
d. connective
e. endoderm

76. Which type of tissue utilizes the inherent capability to synthesize digestive enzymes?
a. epithelial
b. muscle
c. nervous
d. connective
e. glandular

77. During the minute that it will take to answer this question, all but one of these activities will occur.
a. Your body will use about two calories of energy.
b. Your body will exchange six liters of air between the atmosphere and your lungs.
c. More than 10 liters of blood will flow through your kidneys.
d. Your heart will beat 70 times, pumping 5 liters of blood to your lungs and another 5 liters to the rest of your body.
e. Your brain will process 325 afferent impulses.

78. Which of the following best describes embryonic stem cells?
a. The "mother cells" resulting from the early division of the egg.
b. All of the cells in the embryo?
c. The last cells to develop in an embryo.
d. The implantation of cells into embryos.

79. Factors of an internal environment that must be homeostatically maintained include:
a. its pH and temperature
b. its concentration of water, salt, and other electrolytes
c. its volume and pressure
d. all of the above
e. none of the above

80. Which of the following is a component of the immune system?
a. appendix
b. adrenals
c. bulbourethral glands
d. hypothalamus
e. none of the above

81. Which system controls and coordinates bodily activities that require swift responses?
a. skeletal system
b. muscular system
c. reproductive system
d. nervous system
e. endocrine system

82. What feedback continually enhances the output so that the controlled variable continues to be moved in the direction of the initial change?
a. negative
b. positive
c. feed-forward
d. biofeedback
e. cybernetic

83. What controls are built in or inherent to an organ?
a. intrinsic
b. extrinsic
c. endocrine
d. exocrine
e. holocrine

84. What controls are regulatory mechanisms initiated outside of an organ to alter the activity of the organ?
a. intrinsic
b. extrinsic
c. mesocrine
d. epicrine
e. holocrine

85. What controls use the nervous and endocrine systems to control various organs and systems?
a. intrinsic
b. extrinsic
c. mesocrine
d. holocrine
e. exocrine

86. The medical branch that studies the structure of the human body is called:
a. physiology
b. organology
c. histology
d. cytology
e. anatomy

87. The thoracic cavity contains:
a. the heart and lungs
b. the testes, ovaries, and other reproductive organs
c. just the lungs
d. the diaphragm and heart
e. all of the above

88. The smallest functional unit in the human body from a physiologic perspective is the
a. nucleus.
b. atom.
c. cell.
d. tissue.
e. organ.

89. In a negative feedback system, when blood glucose levels fall, the body will respond by causing a change that:
a. increases blood glucose only
b. exhibits positive feedback only
c. exhibits negative feedback only
d. both a and b only
e. both a and c only

90. Which of the following systems does not play a role in protecting the human body?
a. integumentary
b. lymphatic
c. circulatory
d. digestive
e. nervous

91. All of the following are examples of membranes that line body cavities except:
a. mucous
b. serous
c. synovial
d. pleura
e. viscera

92. Which of the following is not a type of connective tissue?
a. bone
b. cartilage
c. blood
d. tendon
e. muscle

93. Which of the following is a major transport system in the human body?
a. muscular
b. nervous
c. circulatory
d. integumentary
e. urinary

94. Which of the following systems plays a role in integration and coordination?
a. nervous
b. skeletal
c. urinary
d. respiratory
e. integumentary

95 Which of the following systems sends electrical signals?
a. nervous
b. respiratory
c. circulatory
d. muscular
e. digestive

96. Which of the following describes phenomena that occur in the body with respect to filling a body need?
a. the mechanistic view of life
b. the teleological view of life
c. both a and b
d. none of the above
e. vitalism

97. All glands are composed of which of the following tissue types?
a. connective
b. muscular
c. epithelial
d. a and b
e. b and c

98. How many body systems are present in your body?
 a. five
 b. six
 c. seven
 d. ten
 e. eleven

99. In your body, you would find a loss of elastin in:
 a. connective tissue
 b. muscle tissue
 c. nervous tissue
 d. skeletal tissue
 e. all of the above

100. The respiratory system contains all of the following except:
 a. nose
 b. pharynx
 c. larynx
 d. bronchi
 e. hair

## *Points to Ponder*

1. Why does the heart rate of a well-trained athlete begin to increase a few moments before the competitive event occurs?

2. As you sit in class you start yawning. Looking around, you see other students doing the same. What do you think is happening, aside from the obvious?

3. How would you rate homeostasis as a diagnostic tool?

4. After studying this chapter on homeostasis why do you think it is important for people to have thorough examinations by their physicians?

5. How are the different structural levels of the human body related?

6. If a bullet entered the right side of the body just below the armpit and exited at the same point on the other side of the body, which cavities would be pierced (in order) and which organs would probably be damaged?

7. Using insulin as an example, explain how the secretion of this hormone is controlled by the effects of insulin's action in various places in the body.

8. What are some reasons artificial organs are so difficult to develop?

9. What does exercise physiology tell us about normal body functions?

10. Do illness and aging involve the same pathophysiological process? Explain your answer.

11. Why are strokes so damaging to the human body?

12. What are some similarities and differences between endocrine and exocrine glands?

13. How does each cell in the body contribute to homeostasis?

14. What are some of the major hurdles to the use of embryonic stem cells?

15. What are some pros and cons with respect to tissue engineering?

## Clinical Perspectives

1. What does your heart rate (in beats per minute) indicate when you are resting versus when you are exercising?

2. Why do health personnel take one's temperature? What does this tell you about normal body function?

3. Why is positive feedback necessary for childbirth to occur? What mechanisms, hormones, and systems are involved?

4. What is one of the first effects of an exercise program on the heart?

## PhysioEdge Activities

*None for this chapter.*

## Media Resources

**PhysioEdge CD-ROM**
PhysioEdge, the CD-ROM packaged with your text, focuses on the concepts students find most difficult to learn. Figures marked with the Physioedge icon have associated activities on the CD. A diagnostic quiz component allows you to receive immediate feedback on your understanding of the concept and to advance through various levels of difficulty.

**Book Companion Website**
The website for this book contains a wealth of helpful study aids, as well as many ideas for further reading and research. Log on to:

**http://www.brookscole.com/sherwoodhp6**

Select a chapter from the drop-down menu, or click on one of these resource areas:

- **Case Histories** introduce the clinical aspects of human physiology.

- For two- and three-dimensional (2-D and 3-D) graphic illustrations and animations of physiological concepts, visit our **Visual Learning Resource**.

- Review the **Chapter Outline**, **Learning Objectives**, and **Chapter Summary**. To test your mastery of important terminology for this chapter, you can use the **Glossary** or the electronic **Flash Cards**, which you can sort by definition or by term.

- For testing your knowledge and preparing for in-class examinations, look for **Fill-in-the-Blank**, **Matching**, **Multiple-Choice**, and **True-False Quizzes** based on each chapter.

- Clicking on Web Links takes you to an extensive list of current links to Internet sites with news, research, and images related to individual topics in the chapter.

- Internet Exercises are critical-thinking questions that involve research on the Internet with starter URLs (web addresses) provided.

- For InfoTrac® Exercises, projects that use InfoTrac® College Edition as a research tool, go to InfoTrac® College Edition/Research.

    - For Suggested Readings, consult **InfoTrac College Edition/Research** on the website or go directly to InfoTrac® College Edition, your online research library, at:

      http://infotrac.thomsonlearning.com

# Cell Physiology

## Chapter Overview

The basic processes associated with life require cellular functions at the molecular level. These molecular activities are performed by various organelles of the cell. Some organelles such as mitochondria and lysosomes act directly in the basic life processes while cytoskeletal structures, polyribosomal clusters, and others function in areas of structural integrity, maintenance, and movement. The molecular activity of the organelles and other cellular components is essential for the proper functioning of tissues, organs, and body systems in the maintenance of homeostasis. Energy produced in organelles drives all the body systems. The role of a body system in homeostasis is inseparable from the basic molecular functions of its cells.

## Chapter Outline

### INTRODUCTION

- Cells are the bridge between molecules and humans.
- Even though the chemicals comprising a cell have been thoroughly analyzed, researchers have not yet organized these chemicals into a cell in the laboratory.
- Cells are the living building blocks for the whole body.

### OBSERVATIONS OF CELLS

- Human cells average about 10 to 20 micrometers in diameter. They are too small to be seen by the unaided eye.
- Lined up side by side, about 100 average-sized cells would occupy a distance of only 1 millimeter.
- Cells were unknown until the middle of the seventeenth century.
- The electron microscope has revealed the great diversity and complexity of the internal structure of the cell.
- Through the availability of better microscopy, biochemical techniques, cell-culture technology, and genetic engineering, the cell has emerged as a complex highly organized, compartmentalized structure.

### AN OVERVIEW OF CELL STRUCTURE

- Most cells are subdivided into the plasma membrane, nucleus and cytoplasm.

*The plasma membrane bounds the cell.*

- The thin membranous structure that encloses the cell is the plasma membrane, or cell membrane, which serves as the boundary of the cell.
- The fluid within the cell is the intracellular fluid (ICF) while fluids outside the cell are referred to as extracellular fluid (ECF).
- The plasma membrane selectively controls the movement of molecules between the ICF and ECF.

*The nucleus contains the DNA.*

- The nucleus, surrounded by the double-layered nuclear membrane, is the largest single organized cellular component.
- Deoxyribonucleic acid (DNA) is found in the nucleus.
- DNA controls protein synthesis.
- Since most cellular activities involve structural or enzymatic proteins, DNA provides the instructions and thus the control of these activities.
- As a genetic blueprint DNA ensures cellular reproduction and organismal reproduction.
- Three types of ribonucleic acid (RNA) play a role in protein synthesis.
- The DNAs genetic code for a particular protein is transcribe into a messenger RNA molecule.
- Within the cytoplasm, messenger RNA delivers the coded message to ribosomal RNA.
- Ribosomal RNA "reads" the code and translates it into the appropriate amino acid sequence.
- Transfer RNA transfers the appropriate amino acid within the cytoplasm to their designated site in the protein under construction.

*The cytoplasm consists of various organelles and the cytosol.*

- The part of the cell outside the nucleus is called the cytoplasm and is composed of a gel-like mass known as cytosol.
- There are six structures or organelles dispersed in the cytosol: endoplasmic reticulum, Golgi complex, lysosomes, peroxisomes, mitochondria and vaults.
- The cytosol is laced with an elaborate protein network which constitutes the cytoskeleton.
- The cytoskeleton provides shape to the cell, internal organization, and internal movement.

## ENDOPLASMIC RETICULUM AND SEGREGATED SYNTHESIS

- There are two types of endoplasmic reticulum (ER): smooth and rough.
- The ER is one continuous membranous organelle with many interconnected channels.
- The rough ER derives its name from dark-staining particles, ribosomes, which are attached to the ER membrane.
- Ribosomes are composed of ribosomal RNA-protein complexes.

*The rough endoplasmic reticulum synthesizes proteins for secretion and membrane construction.*

- The rough ER and its associated ribosome synthesize proteins and release these products into the ER lumen.
- Some of these new proteins are destined for export to the exterior of the cell as a secretion.
- Other new proteins are used within the cell as structural proteins.
- The ribosome synthesizes the leader sequence, which binds to a specific signal-recognition protein in the cytosol.
- This signal-recognition protein facilitates the binding of the ribosome to the rough ER membrane at a binding site known as the ribophorin.
- The new protein is passed into the ER lumen through the ribophorin.

*The smooth endoplasmic reticulum packages new proteins in transport vesicles.*

- Newly synthesized proteins are passed to the smooth ER where they "bud off" and become membrane enclosed transport vesicles.
- These transport vesicles move to the Golgi complex for further processing.
- Smooth and rough ER membranes contain enzymes for lipid synthesis.
- In liver cells the smooth ER has detoxifying enzymes.
- In muscles the smooth ER becomes modified into the sarcoplasmic reticulum and stores calcium.

## GOLGI COMPLEX AND EXOCYTOSIS

- The Golgi complex is a refining plant and directs molecular traffic.
- The Golgi complex consists of membranous, flattened, slightly curved sacs stacked in layers called Golgi stacks.

*Transport vesicles carry their cargo to the Golgi complex for further processing.*

- Transport vesicles enter the Golgi stacks at a point nearest the center of the cell.
- The Golgi complex performs two important interrelated functions: processing the raw materials into finished products and sorting and directing the finished products to their final destination as coated vesicles or secretory vesicles.

*The Golgi complex packages secretory vesicles for release by exocytosis.*

- Coated vesicles are covered with the protein clathrin and accessory proteins.
- The clathrin coat serves as a recognition marker.
- Secretory vesicles lack clathrin and are destined to be released to the exterior of the cell in a process called exocytosis.
- Secretory vesicles fuse only with the plasma membrane.
- Every vesicle bears a protein marker known as v-SNARE that can link lock-and-key fashion with another protein marker, a t-SNARE, found only on the targeted membrane.

## LYSOSOMES AND ENDOCYTOSIS

- Lysosomes serve as the intracellular digestive system.
- Lysosomes are membrane-enclosed sacs containing powerful hydrolytic enzymes.
- Both the membrane and enzyme are derived from the Golgi complex.

*Extracellular material is brought into the cell by endocytosis for attack by lysosomal enzymes.*

- Extracellular material to be attacked by lysosomal enzymes is brought to the interior of the cell through the process of endocytosis.
- Endocytosis can be accomplished in one of two ways: pinocytosis and phagocytosis.

- If fluid is internalized by endocytosis, the process is termed pinocytosis.
- The plasma membrane invaginates, forming a pouch that contains a small bit of extracellular fluid.
- The plasma membrane then seals at the surface of the pouch, forming a small, intracellular, membrane-enclosed vesicle.
- Dynamin has been identified as the molecule responsible for severing the vesicle from the surface membrane.
- White blood cells use a special form of endocytosis called phagocytosis to trap multi-molecular particles.
- Lysosomes fuse to the internalized vesicle and release their hydrolytic enzymes into the vesicle.
- Lysosomes become residual bodies upon completion of their digestive activities and are eliminated by exocytosis.

*Lysosomes remove useless parts of the cell.*
- In specific instances, lysosomes cause intentional self destruction of healthy cells.
- This happens as a normal part of embryonic development when certain unwanted tissues that form are programmed for destruction.
- Such programmed cell death is termed apoptosis.
- Storage diseases may occur if an enzyme is missing in the lysosome.
- The result is massive accumulation within the lysosomes of the specific compound that is normally digested by the missing enzyme.
- Among these storage diseases is Tay-Sachs disease.
- It is characterized by abnormal accumulations of gangliosides, which bring about progressive nervous system degeneration.
- Receptor-mediated endocytosis is a highly selective process that enables cells to import large molecules that it needs from the environment.
- During phagocytosis large multimolecular particles are internalized.

## Peroxisomes and Detoxification

*Peroxisomes house oxidative enzymes that detoxify various wastes.*
- About one-third to one-half the size of lysosomes, peroxisomes are membrane-bound sacs containing oxidative enzymes that strip hydrogen from toxic wastes.
- Peroxisomes also contain catalase, an antioxidant, which breaks down hydrogen peroxide, the product of oxidative detoxification.

## Mitochondria and ATP Production

*Mitochondria, the energy organelles, are enclosed by a double membrane.*
- The rod-shaped mitochondria are the energy organelles of the cell.
- Being descendants of bacteria, mitochondria are rod or oval-shaped organelles living symbiotically with the cell.
- Mitochondria have smooth outer membranes and inner membranes that have infoldings, or cristae, into the gel-like matrix.
- Cristae are the site of the electron transport chain activity.
- The citric acid cycle occurs in the matrix.

*Mitochondria play a major role in generating ATP.*
- To obtain immediate usable energy, cells split the terminal phosphate bond of adenosine triphosphate (ATP), which yields adenosine diphosphate (ADP) plus inorganic phosphate plus energy.
- Glucose from the digestive system is transported across the plasma membrane into the cytosol where it is converted to pyruvic acid and two molecules of ATP through the chemical process glycolysis.
- Pyruvic acid is converted to acetyl coenzyme A in the matrix of the mitochondria. Acetyl coenzyme A joins oxaloacetic acid in the citric acid cycle, which produces two molecules of carbon dioxide, oxaloacetic acid, and one molecule of ATP through a subsequent reaction with guanosine triphosphate.
- Hydrogen is removed using carrier molecules nicotinamide adenine dinucleotide (NAD) and flavine adenine dinucleotide (FAD).
- One molecule of glucose produces two acetic acid molecules, and two turns of the citric acid cycle yielding two molecules of ATP.
- The hydrogen carrier molecules enter the electron transport chain, which produces far more energy than the citric acid cycle itself.
- The electron transport chain is called the respiratory chain.
- High energy electrons are extracted from hydrogen bonds with NAD and FAD producing 32 molecules of ATP.
- At three sites in the electron transport chain ATP is produced using oxygen in a process known as oxidative phosphorylation.

*The cell generates more energy in aerobic than in anaerobic conditions.*
- Glycolysis is anaerobic.
- The citric acid cycle and the electron transport chain are aerobic.

- An oxygen insufficiency will not impede glycolysis thereby producing excessive pyruvic acid.
- Excess pyruvic acid is converted to lactic acid through a reversible reaction.

*The energy stored within ATP is used for synthesis, transport, and mechanical work.*

- Cellular activities that require energy expenditure fall into three main categories:
  (1) Synthesis of new compounds such as protein synthesis by the ER.
  (2) Membrane transport such as selective transport in the kidney.
  (3) Mechanical work such as muscular contraction.

## VAULTS AS CELLULAR TRUCKS

*Vaults may serve as cellular transport vehicles.*

- Vaults are three times as large as ribosomes and are shaped like octagonal barrels.
- These organelles may be used in transport of messenger RNA.
- Since vaults are abundant in areas where actin is assembled they are suspected to be somehow involved in cellular contractile systems.
- Vaults may play an undesirable role in bringing about the multidrug resistance sometimes displayed by cancer cells.

## CYTOSOL: CELL GEL

*The cytosol is important in intermediary metabolism, ribosomal protein synthesis, and nutrient storage.*

- Intermediary metabolism refers to the large set of intracellular chemical reactions that involve degradation, synthesis and transformation of small organic molecules such as simple sugars, amino acids, and fatty acids.
- Glycolysis is such a reaction.
- Throughout the cytosol, free ribosomes are dispersed that synthesize proteins for use in the cytosol.
- Clusters of these free ribosomes are known as polyribosomes.
- Glycogen and fat are stored in the cytosol as inclusions.

## CYTOSKELETON: CELL "BONE AND MUSCLE"

- The cytoskeleton is found in the cytosol and is composed of four distinct elements: microtubules, microfilaments, intermediate filaments, and microtrabecular lattice.

*Microtubules help maintain asymmetric cell shapes and play a role in complex cell movements.*

- Microtubules are long, slender tubes composed of the small globular protein, tubulin.
- A good example of asymmetrical cell shape maintenance is in the axon of a neuron, which may be up to a meter in length.
- Microtubules are essential in the transport of secretory vesicles from one region of a cell to another as in a neuron from the cell body to the axon terminal.
- Protein molecules are moved along the microtubule using transport molecules such as kinesin.
- Microtubules are the dominant structure and functional components of cilia and flagella.
- Cilia are numerous tiny protrusions.
- Flagella are long whiplike structures.
- Both have microtubules arranged in a ring of nine pairs, or doublets, with one pair in the center.
- A basal body within the cell produces the appendages, cilia, and flagella.
- Dynein is a protein that slides adjacent doublets past each other to produce the movement.
- Humans have cilia on the cells lining the trachea, with cilia also lining the oviducts in females.
- Sperm tails are actually flagella.
- Microtubules are produced by the centrioles in the mitotic cells.
- These microtubules form the mitotic spindle.

*Microfilaments are important to cellular contractile systems and as mechanical stiffeners.*

- Microfilaments are composed of actin in most cells and also myosin in muscle cells.
- Actin microfilaments are two twisted strands of the globular protein actin.
- In muscles the contractile force is generated by actin microfilaments sliding past myosin microfilaments.
- Actin microfilaments provide the constricting force to separate the cytoplasm into two daughter cells in mitosis.
- Actin microfilaments are used to produce pseudopoda necessary for locomotion of fibroblasts, skin cells and white blood cells.
- Microfilaments serve as mechanical stiffeners in the microvilli that line the free edge of the epithelial cells lining the intestine and kidney tubules.
- These microvilli increase the absorptive surface area.
- The hair cells of the inner ear are specialized microvilli.

*Intermediate filaments are important in regions of the cell subject to mechanical stress.*

- These filaments derive their name by being between microtubules and microfilaments in size.
- Intermediate filaments are found as neurofilaments in axons and provide strength to these elongated structures.
- In skeletal muscle intermediate filaments hold actin and myosin contractile units in proper alignment.
- Intermediate filaments, made of Keratin, form irregular networks that connect to extracellular filaments to strengthen and waterproof the skin.

*The cytoskeleton functions as an integrated whole and links other parts of the cell together.*

- This lattice of exceedingly fine filaments serves to suspend organelles, microtubules, and free ribosomes.
- These filaments extend from the inner layer of the plasma membrane throughout the cytoplasm.

## Key Terms

actin
adenosine diphosphate (ADP)
adenosine triphosphate (ATP)
adipose tissue
aerobic
aerobic exercise
amoeboid movement
amyotrophic lateral sclerosis (ALS)
anaerobic
anaerobic exercise
ATP synthase
autophagy
basal body
catalase
cell membrane
chemiosmotic mechanism
cilia
cristae
cytoplasm
cytoskeleton
cytosol
deoxyribonucleic acid (DNA)
ectoplasm
electron transport chain
endocytic vesicles
endoplasmic reticulum
exocytosis
flagellum
glycogen
guanosine diphosphate (GDP)
guanosine triphosphate (GTP)
hydrogen peroxide
inclusions
kinesin
Lou Gehrig's disease
lysosomes
matrix
messenger RNA (mRNA)
metabolism
microtubules
microvilli
NAD
nuclear envelope
nucleus
peroxisomes
nuclear pores
oxidative enzymes
oxidative phosphorylation
phagocytosis
pinocytosis
plasma membrane
pseudopod
receptor-mediated endocytosis
ribonucleic acid (RNA)
rough ER
ribosomal RNA
Tay-sachs disease
transport vesicles
smooth ER
transfer RNA
tubulin
vaults

## Review Exercises

*Answers are in the appendix.*

**True/False**

____ 1. Tay-Sachs is a disease in which brain cells cannot perform glycolysis.

____ 2. The electron transport chain in the mitochondrial matrix produces 32 molecules of ATP per molecule of glucose processed.

____ 3. Ketoglutaric acid is part of the citric acid cycle.

____ 4. Glycolysis is a good example of oxidative phosphorylation.

____ 5. The chemiosmotic mechanism is another name for the tricarboxylic acid cycle.

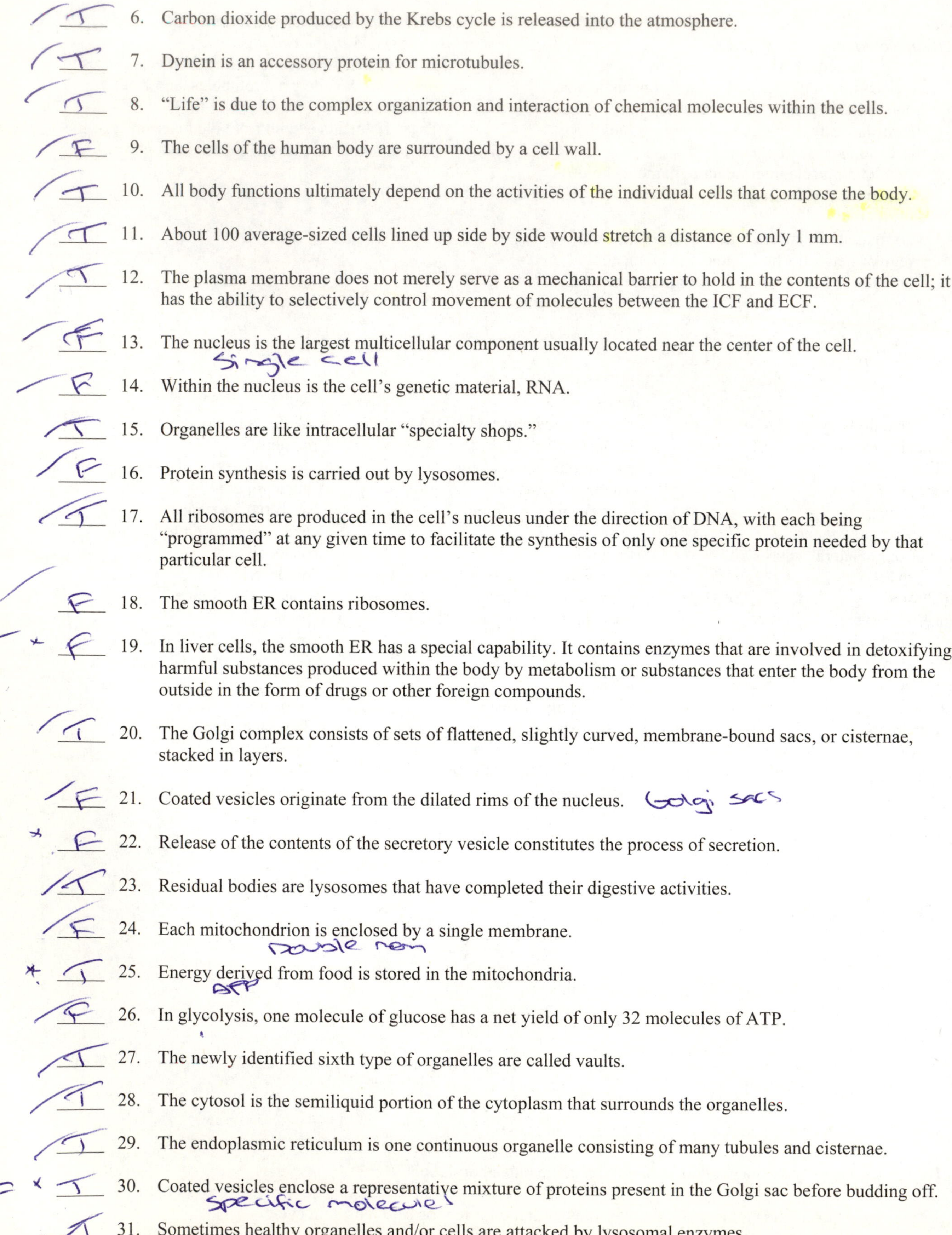

____ 6. Carbon dioxide produced by the Krebs cycle is released into the atmosphere.

____ 7. Dynein is an accessory protein for microtubules.

____ 8. "Life" is due to the complex organization and interaction of chemical molecules within the cells.

____ 9. The cells of the human body are surrounded by a cell wall.

____ 10. All body functions ultimately depend on the activities of the individual cells that compose the body.

____ 11. About 100 average-sized cells lined up side by side would stretch a distance of only 1 mm.

____ 12. The plasma membrane does not merely serve as a mechanical barrier to hold in the contents of the cell; it has the ability to selectively control movement of molecules between the ICF and ECF.

____ 13. The nucleus is the largest multicellular component usually located near the center of the cell.

____ 14. Within the nucleus is the cell's genetic material, RNA.

____ 15. Organelles are like intracellular "specialty shops."

____ 16. Protein synthesis is carried out by lysosomes.

____ 17. All ribosomes are produced in the cell's nucleus under the direction of DNA, with each being "programmed" at any given time to facilitate the synthesis of only one specific protein needed by that particular cell.

____ 18. The smooth ER contains ribosomes.

____ 19. In liver cells, the smooth ER has a special capability. It contains enzymes that are involved in detoxifying harmful substances produced within the body by metabolism or substances that enter the body from the outside in the form of drugs or other foreign compounds.

____ 20. The Golgi complex consists of sets of flattened, slightly curved, membrane-bound sacs, or cisternae, stacked in layers.

____ 21. Coated vesicles originate from the dilated rims of the nucleus.

____ 22. Release of the contents of the secretory vesicle constitutes the process of secretion.

____ 23. Residual bodies are lysosomes that have completed their digestive activities.

____ 24. Each mitochondrion is enclosed by a single membrane.

____ 25. Energy derived from food is stored in the mitochondria.

____ 26. In glycolysis, one molecule of glucose has a net yield of only 32 molecules of ATP.

____ 27. The newly identified sixth type of organelles are called vaults.

____ 28. The cytosol is the semiliquid portion of the cytoplasm that surrounds the organelles.

____ 29. The endoplasmic reticulum is one continuous organelle consisting of many tubules and cisternae.

____ 30. Coated vesicles enclose a representative mixture of proteins present in the Golgi sac before budding off.

____ 31. Sometimes healthy organelles and/or cells are attacked by lysosomal enzymes.

____ 32. Mitochondria are presumably descendants of primitive bacterial cells.

____ 33. Centrioles and basal bodies are identical in structure and may be interconvertible.

____ 34. The protective, waterproof outer layer of skin is formed by the tough skeleton of the microtrabecular lattice, which persists after the surface skin cells die.

____ 35. Each ribosome can synthesize only one specific type of protein.

____ 36. Only one ribosome can translate the mRNA message at any given time.

**Fill in the Blank**

37. Newly formed ribosomes destined to be attached to the rough ER do so at a site on the rough ER that has a specific protein called a ____________.

38. The ____________ "bud off" from the smooth ER.

39. The ____________ appears to suspend microtubules and microfilaments, as well as various organelles.

40. The ____________ occupies about 55 percent of the total cell volume.

41. ____________ is a transport molecule.

42. The ____________ is a specialized cytoplasmic structure that forms cilia.

43. Oxidative enzymes are found in the ____________.

44. A cell is made up of three major subdivisions: (1) ____________, (2) ____________, and (3) ____________.

45. The cytosol is pervaded by a protein scaffolding called ____________, which serves as the "bone and muscle" of the cell.

46. The ribosomes of the rough ER synthesize ____________, whereas its membranous walls contain enzymes essential for the synthesis of ____________.

47. The signal-recognition protein recognizes both the ____________ on the ribosome and the ____________ on the ER to deliver the proper ribosome to the proper site on the rough ER for binding.

48. The ____________ ER is the central packaging and discharge site for molecules to be transported from the ER.

49. The process in which a secretory vesicle fuses with the plasma membrane, then opens up and extrudes its contents to the exterior, is known as ____________.

50. Lysosomes contain (what type of) ____________ enzymes.

51. Foreign material to be attacked by lysosomal enzymes is brought into the cell by the process of ____________.

52. ____________, an enzyme found in peroxisomes, decomposes potentially toxic hydrogen peroxide.

53. The nucleus of the cell contains ____________, the cell's genetic material.

54. Programmed cell death is termed ____________.

55. ________ may play an undesirable role in bringing about the multidrug resistance sometime displayed by cancer.

**Matching**

*Match the component or reaction to the organelle or structure in which it occurs.*

| | | | |
|---|---|---|---|
| ____ 56. | Electron transport chain | a. | cytosol |
| ____ 57. | Catalase | b. | Golgi complex |
| ____ 58. | Microtubules | c. | microvilli |
| ____ 59. | Glycolysis | d. | mitochondrial matrix |
| ____ 60. | Hydrolytic enzymes | e. | ribosome |
| ____ 61. | Clathrin | f. | pseudopodia |
| ____ 62. | Secretory vesicles | g. | mitochondrial inner membrane |
| ____ 63. | Microfilaments | h. | coated vesicle |
| ____ 64. | Plasmasol | i. | lysosomes |
| ____ 65. | Leader sequence | j. | flagella |
| ____ 66. | Citric acid cycle | k. | peroxisomes |

**Multiple Choice**

67. In mitosis the last phase involves the constriction of the cytoplasm to produce two daughter cells. Which of the following is involved in this cytoplasmic constriction?
a. clathrin
b. actin
c. kinesin
d. malic acid
e. dynein

68. Which is the storage form of glucose?
a. citric acid
b. galactose
c. myosin
d. guanosine triphosphate
e. glycogen

69. Tay-Sachs disease is a storage disease involving accumulations of these compounds.
a. gangliosides
b. oxidative enzymes
c. hydrogen peroxide
d. plasma gel
e. ribophorins

70. Which is the correct order for the citric acid cycle?
a. oxaloacetic acid—citric acid—isocitric acid—ketoglutaric acid—fumaric acid—succinyl CoA—succinic acid—malic acid
b. oxaloacetic acid—citric acid—isocitric acid—ketoglutaric acid—succinyl CoA—succinic acid—fumaric acid—malic acid
c. citric acid—isocitric acid—oxaloacetic acid—ketoglutaric acid—succinyl acid—succinic acid—malic acid—fumaric acid
d. oxaloacetic acid—isocitric acid—citric acid—succinyl CoA—succinic acid—ketoglutaric acid—fumaric acid—malic acid
e. oxaloacetic acid—citric acid—isocitric acid—ketoglutaric acid—succinic acid—succinyl CoA—fumaric acid—malic acid

71. Which compound is used as mechanical stiffeners in microvilli?
a. actin
b. kinesin
c. dynein
d. myosin
e. clathrin

72. Which is composed of RNA and proteins?
a. chromosomes
b. peroxisomes
c. lysosomes
d. ribosomes
e. cortisone

73. Which of the following functions as the modifier, packager, and distributor for newly synthesized proteins?
a. vaults
b. Golgi complex
c. secretory vesicles
d. endoplasmic reticulum
e. mitochondria

74. Which are described as octagonal barriers?
a. vaults
b. Golgi complex
c. residual bodies
d. free ribosomes
e. mitochondria

75. Which component holds the actin-myosin units in proper alignment in skeletal muscle?
a. microtubules
b. microfilaments
c. microtrabecular lattice
d. spindle fibers
e. intermediate filaments

76. Dynein is found in which of the following?
a. cilia and flagella
b. cilia and mitotic spindles
c. flagella and centrioles
d. mitotic spindles and flagella
e. none of the above

77. Which of the following processes acetyl CoA?
a. electron transport chain
b. citric acid cycle
c. glycolysis
d. respiratory chain
e. fermentation

78. Which of the following converts glucose into two pyruvic acid molecules?
a. electron transport chain
b. glycolysis
c. citric acid cycle
d. respiratory chain
e. transition reaction

79. Which cellular structure organizes the glycolytic enzymes in sequential alignment?
a. microtubules
b. microfilaments
c. intermediate filaments
d. microtrabecular lattice
e. vaults

80. Which of the following are the smallest elements visible with a conventional electron microscope?
a. microtubules
b. microfilaments
c. intermediate filaments
d. microtrabecular lattice
e. vaults

81. Which ribosome synthesizes proteins used to construct new cell membranes?
a. free ribosome
b. rough ER bound ribosome
c. smooth ER
d. all of the above
e. none of the above

82. Which ribosome synthesizes proteins used intracellularly within the cytosol?
a. free ribosome
b. rough ER bound ribosome
c. neither
d. both a and b
e. none of the above

83. Which of the following takes place in the mitochondrial matrix?
a. glycolysis
b. citric acid cycle
c. electron transport chain
d. fermentation
e. cytolysis

84. Which reaction produces water as a by product?
a. electron transport chain
b. citric acid cycle
c. glycolysis
d. Calvin reaction
e. fermentation

85. Which of the following is not characteristic of the cytoskeleton?
a. Supports the plasma membrane and is responsible for the particular shape, rigidity, and spatial geometry of each different cell type
b. Plays a role in regulating cell growth and division.
c. Elements are all rigid, permanent structures.
d. Is responsible for cell contraction and cell movement.
e. Supports and organizes the ribosomes, mitochondria, and lysosomes.

86. This structure has hairlike motile protrusions.
a. flagella
b. cilia
c. actin bundles
d. sperm
e. fermentation

87. In the plasma membrane, which of the following provides a selective barrier to the movement of molecules and materials?
a. water
b. lipids
c. proteins
d. surface glycoproteins
e. internal carbohydrates

88. The Golgi complex is:
a. located in both the cytosol and nucleoplasm of the cell
b. never associated with the endoplasmic reticulum
c. always associated with ribosomes
d. where some proteins are concentrated before being secreted from the cell
e. involved in cell reproduction and the movement of the chromosome during cell division

89. Which of the following best describes a lysosome?
a. a membrane sac containing hydrolytic enzymes
b. an organelle formed in the smooth endoplasmic reticulum
c. the site of protein synthesis for the cell
d. a type of microfilament found in the cytoplasm
e. a vesicle containing substances for export

90. Which of the following best describes the endoplasmic reticulum of a cell?
a. It is the site of ATP synthesis.
b. It is the site for the attachment of all ribosomes.
c. It is the major store for carbohydrates, proteins, and fats.
d. It helps a variety of materials circulate throughout the cytoplasm.
e. It is physically connected to the plasma membrane but not to the nuclear envelope.

91. Which of the following is the most accurate statement concerning the cell's cytoskeleton?
a. The major structural component is chromatin.
b. It is an inflexible, rigid structure.
c. It is present throughout the cell, including the nucleus.
d. It is a latticed framework that extends throughout the cytoplasm to help organelles in place.
e. Its shape and structure are not altered according to the activities of the cell.

92. The fluid aspect of the fluid-mosaic model of the plasma membrane applies to which of the following?
a. It enables material to be transported across the membrane.
b. The fluid nature is due to water.
c. The fluid nature is due to the lateral movement of proteins and lipids within the bilayer
d. The fluid nature is due to just the lateral movement of lipids.
e. It provides shape to the cell.

93. Which of the following structures provides a clue concerning the physiological relationship between the Golgi apparatus and the endoplasmic reticulum?
a. ribosomes
b. vesicles
c. nuclear pores
d. mitochondria
e. various cytoplasmic inclusions

94. During exocytosis, the membrane of a vacuole fuses with the membrane of:
a. the cell
b. another vacuole
c. the Golgi apparatus
d. ribosomes
e. the nuclear envelope

95. The nucleolus is best described by which of the following statements?
a. It has a thick membrane.
b. It is the site of ribosomal RNA synthesis.
c. It specifies the chemical structure of enzymes.
d. It has fewer secretory cells than nonsecretory cells.
e. It contains ATP.

96. Vaults are the same size and shape as:
a. nuclear pores
b. ribosomes
c. polyribosomes
d. nucleosomes
e. vacuoles

97. Peroxisomes contain:
a. oxidative enzymes
b. DNA
c. RNA
d. mRNA
e. all of the above

98. Which of the following organelles destroys bacteria?
a. peroxisomes
b. lysosomes
c. vacuoles
d. Golgi complex
e. mitochondria

99. How many different types of endoplasmic reticula are found within the cell?
a. one
b. two
c. three
d. four
e. five

## *Points to Ponder*

1. What conditions would prompt primitive bacteria or early mitochondria to invade primitive cells?
2. Aside from energy production, what is the advantage in having the citric acid cycle?
3. How do you think your cells replace nicotinamide adenine dinucleotide?
4. Would it be possible to calculate the ATP production knowing how much carbon dioxide was produced?
5. Why is the plasma membrane considered a semi-permeable structure?
6. Why is the cytoskeleton considered a skeleton?
7. Why are lysosomes and peroxisomes vital to the life of a cell?
8. How does the endoplasmic reticulum differ from the Golgi apparatus?
9. In what way does the meaning of the term "cellular respiration" differ from that of carbohydrate metabolism?

10. Is all the energy liberated from the catabolism of one molecule of sugar trapped in ATP? Explain.

## Clinical Perspectives

1. In carbon monoxide poisoning, the gas carbon monoxide readily combines with hemoglobin. How does this phenomenon produce such fatigue that the victim may not be able to escape the gas?

2. Release of lysosomal enzymes from white blood cells that phagocytize bacteria can contribute to the symptoms of inflammation. Suppose, to alleviate inflammation, you develop a drug that destroys lysosomes. What would be the negative side effects of this drug?

3. A pathologist examines a liver biopsy sample. The cells contain extensive amounts of smooth endoplasmic reticulum. The patient admits to substance abuse. What is the relationship between the drug abuse and the increased ER?

4. Based on what you have learned in this chapter, which is better for a healthy young person: aerobic exercises or lifting weights?

5. What is the relationship between neurofilaments and amyotrophic lateral sclerosis?

## PhysioEdge Activities

**Related to Text:**
Tutorial - Transport Across Membranes
Media Exercise 2.1: Anatomy of a Generic Cell
Media Exercise 2.2: Basic Functions of Organelles

**Related to Figures:**
*Figure 2.1.* For an interaction related to this figure, see Media Exercise 2.1: Anatomy of a Generic Cell.
*Figure 2.2.* For an animation of this figure, click the Exocytosis tab in the Transport Across Membranes tutorial.
*Figure 2.8.* For an animation of this figure, click the Endocytosis tab in the Transport Across Membranes tutorial

## Media Resources

**PhysioEdge CD-ROM**
For a visual review of concepts in this chapter, check out:

Tutorial: Transport across Membranes
Media Exercise 2.1: Anatomy of a Generic Cell
Media Exercise 2.2: Basic Functions of Organelles

**Book Companion Website**
The website for this book contains a wealth of helpful study aids, as well as many ideas for further reading and research. Log on to:

**http://www.brookscole.com/sherwoodhp6**

Select Chapter 2 from the drop-down menu, or click on one of the many resource areas.

For Suggested Readings, consult **InfoTrac® College Edition**, your online research library, at:

http://infotrac.thomsonlearning.com

# 3

# The Plasma Membrane and Membrane Potential

## Chapter Overview

All cells are enveloped by a plasma membrane, a thin, flexible lipid barrier that separates the contents of the cell from its surroundings. The plasma membrane is a lipid bilayer with membrane proteins, membrane carbohydrates and cholesterol arranged as a fluid mosaic model. Some proteins form channels across the lipid bilayer, others serve as carrier molecules, and still others are receptor sites on the outer surface of the cell. Using these proteins and several enzymes molecules can enter the cell. Cells can be held together using fibrous proteins: (1) collagen, (2) elastin, and (3) fibronectin. Cells can be directly linked together using (1) desmosomes, (2) tight junctions, and (3) gap junctions. Two forces are used to transport molecules across a membrane: (1) passive and (2) active. Diffusion is passive down a concentration gradient. Ions may use an electrical gradient for transport. Osmosis is water diffusing down its own concentration gradient. Larger molecules use carrier-mediated transport. The sodium-potassium pump is such a mechanism that performs several functions for the cell. Vesicular transport is a special situation to gain, secrete, rebuild, or destroy various entities. A membrane potential exists due to separated charged ions -- primarily sodium, potassium and large anions. The membrane potential and the various transport mechanisms demonstrate the importance of the plasma membrane in maintaining homeostasis.

## Chapter Outline

### MEMBRANE STRUCTURE AND COMPOSITION

*The plasma membrane is a fluid lipid bilayer embedded with proteins.*

- Each cell type is unique in its intracellular contents.
- The plasma membrane is a thin layer of lipids and proteins forming the outer boundary of the cell.
- This membrane maintains the specific composition of the intracellular fluid by selectively permitting specific substances to pass between the cell and its environment.
- All plasma membranes are composed of lipids, proteins, and carbohydrates.
- The lipids are predominately phospholipid with some cholesterol.
- Phospholipids vary from most lipids by having two fatty acids and a phosphate.
- The polar head is hydrophilic and negatively charged.
- Consistent with the properties of fatty acids, the tails are hydrophobic.
- The two lipid layers are arranged such that the outer polar heads are in contact with the extracellular fluid and the inner polar heads are in contact with the intracellular fluid.
- The hydrophobic tails of each lipid layer are toward the inside of the plasma membrane.
- The fluid nature of the membrane is maintained by the lack of bonding between phospholipids.
- Cholesterol molecules between the phospholipid molecules prevent the fatty acid chains from packing and crystallizing.
- Membrane proteins may extend through the thickness of the membrane, while others are found only on the inner and outer surfaces.
- Short-chain carbohydrates and membrane carbohydrates are bound to the outer membrane proteins and lipids and are named accordingly: glycoproteins and glycolipids.
- The molecular arrangement of plasma membranes is called the fluid mosaic model.

*The lipid bilayer forms the basic structural barrier that encloses the cell.*

- The lipid bilayer provides three important functions: (1) It forms the basic structure of the membrane. (2) It prevents passage of water-soluble substances between the ICF and ECF. (3) It provides fluidity to the membrane.

*The membrane proteins perform a variety of specific membrane functions.*

- Membrane proteins provide seven specialized functions: (1) Proteins spanning the membrane form highly selective channels that allow or prevent passage of small particles such as ions. (2) Other proteins serve as carrier molecules to transport substances between the ECF and ICF.

(3) Proteins on the outer surface may serve as receptor sites for specific molecules in the ECF. (4) Proteins also function as membrane-bound enzymes to control surface chemical reactions. (5) Other proteins on the inner surface form a filamentous meshwork and attach to the cytoskeleton. (6) Additional proteins function on the outer surface as cell adhesion molecules (CAMs) and are used to attach cells and tissues together. (7) Still other proteins, functioning with special carbohydrates, serve to recognize themselves and other cells.

*The membrane carbohydrates serve as self-identity markers.*

- The short sugar chains on the outer-membrane surface serve as self-identity markers that enable cells to identify and interact with each other in the following ways: (1) These carbohydrate chains play an important role in recognition of "self" and in cell-to-cell interactions. Cells are able to recognize other cells of the same type and join together to form tissues. (2) Carbohydrate - containing surface markers also appear to be involved in tissue growth, which is normally held within certain limits of cell density. (3) Some cell adhesion molecules (CAMs) bear carbohydrates on their outermost tip, where these sugary chains participate in cell adhesion activities.

## CELL-TO-CELL ADHESIONS

*The extracellular matrix serves as the biological "glue."*

- The plasma membrane functions in cell-to-cell adhesions that permits groups of cells to bind together into tissues and to be organized further into organs.
- The extracellular matrix is a meshwork of fibrous proteins in a watery gel composed of complex carbohydrates.
- The protein fibers are collagen, elastin and fibronectin.
- The extracellular matrix is secreted by fibroblasts.

*Some cells are directly linked by specialized cell junctions.*

- At a desmosome filaments extend between the plasma membranes of two closely adjacent but nontouching cells to anchor the cells together.
- Sheets of epithelial tissue are joined by tight junctions.
- Adjacent cells are joined together in a tight seal, which is impermeable and thereby prevents materials from passing between cells.
- Selected substances must pass through the cells.
- The third type of junction between cells is a communicating junction known as a gap junction.
- A gap exists between adjacent cells that are linked by small connecting tunnels known as connexons.
- Gap junctions allow particles such as ions to pass between cells, whereas larger molecules are blocked.

## OVERVIEW OF MEMBRANE TRANSPORT

- The membrane is permeable to the substance if the substance can cross the membrane and impermeable if the substance cannot cross.
- Selectively permeable membranes permit some particles to pass but exclude others.
- Two properties of particles determine whether they can cross the membrane without assistance: (1) Particles must be soluble in lipids. (2) Particles must be small.
- Forces are required to move particles across the membrane.
- Passive forces do not require the cell to expend energy to produce movement.
- In conditions where the cell must expend energy, the movement is achieved by active forces.

## UNASSISTED MEMBRANE TRANSPORT

*Particles that can permeate the membrane passively diffuse down their concentration gradient.*

- Diffusion uses passive forces to move particles down a concentration gradient.
- While diffusion involves movement of molecules in all directions the net movement of a substance is from an area where the concentration of the substance is greater to an area of lesser concentration.
- When the molecules of the substance are uniformly distributed, an equilibrium exists.
- Fick's law of diffusion demonstrates the effects of several factors on rate of net diffusion: (1) The greater the difference in concentration the faster the rate of net diffusion. (2) The more permeable the membrane is to a substance, the more rapidly the substance can diffuse down its concentration gradient. (3) The larger the available surface area of the membrane, the greater rate of diffusion it can accommodate. (4) Lighter molecules diffuse more rapidly. (5) The greater the distance through which diffusion must take place, the slower the rate of diffusion.

*Ions that can permeate the membrane also passively move along their electrical gradient.*

- Diffusion also moves substances along an electrical gradient.

- Positively charged ions (cations) move toward negatively charged areas whereas negatively charged ions (anions) move toward positively charged areas.
- When an electrical gradient (difference in charge) exists between the ICF and ECF, and the membrane is permeable to the ions in question then these ions will diffuse along this electrical gradient.

*Osmosis is the net diffusion of water down its own concentration gradient.*

- Osmosis is a special type of diffusion involving a water-soluble solute and water.
- If both solute and water can easily pass through the membrane then both will cross the membrane until each has the same concentration on both sides.
- The resulting volume will be the same as at the onset of the diffusion.
- If the solute cannot pass through the membrane the water will use its own concentration gradient to dilute the solute on the other side.
- In this type of osmosis the volume on the side of the solute increases.
- As water continues to cross the membrane a hydrostatic pressure develops which increases as the volume increases.
- Hydrostatic pressure opposes osmosis.
- Osmotic pressure is equal to the hydrostatic pressure necessary to stop the osmosis.
- The tonicity of a solution refers to the effect the solution has on cell volume.
- The tonicity of a solution is determined by its concentration of nonpenetrating solutes.

## ASSISTED MEMBRANE TRANSPORT

- Carrier-mediated transport is used for large, poorly lipid-soluble molecules such as glucose, proteins and amino acids which cannot cross the plasma membrane.
- This impermeability ensures that large essential molecules cannot escape from the cell.

*Carrier-mediated transport is accomplished by a membrane carrier flipping its shape.*

- The carrier proteins span the thickness of the plasma membrane.
- These proteins change shape to expose the binding sites to each surface of the membrane alternately.
- The substance to be transported attaches to a carrier protein that is specific for one substance or at most a few closely related compounds.
- The transport maximum of a substance is the upper limit that can be transported per unit of time.
- This transport maximum occurs when all the carriers are being used, thus saturation.
- This saturation condition can be overcome by changing the affinity of the binding site for the passenger or by increasing the number of binding sites.
- Closely related chemical compounds compete for binding sites.

*Carrier-mediated transport may be passive or active.*

- Carrier-mediated transport takes two forms: (1) Facilitated diffusion goes down a concentration gradient and is passive. (2) Active transport goes against the concentration gradient.
- In active transport the carrier is phosphorylated on the low concentration side to enhance affinity of the binding site for the passenger.
- The carrier is dephosphorylated on the high concentration side to reduce affinity.
- These active transport mechanisms are called pumps.

*Sodium-potassium pump.*

- The sodium-potassium pump is a mechanism that performs three important roles: (1) It establishes sodium and potassium gradients across the plasma membrane of all cells. (2) It helps regulate cell volume. (3) The energy used to run the sodium potassium pump also indirectly serves as the energy source for the cotransport of glucose and amino acids across intestinal and kidney cells.
- With secondary active transport, energy is required in the entire process but is not directly required to run the pump.

*With vesicular transport, material is moved into or out of the cell wrapped in a membrane.*

- Vesicular transport occurs when either a material is brought into the cell (endocytosis) or a substance leaves the cell (exocytosis) in a membrane-enclosed vesicle.
- Endocytosis is the formation of a vesicle by pinocytosis (cell drinking) or phagocytosis (cell eating).
- Such a vesicle will either be degraded by a lysosome or pass across the cell and be released.
- Exocytosis serves two functions: (1) It provides the mechanism for secreting large polar molecules such as hormones and enzymes. (2) It enables the cell to add specific components to the membrane.

*Caveolae may play roles in membrane transport and signal transduction.*

- The outer surface of the plasma membrane is dimpled with tiny cavelike indentations known as caveolae.
- Caveolae: (1) provide a new route for transport into the cell, and (2) serve as a "switchboard" for relaying signals from extracellular chemical messengers into the cell's interior.
- An abundance of proteins cluster in the caveolae.
- When folic acid binds with its receptors, which are concentrated in the caveolae, the extracellular openings of these tiny caves close off.
- The high concentration of folic acid within the closed caveolar compartment encourages the movement of this vitamin across the caveolar membrane into the cytoplasm.
- Cellular uptake through the caveolae has been termed potocytosis.
- Caveolae also appear to be important sites for signal transduction, a complicated process in which incoming signal from the ECF are conveyed to the cells interior.
- Proteinaceous hormones and other regulatory molecules signal a cell to perform a given response by binding with membrane receptors concentrated in the caveolae.
- Caveolae are thought to be important in cell-to-cell communication.

## Membrane Potential

*Membrane potential is a separation of opposite charges across the plasma membrane.*

- The unequal distribution of a few key ions between the ECF and ICF and their selective movement through the plasma membrane are responsible for the electrical properties of the membrane.
- Membrane potential refers to the separation of charges across the membrane.
- Because separated charges have the "potential" to do work, a separation of charges across the membrane is referred to as a membrane potential.
- The membrane potential is relatively low therefore the unit used is the millivolt.
- These separated charges represent only a small fraction of the total number of charged particles present in the ICF and ECF.
- The magnitude of the potential depends on the degree of separation of the opposite charges: the greater the number of charges separated, the larger the potential.

*Membrane potential is primarily due to differences in the concentration and permeability of key ions.*

- All living cells have a membrane potential, characterized by a slight excess of positive charges outside and a corresponding slight excess of negative charges on the inside.
- The ions primarily responsible for the generation of membrane potential are sodium, potassium, and anionic intracellular proteins.
- These large, negatively charged proteins are found only inside the cell.
- Sodium is in greater concentration in the extracellular fluid and potassium is in much higher concentration in the intracellular fluid.
- These concentration differences are maintained by the sodium-potassium pump at the expense of energy.
- In addition to the active carrier mechanism, sodium and potassium can passively cross the membrane through protein channels specific for them.
- It is usually much easier for potassium than for sodium to get through the membrane because typically more potassium channels than sodium channels are open.
- The concentration gradient for potassium will always be outward and the concentration gradient for sodium will always by inward, because the sodium-potassium pump maintains a higher concentration of potassium inside the cell and a higher concentration of sodium outside the cell.
- Because sodium and potassium are both cations, the electrical gradient for both will always by toward the negatively charged side of the membrane.
- About 20 percent of the membrane potential is directly generated by the sodium-potassium pump.
- The remaining 80 percent is caused by passive diffusion of potassium and sodium down concentration gradients.
- Most of the sodium-potassium pump's role in producing membrane potential is indirect, though its critical contribution to maintaining the concentration gradients is directly responsible for the ion movements that generate most of the potential.
- The resting membrane potential of a typical cell is -70mV.
- By convention, the sign always designates the polarity of the excess charge on the inside of the membrane.
- Because the membrane at rest is more permeable to potassium than to sodium, potassium influences the resting membrane potential to a much greater extent than does sodium.
- At resting potential, neither potassium nor sodium is at equilibrium.
- Both ions leak across the membrane.
- The sodium-potassium pump counterbalances the rate of leakage.

- The pump transports back into the cell essentially the same number of potassium ions that have leaked out and simultaneously transports to the outside the sodium ions that have leaked in.
- Not only is the sodium-potassium pump initially responsible for the sodium and potassium concentration differences across the membrane, but it also maintains these differences.
- Chloride is the principal ECF ion.
- In most cells chloride ions do not influence membrane potential.
- Nerve and muscle cells have developed a specialized use for membrane potential.
- This causes nerve impulses and muscles to contract.

## Key Terms

active forces
active transport
aquaporins
carrier-mediated transport
cell adhesion molecules (CAMs)
channels
chemical gradient
cholesterol
collagen
concentration gradient
cystic fibrosis
desmosome
diffusion
docking marker acceptors
elastin
electrical gradient
electrochemical gradient
endocytosis
exocytosis
extracellular matrix
facilitated diffusion
fibroblasts
fibronectin
Fick's law of diffusion
Fluid mosaic model
gap junction
hydrogen ion pump
hydrostatic pressure
hypertonic solution
hypotonic solution
impermeable
ion concentration gradient
isotonic solution
lipid bilayer
membrane carbohydrates
$Na^+$ equilibrium potential
$Na^+/K^+$ ATPase pump
Nernst equation
net diffusion
osmosis
osmotic pressure
passive forces
permeable
phospholipids
plasma membrane
primary active transport
pumps
receptor sites
resting potential
secondary active transport
selectively permeable
steady state
tight junctions
tonicity
transport maximum
trilaminar structure
vesicular transport

## Review Exercises

*Answers are in the appendix.*

**True/False**

____ 1. Cholesterol is found within the lipid bilayer.

____ 2. Most cells have no active transport for the chloride ion.

____ 3. Large anions are in the ECF.

____ 4. Pinocytosis is the process whereby the white blood cell eats bacteria.

____ 5. The inner and outer layers of the plasma membrane are identical in composition and function.

____ 6. One extracellular messenger molecule can ultimately influence the activity of only one protein molecule within the cell.

____ 7. When equilibrium is achieved and no net diffusion is taking place, there is no movement of molecules.

____ 8. Phosphorylation of a carrier can alter the affinity of its binding sites.

_____ 9. A potential of +30mV is larger than a potential of -70mV.

_____ 10. At resting membrane potential, passive and active forces exactly balance each other, so there is no net movement of ions across the membrane.

_____ 11. The survival of every cell depends on the maintenance of intracellular contents unique for that cell type despite the remarkably different composition of the extracellular fluid surrounding it.

_____ 12. All plasma membranes consist mostly of lipids and proteins plus small amounts of carbohydrates.

_____ 13. The small amount of membrane carbohydrate is located only on the inner surface.

_____ 14. The lipid bilayer forms the basic structure of the membrane.

_____ 15. It is not possible for a given channel to be open or closed to its specific ion as a result of changes in channel shape in response to a controlling mechanism.

_____ 16. Release of chemicals from nerve cells in response to a nerve impulse is an example of second messenger mechanism.

_____ 17. G proteins are so named because they are bound to guanine nucleotides.

_____ 18. Different types of cells have different proteins available for phosphorylation and modification by protein kinase.

_____ 19. Binding of one chemical messenger molecule to a receptor activates a number of adenylate cyclase molecules, each of which activates many cAMP molecules.

_____ 20. Interwoven within extracellular matrix gel, there are three types of protein fibers: collagen, elastin, and fibroblasts.

_____ 21. Elastin is a rubberlike protein fiber most abundant in tissues that must be capable of stretching.

_____ 22. The extracellular matrix is secreted by local cells, most commonly by fibronectin.

_____ 23. Sheets of epithelial tissue are joined by gap junctions.

_____ 24. Tight junctions restrict the movement of membrane proteins to one of two cell surfaces: the luminal border or basolateral border.

_____ 25. Connexons are formed by the joining of proteins that extend outward from each of the adjacent plasma membranes.

_____ 26. Gap junctions are found in skeletal muscle.

_____ 27. Two properties of particles influence whether they can permeate the plasma membrane without any assistance: The relative solubility of the particle in lipid and the polarity of the particle.

_____ 28. If 10 molecules move from A to B while two molecules simultaneously move from B to A, the net diffusion is 12 molecules moving from A to B.

_____ 29. The greater the difference in concentration, the slower the rate of net diffusion.

_____ 30. The more permeable the membrane is to a substance, the more rapidly the substance can diffuse down its concentration gradient.

_____ 31. The larger the surface area available, the greater the rate of diffusion.

_____ 32. Heavier molecules bounce farther upon collision than do lighter molecules.

_____ 33. The greater the distance, the greater the rate of diffusion.

_____ 34. Opposite charges attract each other.

_____ 35. A solution with a high solute concentration exerts greater osmotic pressure than does a solution with a lower solute concentration.

_____ 36. Cells utilize two different mechanisms to accomplish these selective-transport processes: carrier-mediated transport and vesicular transport.

_____ 37. If fluid is internalized by endocytosis, the process is termed pinocytosis.

_____ 38. The separation of like charges across the plasma membrane is referred to as a membrane potential.

_____ 39. All living cells have a membrane potential characterized by a slight excess of positive charges outside and a correspondingly slight excess of negative charges on the inside.

_____ 40. The larger the concentration gradient for an ion, the lesser the ion's equilibrium potential.

**Fill in the Blank**

41. When ________________ occurs, the body's exocrine glands secrete an abnormally thick, sticky mucus.

42. ____________________ transfer the signal to an intracellular chemical messenger, which in turns, triggers a preprogrammed series of biochemical events within the cell.

43. ____________________ brings about the desired intracellular response by opening or closing specific channels in the membrane to regulate the movement of particular ions into or out of the cell.

44. A membrane-bound "middleman," a ____________ acts as an intermediary between the receptor and adenylyl cyclase.

45. ____________________ refers to the transfer of a phosphate group from ATP to the protein at the expense of degrading ATP to ADP.

46. ____________________________________ is suspected of playing a role in postreceptor events in some cells.

47. The ________________ serves as the biological "glue."

48. Once arranged, cells are held together by three different means. They are: (1) ____________________________, (2) __________________________ and (3) __________________________________.

49. ____________________ are adhering junctions that anchor cells together in tissues subject to considerable stretching.

50. List the three major types of protein fibers: (1) __________, (2) __________, and (3) __________

51. The extracellular matrix is secreted by local cells, most commonly by______________ present in the matrix.

52. Some cells are directly linked together by one of three types of specialized cell junctions, which are (1) ______________ (2) ______________, and (3) ______________

53. Tight junctions restrict the movement of membrane proteins to one of two cell surfaces. Name the two cell surfaces. (1) ______________________ and (2) ______________________.

54. Sheets of epithelial tissue are joined by ______________________.

55. ______________ form(s) cablelike fibers or sheets that provide tensile strength.

56. All cells are enveloped by a(n) ______________________, a thin, flexible, lipid barrier that separates the contents of the cell from its surroundings.

57. Cells have a(n) ____________________, which refers to a slight excess of _________ (+/-) charges lined up along the inside of the membrane and separated from a slight excess of positive charges on the outside.

58. ______________ have a polar head containing a negatively charged phosphate group and two nonpolar fatty acid tails.

59. A polar end that interacts with water molecules is referred to as ______________, and a nonpolar end that will not mix with water is called ______________.

60. ______________ contributes to the fluidity as well as the stability of the membrane.

61. The ______________ forms the primary barrier to diffusion, whereas ______________ perform most of the specific membrane functions.

62. Some proteins serves as ____________________ which transfer substances unable to cross the membrane on their own. Many of the proteins on the outer surface serve as ______________ which "recognize" and bind with specific molecules in the environment of the cell.

63. ______________ is a rubberlike protein fiber most abundant in tissues that must be capable of easily stretching and then recoiling after the stretching force is removed.

64. ______________ are communicating junctions consisting of ______________, small connecting tunnels that permit movement of charge-carrying ions between two adjacent cells.

65. ______________ promotes cell adhesion and holds cells in position.

66. Gap junctions are found in ______________ muscle and ______________ muscle.

67. ______________ refers to the difference between two opposing movements.

68. When movement of molecules from A to B are exactly matched by movement of molecules from B to A, this situation is known as a(n)____________________ or ______________.

69. If a substance is unable to pass, the membrane is ______________ to it.

70. All ______________________________ proteins span the thickness of the plasma membrane and are able to undergo reversible changes in shape so that specific binding sites can alternately be exposed at either side of the membrane.

71. ______________ is a disease involving defective cysteine carriers in the kidney membranes.

72. There is a limit to the amount of a substance that can be transported across the membrane via a carrier in a given time. This limit is referred to as ______________.

73. With ____________________, energy is directly required to move a substance uphill.

74. With ____________________, energy is required in the entire process, but it is not directly required to run the pump.

75. In ____________, the plasma membrane surrounds the substance to be ingested, then fuses over the surface, pinching off a membrane-bound vesicle that encloses the engulfed material within the cell.

76. If large multimolecular particles are engulfed, such as bacteria or cellular debris, the process is called ____________.

77. ____________ refers to a separation of charges across the membrane or to a difference in the relative number of cations and anions in the ICF and ECF.

78. If a relative difference in charge exists between two adjacent areas, the positive charged ions, called ____________, tend to move toward the more ____________ charged area, whereas the negatively charged ions, called ____________, tend to move toward the more ____________ charged area.

79. A difference in charge between two adjacent areas thus produces a(n) ____________________ that passively induces the movement of ions.

80. The simultaneous existence of an electrical gradient and concentration (chemical) gradient for a particular ion is referred to as a(n) ____________________.

81. The term ____________ refers to the density of the solute (dissolved substance) in a given volume of water.

82. The net diffusion of water is known as ____________.

83. The magnitude of opposing pressure necessary to completely stop osmosis is equal to the ____________ or hydrostatic pressure.

**Matching**

*Match substance to method of transport.*

____ 84. sodium ions
____ 85. water
____ 86. carbon dioxide
____ 87. calcium ions
____ 88. potassium ions
____ 89. oxygen
____ 90. bacteria
____ 91. chloride ions
____ 92. hormones
____ 93. fatty acids

a. diffusion through lipid bilayer
b. endocytosis
c. exocytosis
d. through protein channel
e. osmosis

**Multiple Choice**

94. Which of the following act as “spot rivets” to anchor cells together?
a. tight junctions
b. gap junctions
c. desmosomes
d. ribosomes
e. archaeosomes

95. Which type of movement is the net diffusion of water down its own concentration gradient?
a. carrier-mediated
b. osmosis
c. facilitated diffusion
d. secondary active transport
e. exocytosis

96. Which process is used to recycle portions of the plasma membrane?
a. primary active transport
b. exocytosis
c. dephosphorylation
d. facilitated diffusion
e. endocytosis

97. Which of the following types of cell junctions involve connexons?
a. tight junctions
b. gap junctions
c. desmosomes
d. all the above
e. none of the above

98. Cholesterol contributes to the fluidity and stability of the membrane. Between which molecules is cholesterol found?
a. membrane proteins
b. membrane carbohydrates
c. cell adhesions molecules
d. phospholipids
e. G proteins

99. Which disorder is characterized by having improperly formed collagen fibers?
a. scurvy
b. diabetes insipidus
c. cystic fibrosis
d. diabetes mellitus
e. cysteinuria

100. Phagocytosis is an example of which of the following?
a. endocytosis
b. mesocytosis
c. ectocytosis
d. hydrocytosis
e. none of the above

101. Membrane potential is measured in these units.
a. volts
b. nanovolts
c. kilovolts
d. millivolts
e. decivolts

102. The negative sign of the membrane potential applies to which of the following?
a. a truly negative potential
b. the polarity of excess charge on the inside of the membrane
c. the polarity of excess charge on the outside of the membrane
d. the concentrations of large anions in the ECF
e. none of the above

103. What is the ratio of sodium pumped out to potassium pumped in by the active transport mechanism?
a. 9:1
b. 3:1
c. 5:2
d. 4:1
e. 3:2

104. Which of the following best describes cadherins?
a. a membrane-bound enzyme
b. a type of CAM
c. a type of receptor site
d. a carrier molecule
e. fibronectin

105. Which of the following describes cellular uptake through the caveolae?
a. phagocytosis
b. exocytosis
c. endocytosis
d. potocytosis
e. pinocytosis

106. Which of the following are though to be used in signal transduction?
a. desmosomes
b. fibronectin
c. caveolae
d. calmodulin
e. membrane potential

107. Which of the following are released into the blood specifically by neurosecretory neurons?
a. neurotransmitters
b. paracrines
c. neurohormones
d. neuromodulators
e. hormones

108. Thyroid cells are the only cells in the body to use:
a. iron
b. iodine
c. sulfur
d. calcium
e. potassium

109. What is the function of carbohydrates in a plasma membrane?
a. form a barrier to diffusion
b. perform specific membrane functions
c. play an important role in "self" regonition
d. both a and b
e. both b and c

110. Cystic fibrosis patients are most likely to become infected with which of the following bacteria?
a. Escherichia coli
b. Pseudomona aeruginosa
c. Staphylococcus aurus
d. Streptococcus pyogenes
e. Beta hemolytic streptococcus

111. Which of the following is a natural antibiotic that kills most airborne bacteria that get into the lungs?
   a. penicillin
   b. streptomyocin
   c. bactin
   d. defensin
   e. none of the above

112. Dense plaque and strong glycoprotein filaments are found in:
   a. tight junctions
   b. desmosomes
   c. gap junctions
   d. both a and b
   e. none of the above

113. Cellular communication occurs most often in which of the following types of junctions?
   a. tight junctions
   b. desmosomes
   c. gap junctions
   d. both a and b
   e. none of the above

114. How many types of gated channels are there?
   a. one
   b. two
   c. three
   d. four
   e. five

115. Programmed cell death is termed:
   a. cellular suicide
   b. apoptosis
   c. euthanasia
   d. Both a and b
   e. None of the above

116. Uncontrolled cell death is termed:
   a. cellular suicide
   b. apoptosis
   c. necrosis
   d. euthanasia
   e. caspases

## *Points to Ponder*

1. After a long, hard day you soak in the tub for several minutes. It feels good to just to relax in the water. As you dry your body you notice that your fingers and toes are very wrinkled. What has happened?

2. As you watch people sitting in the smoking section of a restaurant you notice that shortly after one patron lights up a cigarette others soon follow. What is happening?

3. How would you explain that many different totally man made molecules, such as some drugs, are transported across membranes? Even carcinogenic molecules, both large and small, enter the cell. How is this possible?

4. Why can't humans drink salt water? What happens when they do?

5. If a person has an abnormally low concentration of proteins in his/her plasma, excessive accumulation of fluid in the tissues (edema) will occur. Why?

## Clinical Perspectives

1. Mannitol is a sugar that does not cross the blood-brain barrier. It is a particle that is osmotically active. During head trauma, the brain swells. How might administering mannitol to a head trauma patient prevent this swelling?

2. A physiology student was diagnosed as having hyperkalemia (high potassium) and an abnormal electrocardiogram (EKG). Based on what you know about the membrane potential, what effect would a high potassium level have on the membrane potential of the heart to produce an abnormal EKG?

3. Severe diarrhea is responsible for almost half of all deaths worldwide in children under the age of four. Because intravenous rehydration therapy is not economically feasible, oral rehydration therapy is used. The concept behind oral rehydration therapy is to cause sodium to move into the body and then water follows. Based on the principles of osmosis, what is the rationale for this therapy?

4. If a diabetic patient takes too much insulin, this can lead to hypoglycemia, unconsciousness and death due to a coma. The rate of facilitated diffusion into the brain depends directly on the plasma glucose concentration. Based on what you know about facilitated diffusion, why does a coma occur?

5. Based on the concepts of isotonic, hypertonic, and hypotonic, why is Ringer's lactate or 5 percent dextrose used for IV fluids?

## Experiments of the Day

1. Materials:
   - 2 Glasses
   - 1 Face cloth
   - Water
   - Cooking oil

Place one glass half filled with water next to the other one-fourth filled with cooking oil. Carefully insert one end of the dry face cloth into the water glass and the other end into the glass with oil. Be sure both ends are in the respective fluids. Mark the fluid levels. Explain what happens over the next several days.

2. Take a diet coke and a regular coke. Place them in a tub or bucket of water. One can will sink and one can will float. Based on what you have learned in this chapter, you should be able to explain why this occurs.

## PhysioEdge Activities

**Related to Text:**
Tutorial - Membrane Potential.
Media Exercise 3.1: The Plasma Membrane and Cell-Cell Connections.
Media Exercise 3.2: Means of Transmembrane Exchange.

**Related to Figures:**
*Figure 3.3.* For an interaction related to this figure, see Media Exercise 3.1: The Plasma Membrane and Cell-Cell Connections.
*Figure 3.5.* For an interaction related to this figure, see Media Exercise 3.1: The Plasma Membrane and Cell-Cell Connections.
*Figure 3.7.* For an animation of this figure, click the Diffusion and Membrane Transport tab in the Membrane Potential tutorial.

*Figure 3.8.* For interactions related to this figure, click the Diffusion and Membrane Transport tab in the Membrane Potential tutorial and Media Exercise 3.2: Means of Transmembrane Exchange.
*Figure 3.14.* For interactions related to this figure, click the Diffusion and Membrane Transport tab in the Membrane Potential tutorial and Media Exercise 3.2: Means of Transmembrane Exchange.
*Figure 3.15.* For interactions related to this figure, click the Diffusion and Membrane Transport tab in the Membrane Potential tutorial and Media Exercise 3.2: Means of Transmembrane Exchange.
*Figure 3.16.* For interactions related to this figure, click the Diffusion and Membrane Transport tab in the Membrane Potential tutorial and Media Exercise 3.2: Means of Transmembrane Exchange.
*Figure 3.19.* For an animation of this figure, click the Resting Potential tab in the membrane Potential tutorial.
*Figure 3.20.* For an animation of this figure, click the Resting Potential tab in the membrane Potential tutorial.
*Figure 3.21.* For an animation of this figure, click the Resting Potential tab in the membrane Potential tutorial.
*Figure 3.22.* For an animation of this figure, click the Resting Potential tab in the Membrane Potential tutorial.

## Media Resources

**PhysioEdge CD-ROM**
For a visual review of concepts in this chapter, check out:

Tutorial: Membrane Potential
Media Exercise 3.1: The Plasma Membrane and Cell\-Cell Connections
Media Exercise 3.2: Means of Transmembrane Exchange

**Book Companion Website**
The website for this book contains a wealth of helpful study aids, as well as many ideas for further reading and research. Log on to:

**http://www.brookscole.com/sherwoodhp6**

Select Chapter 3 from the drop-down menu or click on one of the many resource areas.

For Suggested Readings, consult **InfoTrac® College Edition**, your online research library, at:

**http://infotrac.thomsonlearning.com**

# Principles of Neural and Hormonal Communication

## Chapter Overview

Nerve cells, or neurons, make up the nervous system, one of the two major regulatory systems of the body. As changes occur in the external and internal environments of the body, the nerve cells receive this information and transmit it to other cells. Eventually nerve cells initiate the necessary responses to such changes. Therefore, neurons are essential in maintaining homeostasis. The process of transmission of information within a neuron involves electrical signals such as graded and action potentials. Through changes in permeability the nerve cell undergoes depolarization, hyperpolarization, and repolarization. These phenomena transmit the information. The speed of transmission depends on the diameter of the fiber and whether or not the neuron is myelinated. When the action potential reaches the synaptic knob a neurotransmitter is released, which diffuses across the synaptic cleft and changes the permeability of the postsynaptic neuron. Through temporal or spatial summation an action potential is initiated in the axon hillock of the postsynaptic neuron. Synapses may be excitatory or inhibitory. Drugs and diseases may alter synaptic effectiveness.

## Chapter Outline

### INTRODUCTION TO NEURAL COMMUNICATION

*Nerve and muscle are excitable tissues.*

- All cells of the body process a membrane potential related to the distribution and permeability of sodium ions, potassium ions, and large intracellular anions.
- Nerve and muscle are excitable tissues, capable of producing electrical signals when stimulated.
- These fluctuations in membrane potential produce two types of signals: (1) graded potentials (short distance signals) and (2) action potentials (long distance signals).

*Membrane potential decreases during depolarization and increases during hyperpolarization.*

- Charges are separated across the plasma membrane so that the membrane has potential.
- When depolarized, the membrane is less polarized than at resting potential.
- During repolarization, the membrane returns to a resting potential.
- When a membrane is hyperpolarized, it is more polarized (more negative) than at a resting potential.

*Electrical signals are produced by changes in ion movement across the plasma membrane.*

- Changes in ionic movement across the membrane are responsible for the changes in potential.
- Changes in ionic movement in turn are brought about by changes in the permeability of the membrane to specific ions.
- There are two types of channels: leak channels and gated channels.

### GRADED POTENTIALS

*The stronger a triggering event, the larger the resultant graded potential.*

- Graded potentials are local changes in membrane potential that occur in varying grades or degrees of magnitude; the stronger the triggering event, the larger the graded potential.
- The triggering event might be: (1) a stimulus, (2) an interaction of a chemical messenger, or (3) a spontaneous change of potential caused by imbalances in the leak-pump cycle.
- When a graded potential occurs in excitable tissue a different potential exists in this area than in the remainder of the membrane, which is still at resting potential.

*Graded potentials spread by passive current flow.*

- Because opposite charges attract each other, current flows from this active area to the neighboring inactive areas.
- Any reduction in resting potential will produce a current flow.
- Since these areas on excitable tissues are not insulated the current flow leaks into the ECF.

*Graded potentials die out over short distances.*

- The graded potential dies out after a short distance.
- Graded potentials are critically important to the body as: (1) postsynaptic potentials, (2) receptor potentials, (3) end-plate potentials, (4) pacemaker potentials, and (5) slow-wave potentials.

## Action Potentials

- The rapid reversals of membrane potential in excitable tissues are known as action potentials.
- This entire rapid change in potential, from threshold potential to depolarization and back, is called the action potential (also referred to as a spike and firing).

*During an action potential, the membrane potential rapidly, transiently reverses.*

- During an action potential, marked changes in membrane permeability of sodium and potassium take place, permitting rapid fluxes of these ions.
- The rising phase of the action potential (depolarization) is due to a sodium ion influx, induced by an explosive increase in sodium ion permeability at threshold.
- The falling phase (repolarization) is brought about by a potassium ion efflux, caused by a marked increase in potassium ion permeability occurring simultaneously with an inactivation of sodium channels at the peak of the action potential.
- After the action potential, the ion distribution has been altered slightly.
- Sodium entered the cell during the rising phase.
- Potassium exited the cell during the falling phase.
- The sodium-potassium pump restores the normal distribution of these ions.

*Marked changes in membrane permeability and ion movement lead to an action potential.*

- During an action potential, depolarization of the membrane to threshold potential triggers sequential changes in permeability caused by conformational changes in voltage-gated sodium and potassium channels.
- Voltage-gated sodium and potassium channels are the ones involved in action potentials.
- The voltage-gated sodium channel has two gates: an activation gate and an inactivation gate.
- Permeability changes bring about a brief reversal of membrane potential, with sodium influx being responsible for the rising phase, followed by potassium efflux during the falling phase.

*The sodium-potassium pump gradually restores the concentration gradients disrupted by action potentials.*

- At the completion of an action potential, the membrane potential has been restored to its resting condition by the sodium-potassium pump.

*Action potentials are propagated from the axon hillock to the axon terminals.*

- A single action potential involves only a small patch of the total surface membrane of an excitable cell.
- The nerve cell, or neuron, has three basic parts: (1) the cell body, (2) the dendrites, and (3) the axon.
- The nucleus and organelles are in the cell body.
- The dendrites project as antennae to receive signals from other neurons.
- The dendrites carry signals toward the cell body.
- The axon carries the signal away from the cell body.
- Side branches of the axon are called collaterals.

*Once initiated, action potentials are conducted throughout a nerve fiber.*

- The site of initiation of action potentials is the axon hillock.
- The axon hillock is the first part of the axon and the adjacent part of the cell body.
- The impulse carried by an axon ends at the axon terminal where it stimulates the release of chemical messengers, which excite other cells.
- Impulse conduction is by local current flow or saltatory conduction.

*Action potentials occur in an all-or-none fashion.*

- Once an action potential is initiated in one part of a nerve cell membrane, a self-perpetuating cycle is initiated so that the action potential is propagated throughout the rest of the axon automatically.

*The refractory period ensures unidirectional propagation of the action potential and limits the frequency of action potentials.*

- The refractory period has two components: (1) absolute refractory period and (2) the relative refractory period.
- The absolute refractory period is the time when a recently activated patch of membrane is completely incapable of initiating another action potential no matter how strongly it is stimulated.
- The relative refractory period is the time during which a second action potential can be produced only by a stimulus considerably stronger than is usually necessary.
- The refractory period ensures the unidirectional propagation of the action potential down the axon, away from the initial site of activation.
- An excitable membrane either responds to a stimulus with a maximal action potential that

spreads nondecrementally throughout the membrane, or it does not respond with an action potential at all.
- This is known as the all-or-none law.
- Weak and strong stimuli evoke the same action potential, but stronger stimuli initiate more action potentials.

*Myelination increases the speed of conduction of action potentials and conserves energy in the process.*
- The velocity with which an action potential travels down the axon depends on two factors: (1) whether or not the fiber is myelinated and (2) the diameter of the fiber.
- Myelin is composed primarily of lipids.
- Water-soluble ions cannot permeate this thick lipid barrier.
- Myelin-forming cells wrap around the axon in jelly-roll fashion.
- In the central nervous system, oligodendrocytes are the myelin-forming cells, while Schwann cells perform this function in the peripheral nervous system.
- Between the myelinated regions, the axonal membrane is bare and exposed to the ECF.
- These bare areas are known as nodes of Ranvier.
- The distance between nodes (about 1mm) is short enough that local current from an active node can reach an adjacent node.
- This is saltatory conduction.
- Myelinated fibers conduct impulses about 50 times faster than unmyelinated fibers of comparable diameter.
- Since ion fluxes are confined to nodes, energy is conserved.

*Fiber diameter also influences the velocity of action potential propagation.*
- The larger the diameter of the nerve fiber, the faster it can propagate an action potential.
- Speeds vary from 0.7m/sec in small fibers to 120m/sec in large myelinated fibers.
- Multiple sclerosis is a pathophysiological condition in which demyelination of nerve fibers occurs in various locations of the nervous system.

## Regeneration of Nerve Fibers

*Peripheral but not central fibers can regenerate.*
- Loss of myelin slows transmission of impulses in the affected neurons.
- Scarring associated with myelin damage can injure the underlying axon.
- When an axon is cut, the peripheral portion degenerates and this debris is phagocytized by Schwann cells.
- In such an injury the Schwann cells without its axon forms a regeneration tube through which the axon can slowly regenerate.
- Central nervous system axons have no regenerative ability.

## Synapses and Neuronal Integration

*Synapses are junctions between presynaptic and postsynaptic neurons*
- The junction between two neurons is known as the synapse.
- The axon terminal of the presynaptic neuron is referred to as a synaptic knob.
- The synaptic knob contains synaptic vesicles, which store a neurotransmitter.
- The space between the presynaptic and postsynaptic membrane is called the synaptic cleft.
- The portion of the postsynaptic membrane immediately underlying the synaptic knob is referred to as the subsynaptic membrane.

*A neurotransmitter carries the signal across a synapse.*
- The action potential in the presynaptic neuron induces the release of the neurotransmitter by opening calcium ion channels in the synaptic knob.
- The released neurotransmitter binds to specific sites on the subsynaptic membrane.
- This binding triggers the opening of specific ion channels thereby altering the permeability of the postsynaptic neuron.
- Synapses operate in one direction only.

*Some synapses excite whereas others inhibit the postsynaptic neuron.*
- At an excitatory synapse, the permeability of the sodium and potassium ions is increased by the opening of the respective channels in the subsynaptic membrane.
- The result is a net movement of positive ions into the postsynaptic neuron or a small depolarization.
- The potential across the membrane is now closer to the threshold potential.
- This postsynaptic potential change is called an excitatory postsynaptic potential or EPSP.
- At an inhibitory synapse, the permeability of the potassium or chloride ion is increased.
- The resulting ion movements bring about a small hyperpolarization.
- The inside of the postsynaptic neuron becomes more negative.
- The difference between this new potential and the threshold potential is greater.
- This postsynaptic potential change is called an inhibitory postsynaptic potential or IPSP.

- Many chemicals are known or suspected to serve as neurotransmitters.

*Each synapse is either always excitatory or always inhibitory.*

- A given synapse is either always excitatory or always inhibitory.

*Neurotransmitters are quickly removed from the synaptic cleft.*

- The transmitter may be inactivated by specific enzymes within the subsynaptic membrane or may be actively taken back up into the axon terminal.
- Once the transmitter is within the synaptic knob it can be stored or destroyed.

*Some neurotransmitters function through intracellular second messenger systems.*

- Some neurotransmitters, such as serotonin, activate a second messenger.
- Cyclic AMP is such a second messenger and can induce both short- and long-term effects.

*The grand postsynaptic potential depends on the sum of the activities of all presynaptic inputs.*

- The grand postsynaptic potential is the total potential produced by the presynaptic neurons.
- The summing of several EPSPs occurring very close together in time because of successive firing of a single presynaptic neuron is known as temporal summation.
- A second method to elicit a grand postsynaptic potential is through concurrent activation of several excitatory inputs or spatial summation.
- Similarly, IPSPs can undergo temporal and spatial summation.
- The summing of several EPSPs occurring very close together in time because of successive firing of a single presynaptic neuron is called temporal summation.
- The summation of EPSPs originating simultaneously from several different presynaptic inputs is known as spatial summation.
- If excitatory and inhibitory input are simultaneously activated, they cancel each other out.
- The magnitude of GPSP depends on the sum of activity in all presynaptic inputs and in turn determines whether or not the neuron will undergo an action potential to pass information on to the cells at which the neuron terminates.

*Action potentials are initiated at the axon hillock because it has the lowest threshold.*

- When the summation of EPSPs takes place, the lower threshold of the axon hillock is reached first.

*Neuropeptides act primarily as neuromodulators.*

- Neuropeptides are larger than most neurotransmitters.
- Neuropeptides bind to neuronal receptors at nonsynaptic sites and bring about long-term changes that subtly depress or enhance synaptic effectiveness.
- Whereas neurotransmitters serve as chemical messengers for rapid communication between neurons, neuromodulators are involved with more long-term neural events such as learning.

*Presynaptic inhibition or facilitation can selectively alter the effectiveness of a given presynaptic input.*

- A presynaptic axon terminal may itself be innervated by another axon terminal.
- In such an arrangement the amount of neurotransmitter released is altered.
- If the amount of transmitter released is reduced, the phenomenon is known as presynaptic inhibition.
- If the release of a transmitter is enhanced, the effect is called presynaptic facilitation.

*Drugs and disease can modify synaptic transmission.*

- Synaptic drugs may block an undesirable effect or enhance a desirable effect.
- Possible drug actions include (1) altering the synthesis, axonal transport, storage, or release of a neurotransmitter; (2) modifying neurotransmitter receptor; (3) influencing neurotransmitter reuptake or destruction; or (4) replacing a deficient neurotransmitter with a substitute transmitter.

*Neurons are linked through complex converging and diverging pathways.*

- Through convergence, a single cell is influenced by thousands of other cells.
- Divergence refers to the branching of axon terminals so that a single cell synapses with many other cells.

## INTERCELLULAR COMMUNICATION AND SIGNAL TRANSDUCTION

*Communication between cells is largely orchestrated by four types of extracellular chemical messengers.*

- There are three types of intercellular communication: (1) The most intimate means of intercellular communication is through gap junctions. (2) The presence of signaling molecules on the surface membrane of some cells permits

them to directly link up and interact with certain other cells in a specialized way. (3) The most common means by which cells communicate with each other is through intercellular chemical messengers, of which there are four types: paracrines, neurotransmitters, hormones, and neurohormones.
- These four types of chemical messengers differ in their source and the distance and means by which they get to their site of action.
- Paracrines are local chemical transmitters whose effect is exerted only on neighboring cells in the immediate environment of their site of secretion.
- Nerve cells communicate directly with the cells they innervate by releasing short-range chemical messengers, known as neurotransmitters in response to action potentials.
- Hormones are long-range chemical messengers that are specifically secreted into the blood by endocrine glands in response to an appropriate signal.
- Neurohormones are hormones released into the blood by neurosecretory neurons.

*Extracellular chemical messengers bring about cell responses primarily by signal transduction.*
- Dispersed within the outer surface of the plasma membrane are specialized protein receptors that bind with the selected chemical messengers that come into contact with the cell.

*Some extracellular chemical messengers open chemically gated channels.*
- Membrane channels may be either leak channels or gated channels.
- There are two methods by which the desired intracellular responses are achieved using the binding of the protein receptor to the extracellular chemical messenger or first messenger: (1) by opening or closing channels through the membrane or (2) by transferring the signal to an intracellular chemical messenger or second messenger.
- If the first messenger opens or closes the sodium and potassium channels, the electrical activity of cells that generate electrical signals is altered.
- Another intracellular response occurs if the calcium channels are opened.
- Cytosolic calcium increases, triggering the release of secretion in many gland cells.
- Another method is to alter sodium and potassium channels which will set up an electrical impulse within the cell and thereby open the calcium channels.
- The release of chemicals from the nerve cells in response to a nerve impulse is such an example.
- A rise in cytosolic calcium can be brought about by release of calcium from intracellular stores within a modified endoplasmic reticulum in skeletal muscle cells.
- This increase in cytosolic calcium sets in motion events leading to a muscle contraction.

*Many extracellular chemical messengers activate second messenger pathways.*
- There are two second messenger pathways.
- The extracellular messenger binds to a protein receptor which causes the activation of an enzyme on the inner surface which in turn activates the intracellular messengers.
- The intracellular second messengers can be cyclic adenosine monophosphate (cyclic AMP) or calcium.
- Using the cyclic AMP pathway, the extracellular messenger uses the membrane-bound G protein to activate the enzyme adenylyl cyclase, which converts ATP to cyclic AMP by splitting off two phosphates.
- Cyclic AMP triggers the intracellular event by activating cyclic AMP-dependent protein kinase, which phosphorylates the specific intracellular protein.
- The altered enzyme performs its function thus completing the desired cellular response.
- Some cells use calcium as a second messenger.
- The first messenger binds on the outer surface and eventually phospholipase C is activated on the inner surface.
- Phospholipase C breaks down phospholipid tails in the membrane itself, which increases cytosolic calcium.
- The increase in calcium activates calmodulin which alters cellular proteins to bring about the desired cellular response.
- In a few cells cyclic guanosine monophosphate (cyclic GMP) serves as a second messenger in a system analogous to the cyclic AMP system.
- Many disease processes, such as myasthenia gravis and cholera, can be linked to malfunctioning receptors or defects in one of the components of the various pathways.
- Cascading can greatly multiply or amplify the initial signal.
- Receptors themselves can change with respect to affinity and number.
- Many disease processes can be linked to malfunctioning receptors

## Key Terms

action potential
adenyl cyclase
all-or-none law
apoptosis
axon
axon hillock
cascading effect
classical neurotransmitters
collaterals
convergence
current
depolarization
dense core vessicles
endocrinology
EPSP
excitable tissues
excitatory synapse
G-proteins
fast synapses
fire
grand postsynaptic potential
hormone response element
hormones
hydrophilic
hyperpolarization
inhibitory synapse
IPSP
lipophilic hormone
multiple sclerosis
myelin
myelinated fibers
necrosis
nerve fibers
nervous system
neuromodulators
neuropeptides
neurotransmitters
oligodendrocyte
overshoot
paracrines
Parkinson's disease
polarization
postsynaptic neuron
presynaptic facilitation
presynaptic inhibition
presynaptic neuron
protein kinase
refractory period
repolarization
regeneration tube
resistance
resting potential
saltatory conduction
Schwann cell
second messenger
signal transduction
spike
strychnine
synapse
synaptic cleft
synaptic delay
synaptic knob
synaptic vesicle
temporal summation
threshold potential

## Review Exercises

*Answers are in the appendix.*

**True/False**

_____ 1. The only way by which a neurotransmitter receptor combination can influence the postsynaptic cell is to directly alter its permeability to specific ions.

_____ 2. Synapses permit two-way transmission of signals between two neurons.

_____ 3. A given synapse may produce EPSPs at one time and IPSPs at another time.

_____ 4. A balance of IPSPs and EPSPs will negate each other so that the grand postsynaptic potential is essentially unaltered.

_____ 5. Neurons are specialized for rapid electrical and chemical signaling.

_____ 6. The greater the magnitude of a triggering event such as a stimulus, the smaller the graded potential.

_____ 7. Hyperpolarization means that the membrane returns to resting potential after having been depolarized.

_____ 8. On the device used for recording rapid changes in potential, a decrease in potential is represented as an upward deflection whereas an increase in potential is represented as a downward deflection.

_____ 9. The threshold potential is reached typically between-55mV and-75mV.

_____ 10. The terms action potential, spike, and firing all refer to the same phenomenon of rapid potential reversal.

_____ 11. Membrane channels are formed by proteins that span the thickness of the membrane.

_____ 12. Only about 1 out of 100,000 potassium ions present in the cell leaves during an action potential.

_____ 13. Once an action potential is initiated in one part of a nerve cell membrane, a self-perpetuating cycle is initiated so that the action potential is propagated throughout the rest of the fiber automatically.

_____ 14. Myelinated fibers conduct impulses about 80 times faster than nonmyelinated fibers of comparable size.

_____ 15. As a general rule, the most urgent types of information are transmitted via unmyelinated fibers, whereas the nervous pathways carrying less urgent information are myelinated.

_____ 16. Damaged neuronal fibers in the brain and spinal cord often regenerate.

_____ 17. The refractory period ensures the unidirectional propagation of the action potential down the axon from the initial site of activation.

_____ 18. The more complex the pathways, the more synaptic delays, and the longer the total reaction time.

_____ 19. Some neurotransmitters function through intracellular second messenger systems rather than by directly altering membrane permeability.

_____ 20. The highest threshold is present at the axon hillock because this region has an abundance of voltage-gated sodium channels.

_____ 21. The hormone CCK has been found in synaptic vesicles in diffuse areas of the brain.

_____ 22. Parkinson's disease is attributable to an excess of dopamine in a particular region of the brain involved in controlling complex movements.

_____ 23. There are an estimated 100 billion neurons in the brain alone.

**Fill in the Blank**

24. The constant membrane potential that exists when an excitable tissue cell is not displaying rapid changes in potential is referred to as the __________ __________ __________.

25. _______________ carry signals toward the cell body while ________ or ________ carry action potentials away from the cell body.

26. The intervening unmyelinated regions are known as _____________________.

27. Name three neuropeptides that start with the letter G: (1) _________________, (2) ________________, and (3) _____________________________________.

28. The site of synthesis in neurotransmitters is ____________________ and in neuromodulators is _______________________________.

29. _________________ means the membrane has potential; there is a separation of opposite charges.

30. A faster method of propagation, _____________________, takes place in myelinated fibers.

31. The space between the presynaptic and postsynaptic neurons is called the _________.

32. The synaptic delay is usually about ______________.

33. The postsynaptic neuron can be brought to threshold in two ways: (1) ____________________ and (2) ____________________.

34. There are three kinds of channels: (1) ____________________, (2) ____________________, and (3) ____________________.

35. The first portion of the axon plus the region of the cell body from which the axon leaves is known as the ______________.

36. These myelin-forming cells are ____________________ in the central nervous system and ______________ in the peripheral nervous system.

37. The ______________ form a(n) ____________________ to guide the regenerating nerve fiber to its proper destination.

38. Name three neurotransmitters that start with the letter G: (1) ____________________, (2) ____________________, and (3) ____________________.

39. Often an action potential is referred to as an __________.

40. The nucleus and organelles are housed in the ______________, from which numerous extensions known as ___________ typically project like antennae to increase the surface area available for receiving signals from other nerve cells.

41. Myelin is composed primarily of __________.

42. An excitable membrane either responds to a stimulus with a maximal action potential that spreads nondecrementally throughout the membrane, or it does not respond with an action potential at all. This is called the ____________________.

43. The portion of the postsynaptic membrane immediately underlying the synaptic knob is referred to as the ____________________.

44. ____________________ is a well known hormone released from the small intestine that causes the gallbladder to contract and release bile into the intestine.

45. ______________ refers to the branching of axon terminals so that a single cell synapses with many other cells.

**Matching**

_____ 46. Gastrin
_____ 47. Epinephrine
_____ 48. Histamine
_____ 49. Neurotensin
_____ 50. Insulin
_____ 51. Somatostatin
_____ 52. Glutamate
_____ 53. Glucagon
_____ 54. Oxytocin
_____ 55. Aspartate
_____ 56. Small, rapid-acting molecules
_____ 57. Large, slow-acting molecules
_____ 58. CCK

a. Neurotransmitters
b. Neuropeptides

**Multiple Choice**

59. Multiple sclerosis is a pathophysiological condition in which:
a. The release of the neurotransmitter is blocked.
b. The demyelination of nerve fibers occurs.
c. There is a deficiency of dopamine in a particular region of the brain.
d. The release of gamma-aminobutyric acid is prevented.
e. The viscosity of mucus increases.

60. Which of the following neurotransmitters is not a small, rapid-acting molecule?
a. insulin
b. dopamine
c. acetylcholine
d. epinephrine
e. serotonin

61. Which sequence best describes the events in an action potential?
a. polarization—hyperpolarization—depolarization—repolarization
b. hyperpolarization—repolarization—depolarization—polarization
c. polarization—hyperpolarization—repolarization—depolarization
d. polarization—depolarization—hyperpolarization—repolarization
e. depolarization—hyperpolarization—polarization—repolarization

62. Which of the following releases the neurotransmitters?
a. axon hillock
b. dendrite
c. synaptic vesicles
d. subsynaptic membrane
e. synaptic cleft

63. Which of the following are the myelin-forming cells in the central nervous system?
a. presynaptic neurons
b. postsynaptic cells
c. oligodendrocytes
d. Renshaw cells
e. Schwann cells

64. Which part of the neuron has the lowest threshold potential?
a. cell body
b. axon hillock
c. collaterals
d. dendrites
e. axon terminal

65. Which part of the neuron directs the signal toward the cell body?
a. nodes of Ranvier
b. axon hillock
c. collaterals
d. dendrites
e. axon

66. Which best describes a postsynaptic neuron? The neuron whose:
a. Action potentials are propagated toward the synapse.
b. Myelin is formed by oligodendrocytes.
c. Action potentials are propagated away from the synapse.
d. Axons are unmyelinated.
e. Dendrites are unmyelinated.

67. Which cells form a regeneration tube in peripheral nerves?
a. neurons
b. Schwann cells
c. oligodendrocytes
d. nodes of Ranvier
e. Golgi cells

68. Which new drug holds promise in the treatment of Parkinson's disease?
a. paraquat
b. meperidine
c. serotonin
d. neurotensin
e. selegiline

69. Which of the following is not a functionally distinct segment of a neuron?
a. trophic segment
b. initial segment
c. receptive segment
d. conductive segment
e. transmissive segment

70. Which of the following moves by way of an ATP dependent pump?
a. only sodium ions
b. only potassium ions
c. only chloride ions
d. both a and b above
e. all of the above

71. All of the following are a major group of neurotransmitters except one. Which one isn't a transmitter?
a. acetylcholine
b. amino acids
c. monoamines
d. neuropeptides
e. nucleic acids

72. A resting membrane potential along a neuron is maintained by:
a. diffusion
b. ion channels
c. the sodium-potassium pump
d. both a and b
e. both b and c

73. The depolarization of the neuronal membrane is produced by any factor that does one of the following:
a. increases the membrane's permeability to sodium ions
b. inhibits the membrane's permeability to sodium ions
c. stimulates the sodium-potassium pump
d. increases the membrane's permeability to potassium ions
e. decreases the membrane's permeability to calcium ion

74. A progressive degeneration of myelin sheaths in neurons of the brain and spinal cord is called:
a. senility
b. Huntington's chorea
c. Parkinson's disease
d. multiple sclerosis
e. myasthenia gravis

75. An acetylcholine blocking drug that acts on the postsynaptic membrane is:
a. GABA
b. homatropine
c. curare
d. propanonol
e. chlorpromazine

76. Which of the following describe the function of graded potentials?
a. are local changes in membrane potential that occur in varying degrees of magnitude
b. serve as short distance signals
c. serve as long-distance signals
d. both a and b are correct
e. both a and c are correct

77. The cells of excitable and nonexcitable tissue share which of the following properties?
a. a threshold potential
b. a resting membrane potential
c. an ability to open gated channels
d. all of the above (a-c) are correct
e. none of the above (a-c) are correct

78. Which of the following occurs during the rising phase of the nerve action potential?
a. The potassium potential is greater than the sodium potential.
b. The sodium potential is greater than the potassium potential.
c. Both sodium and potassium potentials are the same.
d. Sodium efflux occurs.
e. Both a and b are correct.

79. Which of the following occurs at the peak of an action potential?
a. The electrical gradient for potassium tends to move this ion outward.
b. The concentration gradient for potassium tends to move this ion outward.
c. The potassium permeability greatly increases.
d. Both a and b are correct.
e. All of the above are correct.

80. Which of the following is responsible for the falling phase of an action potential?
a. Sodium gates open
b. The sodium-potassium ATPase pump restores ions to their original location.
c. The increased permeability to sodium.
d. All of the above are correct.
e. None of the above (a-c) are correct.

81. When is an excitable membrane more permeable to potassium than sodium?
a. at a resting potential
b. during the rising phase of the action potential
c. during the falling phase of the action potential
d. both a and b are correct
e. both a and c are correct

82. Where does saltatory conduction occur?
a. in non myelinated nerve fibers
b. in myelinated nerve fibers
c. in all nerve fibers
d. in all excitable tissues
e. only in heart tissue

83. Which of the following statements concerning the propagation of an action potential is incorrect?
a. Saltatory conduction occurs in myelinated nerve fibers.
b. During conduction by local current flow, there is a flow of current between the active and adjacent inactive areas of the cell membrane.
c. The action potential jumps from one Schwann cell to an adjacent Schwann cell in a myelinated fiber.
d. Saltatory conduction is faster than conduction by local current flow.
e. Conduction by local current flow is the method of propagation in unmyelinated fibers.

84. If an action potential were initiated in the center of an axon, in which direction would it travel?
a. toward the cell body
b. away from the cell body
c. in both directions
d. in neither direction
e. none of the above

85. Which of the following is not a graded potential?
a. an action potential
b. an excitatory postsynaptic potential
c. an inhibitory postsynaptic potential
d. a grand postsynaptic potential
e. a summation potential

86. When does spatial summation occur in a postsynaptic neuron?
a. when several EPSPs from a single presynaptic input sum to reach threshold
b. when EPSPs from several presynaptic inputs sum to reach threshold
c. upon simultaneous interaction of an EPSP and a IPSP
d. when several IPSPs from a single presynaptic input sum to hyperpolarize the membrane
e. none of the above are correct

87. At an excitatory synapse:
a. There is increased permeability of the subsynaptic membrane to both sodium and potassium.
b. A small hyperpolarization occurs.
c. An action potential in the presynaptic neuron always causes an action potential in the postsynaptic neuron.
d. Two of the above are correct.
e. All of the above (a-c) are correct.

88. An IPSP:
a. is produced by increased sodium and potassium potentials
b. is produced by increased potassium or increased chloride potentials
c. is a small depolarization of the postsynaptic cell
d. both a and c are correct
e. both b and c are correct

89. If presynaptic neurons "X" and "Z" are stimulated simultaneously, what changes would you expect to occur in the postsynaptic neuron?
a. A single EPSP
b. A single IPSP
c. Temporal summation of EPSPs
d. Spatial summation of EPSPs
e. An IPSP and EPSP would cancel each other out, so there would be no change

90. If presynaptic neuron "Z" is repetitively stimulated very rapidly, what change would you expect to occur in the postsynaptic neuron?
a. a single EPSP.
b. a single IPSP.
c. spatial summation of EPSPs.
d. none of the above (a-c)
e. all of the above (a-c)

**Modified Multiple Choice**

*The following questions refer to comparative concentrations, permeabilities, and potentials under various circumstances. Indicate the relationship between the two items listed in each situation by filling in the appropriate letter in the blank using the following answer code:*

(a) = A is greater than B
(b) = B is greater than A
(c) = A and B are equal

91. _____ A. permeability of resting nerve cell membrane to potassium
B. permeability of a resting nerve cell membrane to sodium

92. _____ A. permeability of a nerve cell membrane to sodium during the rising phase of the action potential
B. permeability of the nerve cell membrane to potassium during the rising phase of the action potential

93. _____ A. permeability of a resting nerve cell membrane to potassium
B. permeability of a nerve cell membrane to sodium during the rising phase of an action potential

94. _____ A. permeability of a resting nerve cell membrane to potassium
B. permeability of a nerve cell membrane to potassium during the falling phase of an action potential

95. _____ A. permeability of a nerve cell membrane to sodium during the falling phase of an action potential
B. permeability of a nerve cell membrane to potassium during the falling phase of an action potential

96. _____ A. concentration of sodium in the intracellular fluid of a nerve cell immediately before an action potential
B. concentration of sodium in the intracellular fluid of a nerve cell immediately following an action potential

97. _____ A. concentration of potassium in the intracellular fluid of a nerve cell immediately before an action potential
B concentration of potassium in the intracellular fluid of a nerve cell immediately following an action potential

98. _____ A. concentration of sodium in the extracellular fluid
B. concentration of sodium in the intracellular fluid of a nerve cell immediately following an action potential

99. _____ A. membrane potential at rest
B. the potential of the same membrane when it is hyperpolarized

100. ____ A. membrane potential at rest
B. the potential of the same membrane when it is depolarized

**True/False**

_____ 101. There are three major types of intercellular communication.

_____ 102. Paracrines are local chemical messengers whose effect is exerted on neighboring cells in the immediate environment.

_____ 103. Neurohormones are hormones released into the blood by neurosecretory cells.

_____ 104. The term "signal transduction" refers to the process by which incoming signals are conveyed to the target cell's interior for execution.

_____ 105. Hydrophobic water soluble hormones include the peptides and catecholamines.

_____ 106. Hydrophilic hormones bind to receptors on the surface of the cell.

_____ 107. Lipophilic hormones bind to receptors on the surface of the cell.

_____ 108. Cyclic AMP is an important second messenger.

_____ 109. Nerves go to target cells whereas hormones are carried by the blood.

_____ 110. In general, the endocrine system coordinates rapid responses.

## *Points to Ponder*

1. Which ion (sodium or potassium) do you think is more important in neuronal physiology?

2. If yelling and grunting enhance an athlete's performance, what do you think happens when the game official tells them to stop yelling and grunting?

3. How do humans perceive differences in the intensity of external stimuli?

4. What is the physiological importance of inhibition?

5. In common usage how does the definition of a nerve differ from its physiological definition?

6. What are five basic properties of neurons?

7. Why are some basic principles of electricity needed to understand nerve physiology?

8. How many synapses does a neuron have?

9. How do pre- and post-synaptic neurons differ?

10. What is a neuromodulator?

## Clinical Perspectives

1. Why are people with multiple sclerosis likely to lose their ability to control skeletal muscles?

2. Suppose that your doctor is performing neurological tests on your arm using tiny needle electrodes. In doing so he places an electrode adjacent to an axon halfway between the cell body and the axon terminal. When the electrical stimulus is applied, which way does the signal travel in the axon?

3. Nerve gas exerts its odious effects by inhibiting acetylcholine esterase in skeletal muscles. Since acetylcholine is not degraded, what will the results be in this clinical situation?

4. Muscle weakness in the disease myasthenia gravis is due to the fact the acetylcholine receptors are blocked and destroyed by antibodies secreted by the immune system of the affected person. What clinical results will become apparent in the affected person?

5. Patient A has clinical depression. After eating a meal of shellfish, she falls out of her chair and experiences difficulty in breathing. The paramedics find a prescription for a monoamine oxidase inhibitor in her pocket. What may have happened to patient A?

6. Monoamine oxidase (MAO) is an enzyme in the endings of presynaptic axons that breaks down catecholamines and seotonin after they have been taken up from the synaptic cleft. For what clinical situations would you use MAO inhibitors?

## Experiment of the Day

Demonstrate the following using a set of dominos.

a. collateral axon
b. convergence
c. divergence
d. threshold

## PhysioEdge Activities

**Related to Text:**
Tutorial - Neuronal Physiology and Hormonal Communication.
Media Exercise 4.1: Basis of a Neuron.
Media Exercise 4.2: Graded Potentials and Action Potentials.
Media Exercise 4.3: Synapses and Neuronal Integration.
Media Exercise 3.3: Signaling at Cell Membranes and Membrane Potential.

**Related to Figures:**
*Figure 4.3.* For an interaction related to this figure, see Media Exercise 4.2: Graded Potentials and Action Potentials.
*Figure 4.5.* For an animation related to this figure, click the Action Potential tab in the Neuronal Physiology and Hormonal Communication tutorial.
*Figure 4.9.* For an animation related to this figure, click the Action Potential tab in the Neuronal Physiology and Hormonal Communication tutorial.
*Figure 4.10.* For an animation of this figure, click on the Conduction of the Action Potential tab in the Neuronal Physiology and Hormonal Communication tutorial. Also review Media Exercise 4.1: Basics of a Neuron.
*Figure 4.11.* For an animation of this figure, click on the Conduction of the Action Potential tab in the Neuronal Physiology and Hormonal Communication tutorial.
*Figure 4.16.* For an animation of this figure, click on the Synapses tab (p.2) in the Neuronal Physiology and Hormonal tutorial.
*Figure 4.17.* For an animation of this figure, click on the Synapses tab (p.4) in the Neuronal Physiology and Hormonal tutorial.

*Figure 4.18.* For an animation of this figure, click on the Synapses and Neural Integration tab in the Neuronal Physiology and Hormonal Communication tutorial.

## Media Resources

**PhysioEdge CD-ROM**
For a visual review of concepts in this chapter, check out:

- Tutorial: Neural and Hormonal Communication
- Media Exercise 3.3: Signaling at Cell Membranes and Membrane Potential
- Media Exercise 4.1: Basics of a Neuron
- Media Exercise 4.2: Graded Potentials and Action Potentials.
- Media Exercise 4.3: Synapses and Neural Integration

**Book Companion Website**
The website for this book contains a wealth of helpful study aids, as well as many ideas for further reading and research. Log on to:

**http://www.brookscole.com/sherwoodhp6**

Select Chapter 4 from the drop-down menu, or click on one of the many resource areas, including **Case Histories**, which introduce clinical aspects of human physiology. For this chapter check out: Case History 15: A Stiff Baby, and Case History 16: And a Limp Baby.

For Suggested Readings, consult **InfoTrac® College Edition**, your online research library, at:

**http://infotrac.thomsonlearning.com**

# 5

# The Central Nervous System

## Chapter Overview

The nervous system is one of the two major regulatory systems of the body, the other being the endocrine system. In order to maintain homeostasis, the body, with its many cells, tissues, organs, and systems must be completely aware of any and all changes in the internal and external environment. The nervous system provides this information using affectors and afferent neurons. The central nervous system, utilizing various components of the brain, processes this information and initiates a response. Rapid responses are performed by the nervous system through efferent neurons and effectors. The fastest response is the reflex. The endocrine system is used when the response requires duration. To perform these essential functions the brain is divided, and subdivided, and these divisions are interconnected and integrated in such a manner that all parts of the body are controlled. Thus, the brain controls the responses through coordination of the various body systems. In this perspective the nervous system controls homeostasis to preserve the cells of the body.

## Chapter Outline

INTRODUCTION

- Many of the basic life-supporting neuronal patterns, such as those controlling respiration and circulation, are similar in all individuals.
- Some differences in the nervous systems between individuals are genetically endowed.
- Other differences, however, are due to environmental encounters and experiences.
- Depending on external stimuli and the extent these pathways are used, some are retained, firmly established, and even enhanced, whereas others are eliminated.

ORGANIZATION OF THE NERVOUS SYSTEM

*The nervous system is organized into the central nervous system and the peripheral nervous system.*

- The nervous system is organized into the central nervous system (CNS), consisting of the brain and spinal cord, and the peripheral nervous system (PNS), consisting of nerve fibers that carry information between the CNS and other parts of the body.
- The afferent division of the PNS carries information to the CNS.
- Instructions from the CNS are transmitted via the efferent division to effector organs (muscles or glands).
- The efferent division is divided into the somatic nervous system, which consists of motor neurons that supply the skeletal muscles and the autonomic nervous system fibers which innervate smooth muscle, cardiac muscle, and glands.
- The latter system is subdivided into the sympathetic nervous system and the parasympathetic nervous system, both of which innervate most of the organs supplied by the autonomic nervous system.

*The three classes of neurons are afferent neurons, efferent neurons, and interneurons.*

- Three classes of neurons make up the nervous system: afferent neurons, efferent neurons, and interneurons.
- At its peripheral ending, an afferent neuron has a sensory receptor that generates action potentials in response to a particular type of stimulus.
- The afferent neuron cell body is located adjacent to the spinal cord.
- Long peripheral axon extends from the receptor to the cell body, and a short central axon passes from the cell body into the spinal cord.
- The cell bodies of efferent neurons originate in the CNS.
- Efferent axons leave the CNS to course their way to the effector organs they innervate.
- Interneurons lie entirely within the CNS.
- They serve two main roles: (1) Interneurons lie between the afferent and efferent neurons and are important in the integration of peripheral responses to peripheral information. (2) Interconnections between interneurons themselves

are responsible for abstract phenomena associated with the "mind."

## PROTECTION AND NOURISHMENT OF THE BRAIN

*Glial cells support the interneurons physically, metabolically and functionally.*

- About 90 percent of the cells within the CNS are not neurons but glial cells or neuroglia.
- Glial cells do not initiate or conduct nerve impulses.
- The glial cells serve as connective tissue of the CNS and as such help support the neurons both physically and metabolically.
- The four major types of glial cells are astrocytes, oligodendrocytes, ependymal cells, and microglia.

*The delicate central nervous system is well protected.*

- Astrocytes provide a number of critical functions: (1) As the main "glue" they hold the neurons together in proper spatial relationships. (2) Astrocytes serve as a scaffold to guide neurons to their proper final destination during fetal brain development. (3) Astrocytes induce the small blood vessels of the brain to undergo the anatomical and functional changes that are responsible for the establishment of the blood - brain barrier. (4) They are important in the repair of brain injuries. (5) Astrocytes play a role in neurotransmitter activity. (6) They take up excess potassium ions from the ECF. (7) Finally, astrocytes have receptors for the same neurotransmitters as do neurons.
- Oligodendrocytes form the insulative myelin sheaths around axons in the CNS.
- Ependymal cells line the internal cavities of the CNS.
- The microglia are the phagocytic scavengers of the CNS.
- Unlike neurons, glial cells do not lose the ability to undergo cell division.

*Three meningeal membranes wrap, protect, and nourish the central nervous system.*

- Four major features help protect the CNS: (1) The CNS is enclosed in hard bony structures. (2) Three protective and nourishing membranes, the meninges, lie between the bony covering and the nervous tissue. (3) The brain floats in a special cushioning fluid, the cerebrospinal fluid (CSF). (4) A highly selective blood-brain barrier limits access of blood-borne materials into the brain tissue.
- Three meningeal "mothers" provide protection and nourishment for the central nervous system.
- The outermost layer, the dura mater, is a tough inelastic covering.
- The arachnoid mater is a delicate, richly vascularized layer with a "cobwebby" appearance.
- The innermost meningeal layer, the pia mater, is the most fragile.
- It is highly vascular and closely adheres to the surfaces of the brain and spinal cord.

*The brain floats in its own cerebrospinal fluid.*

- The cerebrospinal fluid is formed as a result of selective transport mechanisms across the membranes of the choroid plexuses.
- The major function of CSF is to serve as a shock-absorbing fluid to prevent the brain from bumping against the interior of the hard skull.
- The CSF performs another important role related to the exchange of materials between the body fluids and the brain.

*A highly selective blood-brain barrier carefully regulates exchanges between the blood and brain.*

- The brain is carefully shielded from harmful changes in the blood by the blood-brain barrier.
- Unlike the rather free exchange across capillaries elsewhere in the body, there are strict limitations on permissible exchanges across brain capillaries.
- The blood-brain barrier consists of both anatomical and physiological factors.
- In brain capillaries the cells are joined by tight junctions, which completely seal the capillary wall.
- The only possible exchanges are through the capillary cells themselves.
- Certain areas of the brain are not subject to the blood-brain barrier, most notably a portion of the hypothalamus.
- Functioning of the hypothalamus depends on its "sampling" the blood and adjusting its controlling output accordingly in order to maintain homeostasis.

*The brain depends on constant delivery of oxygen and glucose by the blood.*

- The brain cannot produce ATP in the absence of oxygen.
- Unlike other tissues, the brain uses only glucose in energy production but does not store any of this nutrient.
- Therefore, the brain is absolutely dependent on a continuous, adequate blood supply of oxygen and glucose.
- Brain damage may occur in spite of protective mechanisms.
- Direct brain damage occurs if the brain is violently shaken or jarred by a forceful impact.

- Further indirect brain damage may occur following a head injury as a consequence of swelling or hemorrhaging within the enclosed confines of the cranium.
- The most common cause of brain damage is not traumatic head injuries, however, but cerebrovascular accidents (strokes).
- When a brain blood vessel ruptures or is blocked by a clot, the brain tissue being supplied by that vessel is deprived of its oxygen and glucose.
- Brain damage may also occur as a result of infectious or degenerative neural disorders or brain tumors.
- The nature of the ensuing loss of neurological function depends on the area of the brain involved and the extent of permanent damage.

## OVERVIEW OF THE CENTRAL NERVOUS SYSTEM

- Newer, more sophisticated regions of the brain are piled on top of older, more primitive regions.
- The brain stem, the oldest and smallest region of the brain, is continuous with the spinal cord.
- It controls many of the life-sustaining activities, such as breathing, circulation, and digestion.
- Attached at the top rear portion of the brain stem is the cerebellum, which is concerned with maintaining proper position of the body in space and subconscious coordination of motor activity.
- On top of the brain stem, tucked within the interior of the cerebrum, is the diencephalon.
- It houses the hypothalamus, which controls many homeostasis functions, and the thalamus, which performs some primitive sensory processing.
- On top of this "cone" of lower brain regions is the cerebrum.
- The outer layer of the cerebrum is the cerebral cortex, which caps an inner core that houses the basal nuclei.

## CEREBRAL CORTEX

*The cerebral cortex is an outer shell of gray matter covering an inner core of white matter.*

- The cerebral cortex plays a key role in the most sophisticated neural functions, such as voluntary initiation of movement, final sensory perception, conscious thought, language, personality traits, and other factors we associate with the mind or intellect.
- The cerebrum is divided into two halves, the right and left cerebral hemispheres.
- They are connected to each other by the corpus callosum.
- Each hemisphere is composed of a thin outer shell of gray matter, the cerebral cortex, covering a thick central core of white matter.
- Located deep within the white matter is another region of gray matter, the basal nuclei.
- Gray matter consists predominately of densely packaged cell bodies and their dendrites as well as glial cells.
- Bundles or tracts of myelinated nerve fibers constitute the white matter.
- The fiber tracts in the white matter transmit signals from one part of the cerebral cortex to another or between the cortex and other regions of the CNS.
- Such communication enables integration between different areas of the cortex and elsewhere.

*The cerebral cortex is organized into layers and functional columns.*

- The cerebral cortex is organized into six well-defined layers.
- These layers are organized into functional vertical columns that extend perpendicularly from the surface down through the depths of the cortex to the underlying white matter.
- The functional differences between various areas of the cortex result from different layering patterns within the columns and from different input-output connections.

*The four pairs of lobes in the cerebral cortex are specialized for different activities.*

- Each half of the cortex is divided into four major lobes: the occipital, temporal, parietal, and frontal lobes.
- The occipital lobes, which are located posteriorly are responsible for initially processing visual input.
- Sound sensation is initially received by the temporal lobes, located laterally.
- The parietal lobes lie to the rear of the central sulcus on each side.

*The parietal lobes are responsible for somatosensory processing.*

- The parietal lobes are primarily responsible for receiving and processing sensory input such as touch, pressure, heat, cold, and pain from the surface of the body.
- These sensations are collectively known as somesthetic sensations.
- The parietal lobes also perceive awareness of body position, a phenomenon referred to as proprioception.
- The somatosensory cortex, the site for initial cortical processing of this somesthetic and

proprioceptive input, is located at the front of each parietal lobe immediately behind the central sulcus.

*The primary motor cortex is located in the frontal lobes.*

- The frontal lobes, lying at the front of the cortex, are responsible for three main functions: (1) voluntary motor activity, (2) speaking ability, and (3) elaboration of thought.
- The area at the rear of the frontal lobe immediately in front of the central sulcus and adjacent to the somatosensory cortex is the primary motor cortex.
- As in sensory processing, the motor cortex on each side of the brain primarily controls muscles on the opposite side of the body.

*Other regions of the nervous system besides the primary motor cortex are important in motor control.*

- Lower brain regions and the spinal cord control involuntary skeletal muscle activity, such as the maintenance of posture.
- Although fibers originating from the motor cortex can activate motor neurons to bring about the muscle contraction, the motor cortex itself does not initiate voluntary movement.
- The higher motor areas of the brain believed to be involved in the voluntary decision-making period include the supplementary motor area, the premotor cortex, and the posterior parietal cortex.
- These higher areas all command the primary motor cortex.
- Furthermore, the cerebellum appears to play an important role in the planning, initiation, and timing of certain kinds of movement by sending input to the motor areas of the cortex.
- The supplementary motor area lies on the medial surface of each hemisphere anterior to the primary motor cortex.
- Stimulation of various regions of this motor area brings about complex patterns of movement, such as the opening or closing of the hand.
- The premotor cortex located on the lateral surface of each hemisphere in front of the primary motor cortex, is believed to be important in orienting the body and arms toward a specific target.
- The premotor cortex is guided by sensory input processed by the posterior parietal cortex, a region that lies posterior to the primary somatosensory cortex.

*Somatotropic maps vary slightly between individuals and are dynamic, not static.*

- The general pattern of sensory and motor somatotropic maps is governed by genetic and developmental processes, but the individual cortical architecture appears capable of being influenced by use-dependent competition for cortical space.

*Because of its plasticity, the brain can be remodeled in response to varying demands.*

- The brain displays a degree of plasticity, that is, an ability to change or be functionally remodeled in response to the demands placed on it.

*Different regions of the cortex control different aspects of language.*

- The areas of the brain responsible for language ability are found in only one hemisphere, the left hemisphere in the vast majority of the population.
- The primary areas of cortical specialization for language are Broca's and Wernicke's. Broca's area, which is responsible for speaking ability, is located in the left frontal lobe in close association with the motor areas of the cortex that control the muscles necessary for articulation.
- Wernicke's area, located in the left cortex at the juncture of the parietal, temporal, and occipital lobes, is concerned with language comprehension.

*The association areas of the cortex are involved in many higher functions.*

- The prefrontal association cortex is the front portion of the frontal lobe just anterior to the premotor cortex.
- The roles attributed to this region are: (1) planning for voluntary activity, (2) weighing consequences of future actions and choosing between different options for various social or physical situations, and (3) personality traits.
- The parietal-temporal-occipital association cortex is found at the interface of the three lobes for which it is named.
- In this location, this cortical area pools and integrates somatic, auditory, and visual sensations projected from these three lobes for complex perceptual processing.
- This region is also involved in the language pathway connecting Wernicke's area to the visual and auditory cortices.
- The limbic association is located on the bottom and adjoining inner portion of each temporal lobe.
- This area is concerned primarily with motivation and emotion and is extensively involved in memory.

*The cerebral hemispheres have some degree of specialization.*

- The left cerebral hemisphere excels in performance of logical, analytical, sequential, and

verbal tasks, such as math, language forms, and philosophy.
- The right cerebral hemisphere excels in nonlanguage skills, especially spatial perception and artistic and musical endeavors.

*An electroencephalogram is a record of postsynaptic activity in cortical neurons.*
- The electroencephalogram (EEG) has three major uses: (1) It is often used as a clinical tool in the diagnosis of cerebral dysfunction. (2) The EEG is also used to distinguish various stages of sleep. (3) The EEG finds further use in the legal determination of brain death.

*Neurons in different regions of the cortex may fire in rhythmic synchrony.*
- Neural information is coded by changes in the frequency of action potentials.
- Neurons within an assembly that fire together may be widely scattered.
- The brain may use different strategies for encoding information in different contexts.

## BASAL NUCLEI, THALAMUS, AND HYPOTHALAMUS

*The basal nuclei play an important inhibitory role in motor control.*
- The subcortical regions include the basal nuclei, located in the cerebrum, and the thalamus and hypothalamus, located in the diencephalon.
- The basal nuclei play a complex role in the control of movement in addition to having nonmotor functions that are less understood.
- The basal nuclei are important in: (1) inhibiting muscle tone throughout the body; (2) selecting and maintaining purposeful motor activity while suppressing useless or unwanted patterns of movement; and (3) helping monitor and coordinate slow sustained contractions, especially those related to posture and support.

*The thalamus is a sensory relay station and is important in motor control.*
- It is speculated that the thalamus positively reinforces voluntary motor behavior initiated by the cortex, whereas the basal nuclei modulate this activity by exerting an inhibitory effect on the thalamus to eliminate antagonistic or unnecessary movements.
- The basal nuclei also exert an inhibitory effect on motor activity by acting through the neurons in the brain stem.
- The thalamus is a sensory relay station and is important in motor control.
- The diencephalon consists of two main parts, the thalamus and the hypothalamus.
- The thalamus serves as a "relay station" and synaptic integrating center for preliminary processing of all sensory input on its way to the cortex.
- The thalamus, along with the brain stem and cortical association areas, is important in our ability to direct attention to stimuli of interest.

*The hypothalamus regulates many homeostatic functions.*
- The hypothalamus is the area of the brain most notably involved in the direct regulation of the internal environment.
- The hypothalamus: (1) controls body temperature; (2) controls thirst and urine output; (3) controls food intake; (4) controls anterior pituitary hormone secretion; (5) produces posterior pituitary hormones; (6) controls uterine contractions and milk ejection; (7) serves as a major autonomic nervous system coordinating center, which in turn affects all smooth muscle, cardiac muscle, and exocrine glands; and (8) plays a role in emotional and behavioral patterns.

## THE LIMBIC SYSTEM AND ITS FUNCTIONAL RELATIONS WITH THE HIGHER CORTEX

*The limbic system plays a key role in emotion and behavior.*
- The limbic system includes portions of each of the following: the lobes of the cerebral cortex, the basal nuclei, the thalamus, and the hypothalamus.

*Basic behavioral patterns.*
- This complex interacting network is associated with emotions, basic survival, and sociosexual behavioral patterns, motivation, and learning.
- Homeostasis drives represent the subjective urges associated with specific bodily needs that motivate appropriate behavior to satisfy those needs.

*Norepinephrine, dopamine, and serotonin are neurotransmitters in pathways for emotions and behavior.*
- The underlying neurophysiological mechanisms responsible for the psychological observations of motivated behavior and emotions largely remain a mystery, although the neurotransmitters norepinephrine, dopamine, and serotonin all have been implicated.
- Norepinephrine and dopamine, both chemically classified as catecholamines, are known to be transmitters in the regions that elicit the highest

rates of self-stimulation in animals equipped with do-it-yourself devices.

- Schizophrenia, a mental disorder characterized by delusions and hallucinations, probably results from excess dopamine transmissions.
- Serotonin and norepinephrine are synaptic messengers in the regions of the brain involved in pleasure and motivation.
- A functional deficiency of serotonin or norepinephrine or both is implicated in depression, a disorder characterized by a pervasive unpleasant mood accompanied by a general loss of interest and ability to experience pleasure.

*Learning is the acquisition of knowledge as a result of experiences.*

- Learning is the acquisition of knowledge or skills as a consequence of experience, instruction, or both.
- Learning is a change in behavior that occurs as a result of experiences.

*Memory is laid down in stages.*

- Memory is the storage of acquired knowledge for later recall.
- Learning and memory form the basis by which individuals adapt their behavior to their particular external circumstances.
- The neural change responsible for retention or storage of knowledge is known as a memory trace.
- Concepts, not verbatim information, are generally stored.
- Short-term memory lasts for seconds to hours, whereas long-term memory is retained for days to years.

*Memory traces are present in multiple regions of the brain.*

- The hippocampus, the elongated, medial portion of the temporal lobe that is part of the limbic system, plays a vital role in short-term memory involving the integration of various related stimuli and is also crucial for consolidation into long-term memory. The hippocampus and the surrounding regions play an important role in declarative memories.
- Alzheimer's disease accounts for about two-thirds of the cases of senile dementia, which refers to a generalized age-related diminution of mental abilities. Alzheimer's can only be confirmed at autopsy upon finding the characteristic brain lesions associated with the disease: neuritic plaques and neurofibrillary tangles dispersed throughout the cerebral cortex, and degeneration of cell bodies of certain neurons in the basal brain.
- Three recent discoveries have offered hope for diagnosing Alzheimer's: new brain imaging techniques, a possible skin test for defective potassium channels, and a simple eye-drop test.
- The cerebellum seems to play an essential role in procedural memories involving motor skills gained through repetitive training.

*Short-term memory involves transient changes in synaptic activity.*

- Short-term memory involves transient modifications in the function of preexisting synapses, such as a temporary alteration in the amount of neurotransmitter released in response to stimulation within affected nerve pathways.

*Long-term memory involves formation of new, permanent synaptic connections.*

- Long-term memory involves relatively permanent functional or structural changes between existing neurons in the brain.
- Two forms of short-term memory, habituation and sensitization, are due to modification of different channel proteins in presynaptic terminals of specific afferent neurons.
- This modification in turn brings about changes in transmitter release.
- Habituation is a decreased responsiveness to repetitive presentations of an indifferent stimulus.
- Sensitization refers to increased responsiveness to mild stimuli following a strong or noxious stimulus.
- In habituation, the closing of calcium ion channels reduces calcium entry into the presynaptic terminal, which leads to a decrease in neurotransmitter release.
- The postsynaptic potential is reduced compared to normal, resulting in a decrease or absence of the behavioral response controlled by the postsynaptic efferent neuron.
- In contrast to habituation, calcium entry into the presynaptic terminal is enhanced in sensitization.
- The subsequent increase in neurotransmitter release produces a larger postsynaptic potential.
- Existing synaptic pathways may be functionally interrupted (habituated) or enhanced (sensitized) during simple learning.
- Initial storage of declarative information appears to be accomplished by means of long-term potentiation (LTP), which refers to an increase in strength of existing synaptic connections in the involved pathways following brief periods of stimulation.
- LTP has been shown to last long enough for this short-term memory to be consolidated into more permanent long-term memory.

- LTP is especially prevalent in the hippocampus.
- The development of LTP involves changes in both the presynaptic and postsynaptic neurons at a given synapse.
- The postsynaptic neuron releases a retrograde factor, which diffuses to the presynaptic neuron.
- The retrograde factor activates a second messenger system in the presynaptic neuron to enhance transmitter release from this neuron.
- This enhanced transmitter release maintains LTP.
- The retrograde messenger is believed to be nitric oxide.
- Long-term memory storage appears to involve rather permanent physical changes in the brain.
- Examples of such alterations include the formation of new synaptic connections, permanent changes in presynaptic or postsynaptic membranes, or an increase or decrease in transmitter synthesis.
- Long-term memory may be stored at least in part by a particular pattern of dendritic branching and synaptic contacts.
- Drugs that block protein synthesis have been shown to interfere with long-term retention of learned behavior.
- Serving as molecular switch, CREB is believed to activate genes important in long-term memory storage.
- Immediate early genes (IEGs) have been shown to be activated in response to brief bursts of action potential activity and may be the critical link between short-term memory and the synthesis of proteins that encode long-term memory.
- Numerous hormones and neuropeptides are known to affect learning and memory processes.

### Cerebellum

*The cerebellum is important in balance and in planning and execution of voluntary movement.*

- The cerebellum, which is attached to the back of the upper portion of the brain stem, lies underneath the occipital lobe of the cortex.
- The cerebellum consists of three functionally distinct parts: the vestibulocerebellum, the spinocerebellum, and the cerebrocerebellum.
- The vestibulocerebellum is important for the maintenance of balance and controls eye movement.
- The spinocerebellum enhances muscle tone and coordinates skilled, voluntary movements.
- The cerebrocerebellum plays a role in the planning and initiation of voluntary activity by providing input to the cortical motor areas.
- This is also the cerebellar region, involved in procedural memories.
- The cerebellum and basal nuclei both monitor and adjust motor activity commanded from the motor cortex.
- The cerebellum does not have any direct influence on the efferent motor neurons.
- The cerebellum helps maintain balance; smoothes out fast, phasic motor activity; and enhances muscle tone.

### Brain Stem

*The brain stem is a vital link between the spinal cord and higher brain regions.*

- The brain stem consists of the medulla, pons, and midbrain.
- All incoming and outgoing fibers traversing between the periphery and higher brain centers must pass through the brain stem, with incoming fibers relaying sensory information to the brain and outgoing fibers carrying command signals from the brain.
- The functions of the brain stem include the following:
  1. Except for the vagus, the cranial nerves arise from the brain stem and supply structures in the head and neck with both sensory and motor fibers. Most of the branches of the vagus supply organs in the thoracic and abdominal cavities.
  2. Collected within the brain stem are neuronal clusters that control heart and blood vessel functions, respiration, and many digestive activities.
  3. The brain stem also plays a role in modulating the sense of pain.
  4. The brain stem plays a role in the regulation of muscle reflexes involved with equilibrium and posture.
  5. The reticular formation, found in the brain stem, receives and integrates all synaptic input. The reticular activating system (RAS) controls the overall degree of cortical alertness and is important in the ability to direct attention.
  6. The centers responsible for sleep are also housed within the brain stem.

*Sleep is an active process consisting of alternating periods of slow-wave and paradoxical sleep.*

- Consciousness refers to subjective awareness of the external world and self.
- Even though the final level of awareness resides in the cerebral cortex and a crude sense of awareness is detected by the thalamus, conscious experience depends upon the integrated functioning of many parts of the nervous system.

- The following states of consciousness are listed in decreasing order of level of arousal: maximum alertness, wakefulness, sleep, and coma.
- Sleep is an active process.
- There are two types of sleep, characterized by different EEG patterns and different behaviors: slow-wave sleep and paradoxical, or REM, sleep.
- Slow-wave sleep occurs in four stages, each displaying progressively slower EEG waves of higher amplitude.
- In slow-wave sleep, the person still has considerable muscle tone and frequently shifts body position.
- The muscles are completely relaxed in paradoxical sleep with no movement taking place.
- Paradoxical sleep is further characterized by rapid eye movement (REM).
- The EEG pattern during paradoxical sleep is similar to that of a wide-awake, alert individual.
- A person cyclically alternates between the two types of sleep throughout the night.

*The sleep-wake cycle is probably controlled by interactions among three neural systems.*

- The sleep-wake cycle, as well as the various stages of sleep, is believed to be due to the cyclical interplay of three different neural systems in the brain stem: (1) an arousal system, (2) a slow-wave sleep center, and (3) a paradoxical sleep center.
- Some scientists speculate that paradoxical sleep is necessary to allow the brain to "shift gears" to accomplish the long-term structural and chemical adjustments necessary for learning and memory.
- A specific amount of paradoxical sleep appears to be required.

*The function of sleep is unclear.*

- Even though humans spend about a third of their lives sleeping, the reasons sleep is needed largely remains a mystery.

## Spinal Cord

*The spinal cord extends through the vertebral canal and is connected to the spinal nerves.*

- Exiting through the large hole in the base of the skull, the spinal cord is enclosed by the protective vertebral column as it descends through the vertebral canal.
- Paired spinal nerves emerge from the spinal cord through spaces formed between the bony, winglike arches of adjacent vertebrae.
- There are eight pairs of cervical nerves, twelve thoracic nerves, five lumbar nerves, five sacral nerves, and one coccygeal nerve.
- The spinal cord extends only to the level of the first or second lumbar vertebra.
- The remaining nerves are greatly elongated in order to exit the vertebral column at their appropriate space.

*The white matter of the spinal cord is organized into tracts.*

- Tracts are bundles of nerve fibers with similar functions.
- The gray matter in the spinal cord forms a butterfly-shaped region on the inside and is surrounded by the outer white matter.
- The gray matter of the cord consists primarily of neuronal cell bodies and their dendrites, short interneurons, and glial cells.
- The white matter is organized into tracts.
- Each of these tracts begins or ends within a particular area of the brain.
- Some are ascending tracts that transmit to the brain signals derived from afferent input.

*Spinal nerves carry both afferent and efferent fibers*

- Others are descending tracts that relay messages from the brain to the efferent neurons.
- Each half of the gray matter is arbitrarily divided into a dorsal horn, a ventral horn, and a lateral horn.
- Afferent fibers carrying incoming signals enter the spinal cord through the dorsal root.
- Efferent fibers carrying outgoing signals leave the spinal cord through the ventral root.
- The cell bodies of the efferent neurons are in the gray matter of the cord.
- The dorsal and ventral roots at each level join to form a spinal nerve.
- The spinal nerves and the cranial nerves constitute the peripheral nervous system.

*The spinal cord is responsible for the integration of many basic reflexes.*

- The neural pathway involved in accomplishing reflex activity is known as a reflex arc, which typically includes five basic components: a receptor, an afferent pathway, an integrating center, an efferent pathway, and an effector.
- There are several types of reflexes involving the reflex arc.
- Pathways for unconscious responsiveness digress from the typical reflex arc in two general ways: responses mediated at least in part by hormones and local responses that do not involve either nerves or hormones.

## Key Terms

afferent division
afferent neurons
amnesia
amygdale
aphasia
astrocytes
autonomic nervous system
behavior patterns
blood-brain barrier
brain
brainstem
Broca's area
cauda equine
central nervous system (CNS)
cerebellum
cerebral spinal fluid (CSF)
cerebrum
cognition
consolidation
corpus callosum
depression
diencephalons
dorsal horn
dorsal root
dorsal root ganglion
dyslexia
effector organs
efferent division
efferent neuron
electroencephalogram (EEG)
emotion
ependymal cells
epilepsy
frontal lobes
glial cells
gray matter
hippocampus
homeostatic drives
hypothalamus
interneuron
language
learning
limbic system
long-term memory
medulla
memory
meninges
microglia
midbrain
motor program
nerve
neuroendocrinology
neuroglia
occipital lobes
oligodendrocytes
parasympathetic nervous system
parietal lobes
Parkinson's disease
peripheral nervous system
pons
proprioception
reflex
reflex arc
sleep
sleep attacks
sleep-wake cycle
somatic
somatic nervous system
somatosensory cortex
somethetic sensations
spinal reflexes
spinocerebellum
temporal lobes
thalamus
vestibulocerebellum
Wernicke's area
white matter
withdrawal reflexes
working memory

## Review Exercises

*Answers are in the appendix*

**True/False**

_____ 1. The brain stem, the oldest and largest region of the brain, is continuous with the spinal cord.

_____ 2. The stellate cells send fibers down the spinal cord from the cortex to terminate on the efferent motor neurons that innervate the skeletal muscles.

_____ 3. Simple awareness of touch, pressure, or temperature is detected by the thalamus.

_____ 4. The motor cortex on each side of the brain primarily controls muscles on the same side of the body.

_____ 5. The cerebellum appears to play an important role in the planning, initiation, and timing of certain kinds of movement by sending input to the motor areas of the cortex.

_____ 6. The areas of the brain responsible for language ability are found in only one hemisphere—the right hemisphere.

_____ 7. Epileptic seizures occur when a large collection of neurons abnormally undergo synchronous action potentials that produce stereotypical, involuntary spasms and alterations in behavior.

_____ 8. The central nervous system (CNS) consists of the brain and the spinal cord.

_____ 9. The efferent division carries information to the CNS, apprising it of the external environment and providing status reports on internal activities being regulated by the nervous system.

_____ 10. Afferent neurons are shaped differently than efferent neurons and interneurons.

_____ 11. The efferent neuron cell body is located adjacent to the spinal cord.

_____ 12. The cell bodies of afferent neurons originate in the CNS.

_____ 13. Interneurons lie between the afferent and efferent neurons.

_____ 14. Interconnections between interneurons are responsible for the abstract phenomena associated with the mind.

_____ 15. The spinocerebellum essentially acts as "middle management," comparing the "intentions" or "orders" of the higher centers with the "performance" of the muscles and then correcting any "errors" or deviations from the intended movement.

_____ 16. The cerebellum does not have any direct influence on the efferent motor neurons.

_____ 17. The cerebellum plays a role in the planning and initiation of voluntary activity by providing input to the cortical motor areas.

_____ 18. The vestibulocerebellum helps maintain balance; smoothes out fast, phasic motor activity; and enhances muscle tone.

_____ 19. About 90 percent of the cells within the CNS are neurons.

_____ 20. Glial cells do not initiate or conduct nerve impulses.

_____ 21. Astrocytes form myelin sheaths in CNS.

_____ 22. Ependymal cells line the internal cavities of brain and spinal cord.

_____ 23. The ependymal cell lining of the ventricles contributes to the formation of cerebrospinal fluid.

_____ 24. Cerebrospinal fluid is formed by the astrocytes.

_____ 25. The blood-brain barrier protects the delicate brain and spinal cord from chemical fluctuations in the blood and minimizes the possibility that potentially harmful blood-borne substances might reach the central neural tissue.

_____ 26. The brain can produce ATP in the absence of oxygen.

_____ 27. Damaged brain cells release excessive amounts of glutamate.

_____ 28. The spinal cord is about 28 inches long.

_____ 29. There are eight pairs of cervical nerves: twelve thoracic nerves, seven lumbar nerves, five sacral nerves, and one coccygeal nerve.

_____ 30. Ascending tracts relay messages from the brain to efferent neurons.

_____ 31. Efferent fibers carrying incoming signals enter the spinal cord through the dorsal root.

_____ 32. The cell bodies for the efferent neurons originate in the gray matter and send axons out through the ventral root.

_____ 33. The thirty-one pairs of spinal nerves, along with the twelve pairs of cranial nerves that arise from the brain, constitute the peripheral nervous system.

_____ 34. The skin has different receptors for warmth, cold, light touch, pressure, and pain.

_____ 35. All incoming and outgoing fibers traversing between the periphery and higher brain centers must pass through the brain stem.

_____ 36. Most of the branches of the vagus nerve supply organs in the thoracic and abdominal cavities.

_____ 37. The following states of consciousness are listed in increasing order of level of arousal: coma, sleep, wakefulness, and maximum alertness.

_____ 38. The brains overall level of activity is reduced during sleep.

_____ 39. Paradoxical sleep occupies 80 percent of total sleeping time throughout adolescence and most of adulthood.

_____ 40. Slow-wave sleep is characterized by REM.

_____ 41. The subcortical regions include the basal nuclei located in the cerebrum and the thalamus and hypothalamus located in the diencephalon.

_____ 42. The thalamus exerts an inhibitory effect on motor activity by acting through neurons in the brain.

_____ 43. The hypothalamus is the area of the brain most notably involved in the direct regulation of the internal environment.

_____ 44. Serotonin and norepinephrine are synaptic messengers in the regions of the brain involved in pleasure and motivation.

_____ 45. Short-term memory is believed to involve relatively permanent functional or structural changes between existing neurons in the brain.

_____ 46. The development of LTP (long-term potentiation) involves changes in both the presynaptic and postsynaptic neurons at a given synapse.

_____ 47. Numerous hormones and neuropeptides are known to affect learning and memory processes.

_____ 48. Both the nervous and the endocrine system ultimately alter their target cells by releasing chemical messengers.

_____ 49. Both the nervous and the endocrine systems have an influence on other major control systems.

_____ 50. The nervous system has a very slow speed of response.

_____ 51. Binding of a hormone with target-cell receptors initiates a reaction (or series of reactions) that culminates in the hormone's final effect.

_____ 52. The endocrine system is responsible for coordinating rapid, precise responses.

_____ 53. The nervous system directly or indirectly controls the secretion of many hormones.

_____ 54. In the nervous system, neurotransmitters are released into the synaptic knob.

**Fill in the Blank**

55. The ____________________ regulates muscle tone and coordinates skilled, voluntary movements.

56. The ____________________ is important for the maintenance of balance and control of eye movement.

57. The ____________________ plays a role in the planning and initiation of voluntary activity by providing input to the cortical motor areas.

58. There are two types of reflexes: __________ (__________) reflexes and __________ (__________) reflexes.

59. The spinal nerves are named according to the region of the vertebral column from which they emerge. There are eight pairs of ____________ nerves, twelve ________ nerves, five __________ nerves, five __________ nerves, and one coccygeal nerve.

60. The thick bundle of elongated nerve roots within the lower vertebral canal is known as the _________________.

61. __________ fibers carrying incoming signals enter the spinal cord through the dorsal root; __________ fibers carrying outgoing signals leave through the ventral root.

62. A collection of neuronal cell bodies located outside of the CNS is called a(n) _________, whereas a functional collection of cell bodies within the CNS is referred to as a(n) ____________________.

63. The dorsal and ventral roots at each level join to form a(n) _______________.

64. A(n) __________ is a bundle of peripheral neuronal axons, some afferent and some efferent, which are enclosed by a connective-tissue covering.

65. The diencephalon houses two brain components: (1) __________________ and (2) _______________.

66. The ____________ is the largest portion of the human brain.

67. Each hemisphere is composed of a thin outer shell of gray matter called ____________.

68. _______ matter consists predominantly of densely packaged cell bodies and their dendrites as well as glial cells. Bundles or tracts of myelinated nerve fibers (axons) constitute the ______ matter.

69. The _______________ are primarily responsible for receiving and processing sensory input such as touch, pressure, heat, cold, and pain from the surface of the body. These sensations are collectively known as ______________________.

70. __________ area, which is responsible for speaking ability, is located in the left ______________. ______________ area, located in the left _________ at the juncture of the parietal, temporal, and occipital lobes, is concerned with language comprehension.

71. About 90 percent of the cells within the CNS are _____________ or ___________.

72. _____________ serve as a scaffold to guide neurons to their proper final destination during fetal brain development.

73. ____________________ form the insulative myelin sheaths around axons in the CNS.

74. _______________ phagocytic cells are the scavengers of the CNS.

75. The innermost meningeal layer, the ____________, is the most fragile. It is highly vascular and closely adheres to the surfaces of the brain and spinal cord.

76. Cerebrospinal fluid is formed primarily by the __________________ found in particular regions of the ventricle cavities of the brain.

77. The brain stem consists of the ___________, _______, and _____________.

78. Running throughout the entire brain stem and into the thalamus is a widespread network of interconnected neurons called the ___________________.

79. ________________ refers to subjective awareness of the external world and self.

80. _______ refers to the total unresponsiveness of a living person to external stimuli, caused either by brain stem damage or by widespread depression of the cerebral cortex.

81. _____________ sleep can be considered either the deepest sleep or the lightest sleep.

82. In _____________ sleep a person still has considerable muscle tone and frequently shifts body position.

83. In _____ sleep, muscles are completely relaxed, with no movement taking place.

84. The subcortical regions include the _______________, located in the cerebrum, and the ___________ and ________________, located in the diencephalon.

85. The ____________ serves as a "relay station" and synaptic integrating center for preliminary processing of all sensory input on its way to the cortex.

86. The _______________ controls body temperature, controls food intake, and controls thirst.

87. _______________ is the ability to direct behavior toward specific goals.

88. ____________ is the acquisition of knowledge or skills as a consequence of experience, instruction, or both, and is a change in behavior that occurs as a result of experiences.

89. The process of transferring and fixing short-term memory traces into long-term memory stores is known as ___________________.

90. The study of the relationships between the nervous and endocrine systems is called ____________________________.

91. The nervous system is organized into the central nervous system, consisting of the _________ and _____________, and the _______________________.

92. The __________________ carries information to the CNS.

93. Three classes of neurons make up the nervous system: (1) ____________________, (2) ________________________, and (3) _________________.

94. Interneurons lie entirely within the __________.

**Matching**

*Match function to type of glial cell.*

a. Astrocytes
b. Ependymal cells
c. Microglia
d. Oligodendrocytes

_____ 95. Contribute to formation of cerebrospinal fluid
_____ 96. Form myelin sheaths in CNS
_____ 97. Induce formation of blood-brain barrier
_____ 98. Serve as scaffold during fetal brain development
_____ 99. Take up excess potassium to help maintain proper brain ECF ion concentration and normal neural excitability
_____ 100. Physically support neurons in proper spatial relationship
_____ 101. Play a role in defense of brain as phagocytic scavengers
_____ 102. Form neural scar tissue
_____ 103. Line internal cavities of brain and spinal cord
_____ 104. Take up and degrade released neurotransmitters into raw materials for synthesis of more neurotransmitters by neurons
_____ 105. Possess receptors for neurotransmitters, which may be important in a chemical signaling system

**Multiple Choice**

106. Which of the nerve cells loses its ability to undergo cell division?
a. astrocytes
b. ependymal cells
c. interneurons
d. oligodendrocytes
e. microglia

107. Which is the outermost covering of the brain?
a. pia mater
b. arachnoid mater
c. dura mater
d. perineural sac
e. epineural sac

108. Which type of sleep is characterized by having progressively slower EEG waves of higher amplitude?
a. REM sleep
b. slow-wave sleep
c. paradoxical sleep
d. rhythmic sleep
e. parenthetical sleep

109. This disease is characterized early as only short-term memory loss, but as the disease progresses, even firmly entrenched long-term memories are impaired.
a. cerebrovascular accident
b. dyslexia
c. aphasias
d. Alzheimer's
e. Parkinson's

110. In which form of short-term memory is calcium entry into the presynaptic terminal enhanced?
    a. paradoxical
    b. sensitization
    c. aphasias
    d. habituation
    e. amnesia

111. Which structure controls eye movement?
    a. spino cerebellum
    b. cerebrocerebellum
    c. vestibulocerebellum
    d. medulla oblongata
    e. retinal plexus

112. Which disorder is known as a lazy eye?
    a. amblyopia
    b. dyslexia
    c. aphasia
    d. Parkinson's disease
    e. schizophrenia

113. Which of the following is a language disorder caused by damage to specific cortical areas?
    a. amblyopia
    b. hydrocephalus
    c. Alzheimer's disease
    d. schizophrenia
    e. aphasia

114. This disorder is diagnosed when a large collection of neurons abnormally undergo synchronous action potentials that produce stereotypical involuntary spasms and alterations in behavior.
    a. Alzheimer's disease
    b. schizophrenia
    c. Parkinson's disease
    d. epilepsy
    e. dyslexia

115. Which area of the brain produces posterior pituitary hormones?
    a. hippocampus
    b. basal nuclei
    c. hypothalamus
    d. corpus callosum
    e. thalamus

**Modified Multiple Choice**

*Indicate which brain structure is associated with each function by writing the appropriate letter in the blank using the following answer code:*

(a) = cerebellum
(b) = hypothalamus
(c) = brain stem
(d) = basal nuclei
(e) = cerebral cortex

116. ______ Controls anterior pituitary hormone secretion.
117. ______ Structure that initiates all voluntary movement.
118. ______ Inhibits muscle tone throughout the body.
119. ______ Damage to this structure is associated with a resting tremor.
120. ______ Controls thirst, urine output, food intake, and body temperature.
121. ______ Contains the centers for respiration and heart and blood vessel function.
122. ______ Helps monitor and coordinate slow, sustained contractions with respect to balance
123. ______ Contains the autonomic nervous system coordinating center.
124. ______ Disorders of this structure produce an intention tremor.
125. ______ Plays an important role in emotion and behavior patterns.
126. ______ Coordinates motor activity.
127. ______ The structure that accomplishes final sensory perception.
128. ______ Consists of the medulla and pons.

## *Points to Ponder*

1. Suppose you had a dream last night in which you, David Crockett and Columbus were building a space laboratory on the moon. How did your brain come up with such a story?

2. After playing tennis yesterday you stopped at a fast-food store for something to drink and then decided to have a snack. What were the cues your brain received and how were they processed?

3. It has been demonstrated by experimentation that learning can occur by rewarding the animal if it responds in the desirable manner. If the desirable response is for the rat to run through a maze, and we reward the animal with food upon completion of the task, would it be correct to assume that learning occurs only if the rat is stressed? What happens if you feed the rat first?

4. How do the meninges support and/or protect the brain?

5. What is the functional significance of the gyri and sulci of the cerebral cortex?

6. In what ways are the physiological functions of stimulants and antidepressants similar? How do they differ physiologically?

7. Why do most people dream every night?

8. What happens when the corpus callosum is severed?

9. How are nuclei and ganglia similar? How do they differ?

10. What is the role of the processing center in the spinal cord?

## Clinical Perspectives

1. What would you suggest to a person complaining of frequent severe headaches of long duration?

2. What loss in body function would you expect in a car-crash victim with extensive injury to the cerebellum, the left temporal lobe, and the left occipital lobe?

3. Based on what you know about the brain, why does a person with Parkinson's disease suffer from muscle rigidity, resting tremors, and difficulty in initiating voluntary movements?

4. How do general anesthetics produce unconsciousness by depressing the RAS?

5. Explain how "split-brain" patients have contributed to research on the function of the cerebral hemispheres.

6. Fetal alcohol syndrome, produced by consuming too much alcohol during pregnancy, damages the corpus callosum and basal nuclei. What effects would damage to these areas after the child is born?

7. If a patient is taking a drug that affects protein synthesis, how would the drug affect short- and long-term memory?

8. What would you tell a Parkinson's patient about his/her illness?

9. Based on your knowledge of the brain, what would you tell a patient about the cause of his or her headache?

10. A young college student begins to have lower-back pain. The pain radiates down the student's left leg and is intense when the student coughs and sneezes. In addition, there is some loss of motor function in the left leg. What neurological problem exists? What are the neurological causes of the symptoms? Which spinal nerve is probably involved?

## PhysioEdge Activities

**Related to Text:**
Media Exercise 5.1: The Nervous and Endocrine Systems.
Media Exercise 5.2: The Cerebral Cortex.
Media Exercise 5.3: Subcortical Structures.

**Related to Figures:**
*Figure 5.1.* For an interaction related to this figure, see Media Exercise 5.1: The Nervous and Endocrine Systems.
*Figure 5.2.* For an interaction related to this figure, see Media Exercise 5.1: The Nervous and Endocrine Systems.
*Figure 5.8.* For an interaction related to this figure, see Media Exercise 5.2: The Cerebral Cortex.
*Figure 5.9.* For an interaction related to this figure, see Media Exercise 5.2: The Cerebral Cortex.
*Figure 5.15.* For an interaction related to this figure, see Media Exercise 5.3: Subcortical Structures.
*Figure 5.16.* For an interaction related to this figure, see Media Exercise 5.3: Subcortical Structures.

## *Media Resources*

**PhysioEdge CD-ROM**
For a visual review of concepts in this chapter, check out:

Tutorial: Skeletal Muscle Contraction Muscle Spindles tab
Media Exercise 5.1: The Nervous and Endocrine Systems
Media Exercise 5.2: The Cerebral Cortex
Media Exercise 5.3: Subcortical Structures

**Book Companion Website**
The website for this book contains a wealth of helpful study aids, as well as many ideas for further reading and research. Log on to:

**http://www.brookscole.com/sherwoodhp6**

Select Chapter 5 from the drop-down menu, or click on one of the many resource areas, including **Case Histories**, which introduce clinical aspects of human physiology. For this chapter check out: Case History 13: A Critical Twenty-four Hours and Case History 14: One Disease, Two Outcomes.

For Suggested Readings, consult **InfoTrac® College Edition**, your online research library, at:

**http://infotrac.thomsonlearning.com**

# The Peripheral Nervous System: Afferent Division; Special Senses

## Chapter Overview

The nervous system, one of the two major regulatory systems of the body, consists of the central nervous system (CNS), composed of the brain and spinal cord, and the peripheral nervous system, composed of the afferent and efferent fibers that relay signals between the CNS and the periphery. The afferent division of the peripheral nervous system detects, encodes, and transmits peripheral signals to the CNS for processing. It is the communication link by which the CNS is informed about the internal and external environment. Perception plays a major role in maintaining homeostasis. Any change in the environment must be dealt with by the organism if homeostasis is to be maintained. The necessary internal and external environmental information is picked up at the peripheral endings of afferent neurons. These specialized endings are classified as follows: photoreceptors, mechanoreceptors, thermoreceptors, osmoreceptors, chemoreceptors, and nociceptors. All of these receptors depolarize or hyperpolarize to produce a receptor or generator potential that is a graded potential. Graded potentials initiate action potentials thus the peripheral nervous system delivers the signal to the CNS for interpretation. Each type of receptor is part of a designed pathway going to a specific part of the CNS. With the information reaching the CNS, the nervous system performs its all-important role of maintaining homeostasis. In this chapter special emphasis is placed on pain (nociceptors), vision (photoreceptor), hearing and equilibrium (mechanoreceptors), and taste and smell (chemoreceptors).

## Chapter Outline

INTRODUCTION

- The afferent division of the peripheral nervous system sends information about the internal and external environment to the CNS.

*Visceral afferents carry subconscious input while sensory afferents carry conscious input.*

- The incoming pathway for subconscious information derived from the internal viscera is called a visceral afferent.
- Afferent input that reaches the level of conscious awareness is known as sensory information, and the incoming pathway is a sensory afferent.

*Perception is the conscious awareness of surroundings derived from interpreting sensory input.*

- Perception is our conscious interpretation of the external world as created by the brain from a pattern of nerve impulses delivered to it from sensory receptors.

RECEPTOR PHYSIOLOGY

*Receptors have differential sensitivities to various stimuli.*

- At their peripheral endings afferent neurons have receptors that apprise the CNS of detectable changes in both the external world and the internal environment by generating action potentials in response to the stimuli.
- Receptors must convert other forms of energy into action potentials.
- This energy conversion process is known as transduction.
- Photoreceptors are responsive to light.
- Mechanoreceptors are sensitive to mechanical energy.
- Thermoreceptors are sensitive to heat and cold.
- Osmoreceptors detect changes in the concentration of solutes in the body fluids and the resultant changes in osmotic activity.
- Chemoreceptors are sensitive to specific chemicals.
- Nociceptors, or pain receptors are sensitive to tissue damage such as pinching or burning or to distortion of tissue.
- Intense stimulation of any receptor is also perceived as painful.

*A stimulus alters the receptor's permeability, leading to a graded receptor potential.*

- A receptor may be either a specialized ending of the afferent neuron or a separate cell closely associated with the peripheral ending of the neuron.
- The receptor potential is a graded potential whose amplitude and duration can vary, depending on the strength and the rate of application or removal of the stimulus.
- In the case of a separate receptor, a receptor potential triggers the release of a chemical messenger that diffuses across the small space separating the receptor from the ending of the afferent neuron.
- Binding of the chemical messenger with specific protein receptor sites on the afferent neuron opens chemical messenger-gated sodium channels.
- In the case of a specialized afferent ending, local current flow between the activated receptor ending undergoing a generator potential and the cell membrane adjacent to the receptor brings about the opening of voltage-gated sodium channels in this adjacent region.

*Receptor potentials may initiate action potentials in the afferent neuron.*

- A larger receptor potential cannot bring about a larger action potential (all-or-none law), but it can induce more rapid firing of action potentials.
- Stimulus intensity is distinguished both by the frequency of action potentials generated in the afferent neuron (frequency code) and by the number of receptors activated within the area (population code).

*Receptors may adapt slowly or rapidly to sustained stimulation.*

- There are two types of receptors—tonic receptors and phasic receptors—based on their speed of adaptation.
- Tonic receptors do not adapt at all or adapt slowly.
- Phasic receptors are rapidly adapting receptors.
- Phasic receptors are useful in situations where it is important to signal a change in stimulus intensity rather than to relay status quo information.
- The Pacinian corpuscle is a rapidly adapting receptor that detects pressure and vibration.
- Adaptation in a Pacinian corpuscle is believed to involve both mechanical and electrochemical components.

*Each somatosensory pathway is "labeled" according to modality and location.*

- On reaching the spinal cord, afferent information has two possible destinies: (1) it may become part of a reflex arc, bringing about an appropriate effector response, or (2) it may be relayed upward to the brain via ascending pathways for further processing and possible conscious awareness.
- Somatosensory pathways consist of discrete chains of neurons synaptically interconnected in a particular sequence to accomplish progressively more sophisticated processing of the sensory information.
- The afferent neuron with its peripheral receptor is known as a first-order sensory neuron.
- It synapses on a second-order sensory neuron, either in the spinal cord or the medulla.
- This neuron then synapses on a third-order sensory neuron in the thalamus, and so on.
- A particular sensory input is "projected" to a specific region of the cortex.
- Thus, information is kept separated within specific labeled lines between the periphery and the cortex.
- In this way, even though all information is propagated to the CNS via the same type of signal, the brain can decode the type and location of the stimulus.
- Phantom pain, for example, is pain perceived as originating in the foot by a person whose leg has been amputated at the knee.
- The sensation of phantom pain might arise from irritation of the second endings of afferent pathways or from recently documented extensive remodeling of the brain region that originally handled sensations from the severed limb.

*Acuity is influenced by receptive field size and lateral inhibition.*

- Each sensory neuron responds to stimulus information only within a circumscribed region of the skin surface surrounding it: this region is known as its receptive field.
- The smaller the receptive field in a region, the greater its acuity or discriminative ability.
- More cortical space is allotted for sensory reception from areas with smaller receptive fields and, accordingly, greater tactile discriminative ability.
- The most strongly activated signal pathway originating from the center of the stimulus area inhibits the less excited pathways from the fringe areas.
- This occurs via inhibitory interneurons that pass laterally between ascending fibers serving neighboring receptive fields.

## PAIN

*Stimulation of nociceptors elicits the perception of pain plus motivational and emotional responses.*

- Pain is primarily a protective mechanism meant to bring to conscious awareness to the fact that tissue damage is occurring or is about to occur.
- There are three categories of pain receptors: mechanical nociceptors respond to mechanical damage such as cutting, crushing, or pinching; thermal nociceptors respond to temperature extremes; and polymodal nociceptors respond equally to all kinds of damaging stimuli, including irritating chemicals released from injured tissues.
- None of the nociceptors have specialized receptor structures; they are all naked nerve endings.
- Nociceptors do not adapt to sustained or repetitive stimulation.
- All nociceptors can be sensitized by the presence of prostaglandins.
- Signals arising from mechanical and thermal nociceptors are transmitted over large myelinated A-delta fibers (the fast pain pathway).
- Impulses from polymodal nociceptors are carried by small unmyelinated C-fibers (slow pain pathway).

*The brain has a built-in analgesic system.*

- In addition to the chain of neurons connecting peripheral nociceptors with higher CNS structures for pain perception, the CNS also contains a neuronal system that suppresses pain.
- Electrical stimulation of the periaqueductal gray matter results in profound analgesia, as does stimulation of the reticular formation.
- These two regions are thought to be part of a descending analgesic pathway that blocks the release of substance P from afferent pain fiber terminals.
- Endogenous opiates serve as analgesic neurotransmitters; they are released from a descending analgesic pathway and bind with opiate receptors on the afferent pain fiber terminal.
- This binding suppresses the release of substance P.
- Acupuncture analgesia compares favorably with morphine for treating chronic pain.
- Acupuncture needles activate specific afferent nerve fibers, which send impulses to the CNS.
- The incoming impulses activate three centers (a spinal cord center, a midbrain center, and the hypothalamus/anterior pituitary unit) to cause analgesia.
- All three centers block pain transmission through the use of endorphins and closely related compounds.

## EYE: VISION

*Protective mechanisms help prevent eye injuries.*

- The eyelids and tears help protect the eye from injury.

*The eye is a fluid-filled sphere enclosed by three specialized tissue layers.*

- Most of the eyeball is covered by a tough outer layer of connective tissue, the sclera.
- Anteriorly, the outer layer consists of the transparent cornea through which light rays pass into the interior of the eye.
- The middle layer underneath the sclera is the highly pigmented choroid.
- The choroid layer becomes specialized anteriorly to form the ciliary body and the iris.
- The innermost coat under the choroid is the retina, which consists of an outer pigmented layer and an inner nervous tissue layer.
- The retina contains the rods and cones.
- The anterior cavity between the cornea and lens contains a clear, watery fluid, the aqueous humor, and the larger posterior cavity between the lens and the retina contains a semifluid, jellylike substance, the vitreous humor.
- The aqueous humor carries nutrients for the cornea and lens.
- Aqueous humor is produced by a capillary network within the ciliary body.
- Excess aqueous humor will accumulate in the anterior cavity, causing the intraocular pressure to rise.
- This condition is known as glaucoma.

*The amount of light entering the eye is controlled by the iris.*

- The iris contains two sets of smooth muscle networks, one circular and the other radial.
- The pupil gets smaller when the circular muscle contracts.
- When the radial muscle shortens, the size of the pupil increases.
- Iris muscles are controlled by the autonomic nervous system.
- Parasympathetic nerve fibers innervate the circular muscle, and sympathetic fibers supply the radial muscle.

*The eye refracts the entering light to focus the image on the retina.*

- The bending of a light ray (refraction) occurs when the ray passes from a medium of one density into a medium of a different density.
- A lens with a convex surface converges light rays, bringing them closer together, a requirement for bringing an image to a focal point.

- Refractive surfaces of the eye are convex.
- The two structures most important in the eye's refractive ability are the cornea and the lens.

*Accommodation increases the strength of the lens for near vision.*

- The ability to adjust the strength of the lens so that both near and far sources can be focused on the retina is known as accommodation.
- The strength of the lens depends on its shape, which in turn is regulated by the ciliary muscle.
- The ciliary body has two major components: the ciliary muscle and the capillary network which produces the aqueous humor.
- The ciliary muscle is attached to the lens by suspensory ligaments.
- When the ciliary muscle contracts, the lens assumes a more spherical shape.
- The greater the curvature of the lens increases its strength, causing greater bending of light rays.
- With loss of elasticity, the lens is no longer able to assume the spherical shape required to accommodate for near vision.
- This age-related reduction in accommodative ability, presbyopia, affects most people by middle age.
- Other common vision disorders are myopia and hyperopia.

*Light must pass through several retinal layers before reaching the photoreceptors.*

- The major function of the eye is to focus light rays from the environment on the rods and cones, the photoreceptor cells of the retina.
- The photoreceptors then transform the light energy into electrical signals for transmission to the CNS.
- The neural portion of the retina consists of three layers: (1) the outermost layer containing the rods and cones, whose light-sensitive ends face the choroid; (2) a middle layer of bipolar neurons; and (3) an inner layer of ganglion cells. Axon of the ganglion cells join together to form the optic nerve.
- The point on the retina at which the optic nerve leaves and through which blood vessels pass, is the optic disc.
- In the fovea, which is a pinhead-sized depression located in the exact center of the retina, the bipolar and ganglion cell layers are pulled aside so that light directly strikes the photoreceptors.
- This feature, coupled with the fact that only cones are found here, makes the fovea the point of most distinct vision.

*Phototransduction by retina cells converts light stimuli into neural signals.*

- Photoreceptors consist of three parts: (1) an outer segment, which lies closest to the eye's exterior, facing the choroid, and detects the light stimulus; (2) an inner segment, which contains the metabolic machinery; and (3) a synaptic terminal, which lies closest to the eye's interior, facing the bipolar neurons and transmits the signal generated in the photoreceptor upon light stimulation to the next cells in the visual pathway.
- The outer segment is rod-shaped in rods and cone-shaped in cones.
- The photopigment, which is found in the outer segment, consists of an enzymatic protein called opsin combined with retinene, a derivative of vitamin A.
- There are four different photopigments, one in rods and one in each of three types of cones.
- Rhodopsin, the rod photopigment, cannot discriminate between various wavelengths.
- Rods provide vision only in shades of gray by detecting different intensities.
- The photopigments in the three types of cones—red, green, and blue cones—respond selectively to various wavelengths, making color vision possible.
- Phototransduction, the mechanism of excitation, is basically the same for all photoreceptors.
- When a photopigment molecule absorbs light it dissociates into its retinene and opsin components.
- This light-induced breakdown and subsequent activation of the photopigment bring about a hyperpolarizing receptor potential that influences transmitter release from the synaptic terminal of the photoreceptor.
- The light-induced biochemical change in the photopigment leads to closing of chemical messenger-gated sodium channels in the outer segment's membrane.
- These channels respond to an internal second messenger, cyclic guanosine monophosphate (GMP), which links photopigment light absorption with sodium channel closing.
- The sodium channels of a photoreceptor are open in the absence of stimulation, that is, in the dark.
- Upon exposure to light, the activated photopigment activates transducin, which activates phosphodiesterase.
- This enzyme degrades cyclic GMP.
- During the light excitation processes, the reduction in cyclic GMP brings about closure of sodium channels, which stops the depolarizing sodium leak and causes membrane hyperpolarization.

- This hyperpolarization, which is the receptor potential, passively spreads from the outer segment to the synaptic terminal of the photoreceptor.
- The photoreceptors synapse with bipolar cells.
- These cells terminate on the ganglion cells, whose axons form the optic nerve for transmission of signals to the brain.
- The transmitter released from the photoreceptors' synaptic terminal has an inhibitory action on the bipolar cells.
- The reduction in transmitter release that accompanies light-induced receptor hyperpolarization decreases this inhibitory action on the bipolar cells.
- Bipolar cells display graded potentials similar to the photoreceptors.
- Action potentials do not originate until the ganglion cells, the first neurons in the chain that must propagate the visual message to the brain.
- Researchers are currently working on a microelectronic chip that would serve as a partial substitute retina.
- The chip would benefit persons with macular degeneration.
- This condition is characterized by loss of photoreceptors.
- The "vision chip" would bypass the photoreceptor step.

*Rods provide indistinct gray vision at night, whereas cones provide sharp color vision during the day.*

- Cones are abundant in the center of the retina.
- Rods are most abundant in the periphery.
- Cones provide sharp vision with high resolution for fine detail.
- Rods have low acuity but high sensitivity, so they respond to the dim light of night.
- There is little convergence of neurons in the retinal pathway for cone output.
- In contrast, there is much convergence of neurons in rod pathways.

*The sensitivity of the eyes can vary markedly through dark and light adaptation.*

- When you go from bright sunlight into darkened surroundings, you cannot see anything at first, but gradually you begin to distinguish objects as a result of the process of dark adaptation.
- When you move from the dark to the light, your eyes are very sensitive to the light at first.
- The sensitivity of the eyes decreases and normal contrasts can once again be detected, a process known as light adaptation.

*Color vision depends on the ratios of stimulation of the three cone types.*

- The pigment in various objects selectively absorbs particular wavelengths of light transmitted to them from light-emitting sources, and the unabsorbed wavelengths are reflected from the objects surfaces.
- These reflected light rays enable us to see the objects.
- Our perception of the many colors of the world depends on the three cone types various ratios of stimulation in response to different wavelengths.
- DNA studies have shown that men with normal color vision have a variable number of genes coding for cone pigments.
- This finding will undoubtedly lead to a reevaluation of how the various photo pigments contribute to color vision.

*Visual information is separated, modified, and again separated before reaching the visual cortex.*

- The image detected on the retina at the onset of visual processing is upside down and backward.
- Once it is projected to the brain, the inverted image is interpreted as being in its correct position.
- Before information reaches the brain, the retinal neuronal layers beyond the rods and cones reinforce selected information and suppress other information to enhance contrast.
- Various aspects of visual information such as form, color, depth, and movement are separated and projected in parallel pathways to different regions of the cortex.
- Because of the pattern of wiring between the eyes and the visual cortex, the left half of the cortex receives information only from the right half of the visual field, and the right half receives input only from the left half of the visual field of both eyes.

*The thalamus and visual cortices elaborate the visual message.*

- The first stop in the brain for information in the visual pathway is the lateral geniculate nucleus in the thalamus.
- It separates information received from the eyes and relays it to different zones in the cortex, each of which processes different aspects of the visual stimulus (color, form, depth, and movement).
- Although each half of the visual cortex receives information simultaneously from the same part of the visual field as received by both eyes, the messages from the two eyes are not identical.
- The brain uses the slight disparity in the information received from the two eyes to

estimate distance, allowing us to perceive three-dimensional objects in spatial depth.
- Three types of visual cortical neurons have been identified based on the complexity of stimulus requirements needed for the cell to respond.
- Unlike a retinal cell, which responds to the amount of light, a cortical cell fires only when it receives a particular pattern of illumination for which it is programmed.

*Visual input goes to other areas of the brain not involved in visual perception.*
- Not all fibers in the visual pathway terminate in the visual cortices.
- Following are examples of nonsight activities dependent on input from the retina: (1) control of pupil size, (2) synchronization of biological clocks, (3) contribution to cortical alertness and attention, and (4) control of eye movements.

*Some sensory input may be detected by multiple sensory processing areas in the brain.*
- A growing body of evidence indicates that brain regions actually receive a variety of sensory signals.

## EAR: HEARING AND EQUILIBRIUM

*Sound waves consist of alternate regions of compression and rarefaction of air molecules.*
- Hearing is the neural perception of sound energy.
- Sound waves are traveling vibrations of air that consist of regions of high pressure caused by compression of air molecules alternating with regions of low pressure caused by rarefaction of the molecules.
- Sound waves can also travel through media other than air, such as water.
- Sound is characterized by its pitch, intensity, and timbre.

*The external ear plays a role in sound localization.*
- The pinna collects sound waves and channels them down the external ear canal.
- The tympanic membrane, which is stretched across the entrance to the middle ear, vibrates when struck by sound waves.

*The tympanic membrane vibrates in unison with sound waves in the external ear.*
- The middle ear transfers the vibratory movements of the tympanic membrane to the fluid of the inner ear.

*The middle ear bones convert tympanic membrane vibrations into fluid movements in the inner ear.*
- This transfer is facilitated by a moveable chain of three small bones (the malleus, incus, and stapes) that extend across the middle ear.

*The cochlea contains the organ of Corti, the sense organ for hearing.*
- The stapes is attached to the oval window, the entrance into the fluid-filled cochlea.

*Hair cells in the organ of Corti transduce fluid movements into neural signals.*
- The snail-shaped cochlear portion of the inner ear is a coiled tubular system lying deep within the temporal bone.
- The cochlea is divided throughout most of its length into three fluid-filled longitudinal compartments (the cochlear duct, scala tympani, and scala vestibuli).
- The scala vestibuli follows the inner contours of the spiral.
- The scala vestibuli is sealed from the middle ear cavity by the oval window, to which the stapes is attached.
- The round window seals the scala tympani from the middle ear.
- The basilar membrane forms the floor of the cochlear duct and bears the organ of Corti, the sense organ for hearing.
- The organ of Corti contains hair cells that are the receptors of sound.
- Pressure waves through the basilar membrane cause this membrane to vibrate in synchrony with the pressure wave.
- Since the organ of Corti rides on the basilar membrane, the hair cells also move up and down as the basilar membrane oscillates.
- This back and forth mechanical deformation of the hairs alternately opens and closes mechanically gated ion channels in the hair cell, resulting in alternating depolarizing and hyperpolarizing potential changes—the receptor potential—at the same frequency as the original sound stimulus.
- The inner hair cells are specialized receptor cells that communicate via a chemical synapse with the terminals of afferent nerve fibers making up the auditory nerve.
- The outer hair cells enhance the response of the inner hair cells, the real auditory sensory receptors.

*Pitch discrimination depends on the region of the basilar membrane that vibrates.*
- Different regions of the basilar membranes naturally vibrate maximally at different frequencies; that is, each frequency displays peak

vibration at a different position along the membrane.
- Intensity discrimination depends on the amplitude of the vibration
- The greater tympanic membrane deflection is converted into a greater amplitude of basilar membrane movement in the region of peak responsiveness.

*Loudness discrimination depends on the amplitude of basilar-membrane vibration.*
- The CNS interprets greater basilar membrane oscillation as louder sound.

*The auditory cortex is mapped according to tone.*
- Each region of the basilar membrane is linked to a specific region of the auditory cortex in the temporal lobe, which becomes excited in response to a specific tone.
- The neural pathway between the organ of Corti and the auditory cortex involves several synapses en route, the most notable of which are in the brain stem and median geniculate nucleus of the thalamus.
- The brain stem uses the auditory input for alertness and arousal.
- The thalamus sorts and relays the signal upward.
- Auditory signals from each ear are transmitted to both temporal lobes.
- The primary auditory cortex appears to perceive discrete sounds, while the surrounding higher-order auditory cortex integrates the separate sounds into a coherent, meaningful pattern.

*Deafness is caused by defects either in conduction or neural processing of sound waves.*
- Conductive deafness occurs when sound waves are not adequately conducted through the external and middle portions of the ear to set the fluids in the inner ear in motion.
- In sensorineural deafness, the sound waves are transmitted to the inner ear, but they are not translated into nerve signals that are interpreted by the brain as sound sensations.
- Neural presbycusis is a degenerative, age-related process that occurs as hair cells "wear out" with use.
- Cochlear implants, electronic devices, that are surgically implanted, transduce sound signals that can directly stimulate the auditory nerve, thus bypassing a defective cochlear system.

*The vestibular apparatus is important for equilibrium by detecting position and motion of the head.*
- The vestibular apparatus consists of two sets of structures within the temporal bone near the cochlea, the semicircular canals and the otolith organs.
- The semicircular canals detect rotational or angular acceleration or deceleration of the head.
- Each ear contains three semicircular canals arranged three-dimensionally in planes that lie at right angles to each other.
- The receptive hair cells of each semicircular canal are situated on top of a ridge located in the ampulla.
- The hairs are embedded in the cupula.
- As the head starts to move, the fluid within the semicircular canal does not move in the direction of the rotation but lags behind because of its inertia.
- The hairs of a vestibular hair cell consist of 20 to 50 stereocilia and one kinocilium.
- Each hair cell is oriented so that it depolarizes when its stereocilia are bent toward the kinocilium; bending in the opposite direction hyperpolarizes the cell.
- The hair cells form a chemically mediated synapse with terminal endings of afferent neurons whose axons join with those of the other vestibular structures to form the vestibular nerve.
- This nerve unites with the auditory nerve from the cochlea to form the vestibulocochlear nerve.
- The otolith organs provide information about the position of the head relative to gravity and detect changes in rate of linear motion.
- The utricle and saccule are saclike structures housed within a bony chamber situated between the semicircular canals and the cochlea.
- The hairs of the receptive hair cells in these sense organs also protrude into an overlying gelatinous sheet, whose movement displaces the hairs and results in changes in hair cell potential.
- When a person is in an upright position, the hairs within the utricles are oriented vertically and the saccule hairs are lined up horizontally.
- Signals are carried through the vestibulocochlear nerve to the vestibular nuclei in the brain stem and on to the cerebellum.
- Here the vestibular information is integrated with input from the skin surface, eyes, joints and muscles for: (1) maintaining balance and desired posture, (2) controlling the external eye muscles so that the eyes remain fixed on the same point, and (3) perceiving motion and orientation.

## Chemical Senses: Taste and Smell

*Taste receptor cells are located primarily within tongue taste buds.*
- The chemoreceptors for taste sensation are packaged in taste buds.

- Each taste bud has a small opening, the taste pore.
- Taste receptor cells are modified epithelial cells with microvilli.
- The plasma membrane of the microvilli contains receptor sites that bind selectively with chemical molecules in the environment.

*Taste discrimination is coded by patterns of activity in various taste bud receptors.*

- Binding of a taste-provoking chemical with a receptor cell alters its ionic channels to produce a depolarizing receptor potential.
- The receptor potential initiates an action potential within the terminal endings of afferent nerve fibers with which the receptor cell synapses.
- Signals in these sensory inputs are conveyed via synaptic stops in the brain stem and thalamus to the cortical gustatory area.
- All tastes are varying combinations of four primary tastes: salty, sour, sweet, and bitter.
- A number of sensory physiologists suggest a fifth primary taste—umami, an amino and peptide taste.
- The first G protein in taste—gustducin—has been identified in one of the bitter signaling pathways.
- Taste perception is also influenced by information from other receptors, especially odor.

*The olfactory receptors in the nose are specialized endings of renewable afferent neurons.*

- The olfactory mucosa contains olfactory receptors, supporting cells, and basal cells.

*Various parts of an odor are detected by different olfactory receptors and sorted into "smell files."*

- To be smelled, a substance must be: (1) sufficiently volatile that some of its molecules can enter the nose, and (2) sufficiently water-soluble that it can dissolve in the mucus layer coating the olfactory mucosa that leads to opening of sodium channels.
- The resultant ion movement brings about a receptor potential.
- The receptor potential generates the action potential in the afferent fiber.
- The afferent fibers pass through the bone and immediately synapse in the olfactory bulb.

*Odor discrimination is coded by patterns of activity in the olfactory bulb glomeruli.*

- Each olfactory bulb is lined by small, ball-like neural junctions known as glomeruli.
- Because each glomerulus receives signals only from receptors that detect a particular odor component, the glomeruli serve as "smell files."
- Fibers leaving the olfactory bulb travel in two different routes: (1) a subcortical route going to regions of the limbic system, and (2) a thalamic-cortical route.

*The vomeronasal organ detects pheromones.*

- The vomeronasal organ (VNO) detects pheromones.
- Researchers suspect that VNO is responsible for spontaneous "feelings" between people such as "good chemistry" or "bad vibes."

## Key Terms

accommodation
acuity
A-delta fibers
analgesic system
basilar membrane
C fibers
cataract
chemoreceptors
cochlea
cochlear implants
cones
fast pain pathway
generatory potential
glaucoma
gustation
intensity discrimination
kinocilium
lateral inhibition
mechanoreceptors
mitral cells
morphine
motion sickness
myopia
nociceptors
opiate receptors
optic disc
organ of Corti
osmoreceptors
ossicles
otoliths
oval window
Pacinian corpuscles
pain
perception
pheromones
photoreception
phototransduction
receptive field
receptor potential
receptors
retina
rhodopsin
rods
saccule
scala media
scala vestibule
sclera
semicircular canal
sensory afferent
slow pain pathway
somatic sensation
stereocilia
tastant
taste

taste buds
taste pore
thermoreceptors
tone
tympanic membrane
utricle
visceral afferent
vitreous humor

## *Review Exercises*

*Answers are in the appendix.*

**True/False**

_____ 1. All of the nociceptors have specialized receptor structures; they are all naked nerve endings.

_____ 2. All nociceptors can be sensitized by the presence of prostaglandins.

_____ 3. Impulses from polymodal nociceptors are carried by small myelinated C fibers.

_____ 4. The CNS contains a neuronal system that suppresses pain.

_____ 5. Endogenous opiates serve as analgesic neurotransmitters; they are released from a descending analgesic pathway and bind with opiate receptors on the afferent presynaptic terminal, this suppresses the release of substance Q.

_____ 6. Pain impulses originating at nociceptors are transmitted to the CNS via one of two types of efferent fibers.

_____ 7. Taste receptor cells are modified epithelial cells with microvilli.

_____ 8. Binding of a taste-provoking chemical with a receptor cell alters its ionic channels to produce a depolarizing receptor potential.

_____ 9. Acids cause a salty taste.

_____ 10. Olfactory receptors are specialized endings of afferent neurons.

_____ 11. Humans can distinguish tens of thousands of different odors.

_____ 12. Our sensitivity to a new odor rapidly diminishes after a short period of exposure to it, even though the odor source continues to be present.

_____ 13. Intense stimulation of any receptor is perceived as painful.

_____ 14. A receptor may be either a specialized ending of the afferent neuron or a separate cell closely associated with the peripheral ending of the neuron.

_____ 15. Binding of a chemical messenger with specific protein receptor sites on the afferent neuron closes chemical messenger-gated sodium channels.

_____ 16. A larger receptor potential cannot bring about a larger action potential (all-or-none law), but it can induce more rapid firing of action potentials.

_____ 17. Tonic receptors are useful in situations where it is important to signal a change in stimulus intensity rather than to relay status quo information.

_____ 18. The brain cannot decode the type and location of the stimulus.

_____ 19. Each sensory neuron responds to stimulus information only within a circumscribed region of the skin surface surrounding it.

_____ 20. The pigment in the choroid and retina absorbs light after it strikes the retina to prevent reflection or scattering of light within the eye.

_____ 21. The aqueous humor is important in maintaining the spherical shape of the eyeball.

_____ 22. Iris muscles are controlled by the autonomic nervous system.

_____ 23. The strength of the lens depends on its shape, which in turn is regulated by the ciliary muscle.

_____ 24. Retinene is identical in all four photopigments, but the photoreceptor's opsin vary slightly.

_____ 25. Rhodopsin is a cone-shaped photopigment.

_____ 26. The sodium channels of a photoreceptor are closed in the absence of stimulation, that is, in the dark.

_____ 27. Sound waves can only travel through air.

_____ 28. Sound is characterized by its pitch (tone), intensity (loudness), and timbre (quality).

_____ 29. The tympanic membrane collects sound waves and channels them down the external ear canal.

_____ 30. The middle ear transfers the vibratory movements of the tympanic membrane to the fluid of the inner ear.

_____ 31. The cochlea is divided throughout most of its length into four fluid-filled longitudinal compartments.

_____ 32. The fluid within the cochlear duct is called perilymph.

_____ 33. The vestibular membrane forms the floor of the cochlear duct, separating it from the scala tympani.

_____ 34. Hair cells are specialized receptor cells that communicate via a chemical synapse with the terminals of afferent nerve fibers making up the auditory nerve.

_____ 35. Photoreceptors synapse with bipolar cells.

_____ 36. Rods are most abundant in the center of the retina.

**Fill in the Blank**

37. Most of the eyeball is covered by a tough outer layer of connective tissue called __________.

38. When the _________ muscle shortens, the size of the pupil increases.

39. The ability to adjust the strength of the lens so that both near and far sources can be focused on the retina is known as ______________________.

40. The lens is an elastic structure consisting of transparent fibers. Occasionally these fibers become opaque so that light rays cannot pass through, a condition known as a ___________.

41. The point on the retina at which the optic nerve leaves and through which blood vessels pass is the ________________.

42. A photopigment consist of an enzymatic protein called ________ combined with retinene, a derivative of vitamin A.

43. The reorganized bundles of fibers leaving the optic chiasm are known as __________.

44. Each taste bud has a small opening called the ______________.

45. Signals in these sensory inputs are conveyed via synaptic stops in the brain stem and thalamus to the ___________________________.

46. The axons of the receptor cells collectively form the ___________________.

47. The chemoreceptors for taste sensations are packaged in _____________________.

48. Fibers leaving the olfactory bulb travel in two different routes: (1) __________________________ and (2) _________________________________.

49. The olfactory mucosa contains three cell types: (1) ______________________, (2) _______________________, and (3) _________________________________.

50. Afferent input that does reach the level of conscious awareness is known as ____________________ and includes somatic sensation and special senses.

51. ____________ receptors are rapidly adapting receptors.

52. _____________________ is a skin receptor that detects pressure and vibration.

53. The incoming pathway for subconscious information derived from the internal viscera is called a ______________________.

54. If the information detected by the receptors is propagated to the conscious level of the brain, it is called ______________________.

55. The afferent neuron with its peripheral receptor that first detects the stimulus is known as a(n) ___________________________.

56. Pain perceived as originating in the foot by a person whose leg has been amputated at the knee is known as ________________.

57. The __________________________ houses sensory systems for equilibrium, and provides input essential for maintenance of posture and balance.

58. The skin lining the ear canal contains modified sweat glands that produce ______________, a sticky secretion that traps fine foreign particles.

59. The fluid within the cochlear duct is called _______________.

60. The thin ___________ membrane separates the cochlear duct from the scala vestibuli.

61. Hair cells are specialized receptor cells that communicate via a chemical synapse with the terminals of afferent nerve fibers making up the ______________________.

62. Pitch discrimination depends on the shape and properties of the ______________ membrane.

63. The _________________ provide information about the position of the head relative to gravity and also detect changes in rate of linear motion.

64. Name the three categories of pain receptors: (1) ______________________________, (2) ____________________________, and (3) ______________________________.

65. Signals arising from mechanical and thermal nociceptors are transmitted over large myelinated ______________________.

66. One of the neurotransmitters released from these afferent pain terminals is __________________, which is believed to be unique to pain fibers.

67. All nociceptors can be sensitized by the presence of ______________________.

68. Ascending pain pathways have poorly understood destinations in the ______________________________, the __________________________, and the __________________________________.

69. The built-in analgesic system is dependent on the presence of ____________________.

70. The __________ effect refers to a chemical or technique that brings about a desired response through the power of suggestion or distraction rather than through any direct action.

**Matching**

*Match functions to structure.*

a. oval window
b. organ of Corti
c. round window
d. semicircular canals membrane
e. malleus, incus, staples
f. scala vestibuli
g. utricle
h. basilar membrane
i. saccule
j. scala media

____ 71. Detects changes in head position away from vertical
____ 72. Vibrates in unison with movement of stapes, to which it is attached
____ 73. Contains endolymph; houses the basilar
____ 74. Contains hair cells
____ 75. Dissipates pressure in cochlea
____ 76. Detects changes in head position away
____ 77. Detect angular deceleration
____ 78. Bears the organ of Corti
____ 79. Contains perilymph that is set in motion by oval window movement
____ 80. Oscillate in synchrony with tympanic-membrane vibrations

**Multiple Choice**

81. Which receptor has a life span of about 10 days?
a. Ear
b. Eye
c. Taste
d. Pain
e. None of the above

82. Which receptors relay status quo information?
    a. Photoreceptors
    b. Phasic receptors
    c. Nociceptors
    d. Transduction receptors
    e. Tonic receptors

83. Which eye disorder is characterized by the surface of the cornea being uneven?
    a. Astigmatism
    b. Cataract
    c. Diplopia
    d. Glaucoma
    e. Hyperopia

84. Which part of the CNS is involved in taste?
    a. Cerebellum
    b. Lateral geniculate nucleus
    c. Gustatory nucleus
    d. Cortical gustatory area
    e. Periaqueductal gray matter

85. Which part of the CNS plays a role in the analgesic system?
    a. Lateral geniculate nucleus
    b. Analgesic radiations
    c. Medial geniculate nucleus
    d. Periaquaductal gray matter
    e. Primary analgesic cortex

86. Which type receptor is the naked peripheral end of an afferent neuron?
    a. Nociceptors
    b. Mechanoreceptors
    c. Photoreceptors
    d. Opiate receptors
    e. Vitreous receptors

87. Receptors convert various forms of energy into electrical energy. What is this conversion process called?
    a. Depolarization
    b. Hyperpolarization
    c. Frequency modulation
    d. Somesthetic propagation
    e. Transduction

88. Which type of stimuli is detected by the Pacinian corpuscle?
    a. Specific chemicals
    b. Pressure changes
    c. Temperature changes
    d. Odors
    e. Solute concentration changes

89. Which part of the CNS is involved with sight?
    a. Periaqueductal gray matter
    b. Lateral geniculate nucleus
    c. Reticular function
    d. Olfactory bulb
    e. Third ventricle

90. Which receptors do not adapt at all or adapt slowly?
    a. Tactile
    b. Nociceptors
    c. Phasic
    d. Tonic
    e. Taste

91. Which of the following structures are thought to be used when "getting bad vibes" from someone?
    a. Olfactory bulb
    b. Vomeronasal organ
    c. Macula lutea
    d. Cupula
    e. Mitral cells

92. Which of the following disorders is not related to age?
    a. Neural presbycusis
    b. Manular degeneration
    c. Presbyopia
    d. Meniere's disease
    e. None of the above

93. Which of the primary tastes listed below is suggested to include the taste of amino acids and peptides?
    a. Sweet
    b. Umami
    c. Salty
    d. Sour
    e. Bitter

**Modified Multiple Choice**

*Indicate the proper sequence of involvement in the visual pathway by writing the appropriate letter in the blank, using the answer key below.*

A = bipolar neurons
B = optic nerve
C = latera geniculate nucleus
D = rods and cones
E = visual cortex
F = optic tracts
G = ganglion cells
H = optic radiation

_____ 94. First
_____ 95. Second
_____ 96. Third
_____ 97. Fourth
_____ 98. Fifth
_____ 99. Sixth
_____ 100. Seventh
_____ 101. Eighth

## Points to Ponder

*Since perception is not always reality, consider these questions:*

1. What color is a bluebird? Be careful.

2. As you listen to a recording of a train on your stereo system you hear the approach of the train in one speaker and it seems to travel across in front of you to exit in the other speaker. How do orchestras accomplish this phenomenon in live concerts?

3. On a very hot day in July, you buy a nice cold canned diet drink containing the artificial sweetener aspartame. Upon drinking the beverage you experience a bitter taste instead of the normal sweet taste. How would you explain this?

## Clinical Perspectives

The patient complains of severe dizziness when riding the elevator to her job on the fifteenth floor. Explain how she could solve this problem, without walking up the stairs.

## Experiment of the Day

1. Place your index finger against the corner of your closed eye. Apply short bursts of gentle pressure on the eyeball. Describe and explain this phenomenon.

2. During your travels about campus, locate some people wearing excessive amounts of perfume or cologne. Explain why they do this. (Hint: Check their other habits.)

## PhysioEdge Activities

**From Text:**
Media Exercise 6.1: Pain.
Media Exercise 6.2: The Eye.
Media Exercise 6.3: The Ear.
Media Exercise 6.4: Taste.

**From Figures:**
*Figure 6.9.* For an interaction related to this figure, see Media Exercise 6.2: The Eye.
*Figure 6.10.* For an interaction related to this figure, see Media Exercise 6.2: The Eye.
*Figure 6.17.* For an animation of this figure, click the Sight tab (p.3) in the Special Sense tutorial.
*Figure 6.19.* For an animation of this figure, click the Sight tab (p.4) in the Special Sense tutorial.
*Figure 6.21.* For an interaction related to this figure, see Media Exercise 6.2: The Eye.
*Figure 6.30.* For an interaction related to this figure, see Media Exercise 6.3: The Ear.
*Figure 6.31.* For an animation of this figure, click the Hearing and Equilibrium tab in the Special Sense tutorial.
*Figure 6.34.* For an animation of this figure, click the Hearing and Equilibrium tab in the Special Sense tutorial.
*Figure 6.35.* For an animation of this figure, click the Hearing and Equilibrium tab in the Special Sense tutorial.
*Figure 6.43* For an interaction related to this figure, see Media Exercise 6.4: Taste.

## *Media Resources*

**PhysioEdge CD-ROM**
For a visual review of concepts in this chapter, check out:

Tutorial: Special Senses
Media Exercise 6.1: Pain
Media Exercise 6.2: The Eye
Media Exercise 6.3: The Ear
Media Exercise 6.4: Taste

**Book Companion Website**
The website for this book contains a wealth of helpful study aids, as well as many ideas for further reading and research. Log on to:

**http://www.brookscole.com/sherwoodhp6**

Select Chapter 6 from the drop-down menu, or click on one of the many resource areas.

For Suggested Readings, consult **InfoTrac® College Edition**, your online research library, at:

**http://infotrac.thomsonlearning.com**

# The Peripheral Nervous System: Efferent Division

## Chapter Overview

The nervous system, one of the two major regulatory systems of the body, consists of the central nervous system, composed of the brain and spinal cord, and the peripheral nervous system, composed of the afferent and efferent fibers that relay signals between the CNS and the periphery. Thus, as one of the major control systems of the body, the CNS transmits impulses through the efferent portion of the peripheral nervous system to the effector organs. The transmitted signal is usually the result of a CNS interpretation of conditions that warrant changes in order to maintain homeostasis. Voluntary and involuntary effector organs are controlled using the somatic and autonomic nervous system respectively. The key to this very precise autonomic control is in the neurotransmitters, acetylcholine and norepinephrine, and the dual innervation by the sympathetic and parasympathetic systems. The somatic nervous system acts through the release of a neurotransmitter at the neuromuscular junction to cause skeletal muscle contractions. The efferent division of the peripheral nervous system is the final link in the CNS control of the body. The actions elicited by the efferent neurons are essential for homeostasis.

## Chapter Outline

INTRODUCTION

- The efferent division of the peripheral nervous system is the communication link by which the CNS controls the activities of muscles and glands.
- Cardiac muscle, smooth muscle, most exocrine glands, and some endocrine glands are innervated by the autonomic nervous system.
- Skeletal muscle is innervated by the somatic nervous system.
- Two neurotransmitters—acetylcholine and norepinephrine—are released from efferent neuronal terminals to elicit essentially all the neurally controlled effector organ responses.

AUTONOMIC NERVOUS SYSTEM

*An autonomic nerve pathway consists of a two-neuron chain.*

- Each autonomic nerve pathway consists of a two-neuron chain.
- The cell body of the first neuron is located in the CNS.
- Its axon, the preganglionic fiber, synapses with the cell body of the second neuron, which lies within a ganglion outside the CNS.
- The axon of the second neuron, the postganglionic fiber, innervates the effector organ.
- The autonomic nervous system consists of two divisions—the sympathetic and the parasympathetic nervous system.
- Sympathetic nerve fibers originate in the thoracic and lumbar regions of the spinal cord.
- Most sympathetic preganglionic fibers are very short, synapsing within a sympathetic ganglion chain located along either side of the spinal cord.
- Long postganglionic fibers terminate on the effector organs.
- Parasympathetic preganglionic fibers arise from the cranial and sacral areas of the CNS.
- These fibers are long because they do not end until they reach terminal ganglia, which lie in or near the effector organs.
- Parasympathetic postganglionic fibers are very short.

*Parasympathetic postganglionic fibers release acetylcholine; sympathetic ones release norepinephrine.*

- Sympathetic and parasympathetic preganglionic fibers release the same neurotransmitter, acetylcholine (ACh).
- Parasympathetic postganglionic fibers release acetylcholine and are called cholinergic fibers.
- Most sympathetic postganglionic fibers are called adrenergic fibers because they release norepinephrine.
- The terminal branches of autonomic fibers contain numerous swellings, or varicosities, that simultaneously release neurotransmitters over a large area of the innervated organ.

- This diffuse release means whole organs instead of discrete cells are typically influenced by autonomic activity.

*The autonomic nervous system controls involuntary visceral organ activities.*

- The autonomic nervous system regulates visceral activities normally outside the realm of consciousness and voluntary control, such as circulation, digestion, sweating, and pupillary size.
- With the technique of biofeedback, people are provided with a conscious signal regarding visceral afferent information.

*The sympathetic and parasympathetic nervous systems dually innervate most visceral organs.*

- Most visceral organs are innervated by both sympathetic and parasympathetic nerve fibers.
- Both systems increase the activity of some organs and reduce the activity of others.
- Normally some level of action potential activity exists in both, the sympathetic and the parasympathetic fibers supplying a particular organ.
- This ongoing activity is called sympathetic or parasympathetic tone or tonic activity.
- The sympathetic system promotes responses that prepare the body for strenuous physical activity in the face of emergency or stressful situations.
- This response is typically referred to as a fight-or-flight response.
- The parasympathetic system dominates in quiet, relaxed situations.
- Dual innervation enables precise control over an organ's activity.
- There are several exceptions to the general rule of dual reciprocal innervation; the most notable are the following: (1) Innervated blood vessels receive only sympathetic nerve fibers. The only blood vessels to receive parasympathetic fibers are those supplying the penis and clitoris. (2) Sweat glands are innervated only by sympathetic nerves. The postganglionic fibers of these nerves are unusual because they secrete acetylcholine. (3) Salivary glands are innervated by both autonomic divisions and both stimulate secretion.

*The adrenal medulla is a modified part of the sympathetic nervous system.*

- The adrenal medulla is considered to be a modified sympathetic ganglion that does not give rise to postganglionic fibers.
- Upon stimulation by the preganglionic fiber the adrenal medulla secretes hormones into the blood.
- The hormones are identical or similar to postganglionic sympathetic neurotransmitters: norepinephrine and epinephrine.

*Several different receptor types are available for each autonomic neurotransmitter.*

- Responsive tissue cells possess one or more of several different types of plasma membrane receptor proteins for these chemical messengers.
- Two types of acetylcholine receptors have been identified.
- Nicotinic receptors, found in all autonomic ganglia, respond to acetylcholine released from both sympathetic and parasympathetic preganglionic fibers.
- Muscarinic receptors, found on effector cell membranes, bind with acetylcholine released from parasympathetic postganglionic fibers.
- There are two major classes of adrenergic receptors for norepinephrine and epinephrine, designated as alpha and beta receptors.

*Many regions of the central nervous system are involved in the control of autonomic activities.*

- Some autonomic reflexes, such as urination, defecation, and erection, are integrated at the spinal cord level, but all of these spinal reflexes are subject to control by higher levels of consciousness.
- The medulla within the brain stem is the region most directly responsible for autonomic output.
- Centers for controlling cardiovascular, respiratory, and digestive activity via the autonomic system are located in the medulla.
- The hypothalamus plays an important role in integrating the autonomic, somatic, and endocrine responses that autonomically accompany various emotional and behavioral states.
- Autonomic activity can also be influenced by the frontal association cortex through its involvement with emotional expression characteristics of the individual's personality.

## SOMATIC NERVOUS SYSTEM

*Motor neurons supply skeletal muscle.*

- Skeletal muscle is innervated by motor neurons, the axons of which constitute the somatic nervous system.
- The cell bodies of these motor neurons are located within the ventral horn of the spinal cord.
- The axon of a motor neuron is continuous from its origin in the spinal cord to its termination on skeletal muscle.

- Motor neuron axon terminals release acetylcholine, which brings about excitation and contraction of the innervated muscle fibers.
- Motor neurons can only stimulate skeletal muscles.

*Motor neurons are the final common pathway.*
- The somatic system is considered to be under voluntary control, but much of the skeletal muscle activity involving posture, balance, and stereotypical movements is subconsciously controlled.
- The cell bodies of the crucial motor neurons may be selectively destroyed by polio virus.
- Amyotropic lateral sclerosis, Lou Gehrig's disease, is characterized by progressive degeneration and death of motor neurons.

## NEUROMUSCULAR JUNCTION

*Most neurons and skeletal muscle fibers are chemically linked at neuromuscular junctions.*
- As an axon approaches a muscle, it divides into many terminal branches and loses its myelin sheath.
- Each of these axon terminals forms a special junction, a neuromuscular junction.

*Acetylcholine is the neuromuscular junction neurotransmitter.*
- A single muscle cell, a muscle fiber, is long and cylindrical in shape.
- The axon terminal is enlarged into a knoblike structure, the terminal button.
- The specialized portion of the muscle cell membrane immediately under the terminal button is known as the motor end plate.
- Nerve and muscle cells do not actually come into direct contact at the neuromuscular junction.
- A chemical messenger is used to carry the signal between the neuron terminal and the muscle fiber.
- Propagation of an action potential to an axon terminal triggers the opening of voltage-gated calcium channels in the terminal button.
- Opening the calcium channels permits calcium to diffuse into the terminal button, which in turn causes the release of acetylcholine from several hundred of the vesicles into the cleft.
- The released ACh diffuses across the cleft and binds with specific receptor sites.
- These cholinergic receptors are of the nicotinic type.
- Binding of ACh with these receptor sites induces the opening of chemical messenger-gated channels in the motor end plate.
- When ACh triggers the opening of these channels, considerably more sodium moves inward than potassium outward, bringing about a depolarization of the motor end plate.
- This potential change is known as the end-plate potential (EPP).
- It is similar to an EPSP except that the magnitude of an EPP is much larger because: (1) more transmitter is released from a terminal button than from a presynaptic knob in response to the action potential; (2) the motor end plate has a larger surface area and, accordingly, more transmitter receptor sites than a subsynaptic membrane; and (3) many more ion channels are opened in response to the transmitter receptor complex in the motor end plate.
- An EPP is a graded potential.
- The subsequent action potential triggers contraction of the muscle fiber.

*Acetylcholinesterase ends acetylcholine activity at the neuromuscular junction.*
- The muscle's electrical response is turned off by an enzyme in the motor end plate membrane, acetylcholinesterase (AChE), which inactivates ACh.
- Removal of Ach terminates the EPP so that no more action potentials are initiated.

*The neuromuscular junction is vulnerable to several chemical agents and diseases.*
- The venom of black widow spiders exerts its deadly effect by causing an explosive release of ACh from storage vesicles, the most detrimental consequence of which is respiratory failure.
- Botulinum toxin exerts its lethal blow by blocking the release of ACh from the terminal in the motor neuron.
- Death is due to respiratory failure.
- Curare reversibly binds to the ACh receptor sites on the motor end plate.
- Unlike ACh, however, curare does not alter membrane permeability, nor is it inactivated by AChE.
- When sufficient curare is present the person dies of respiratory paralysis.
- Organophosphates irreversibly inhibit AChE.
- Death from organophosphates is also due to respiratory paralysis.
- In the disease myasthenia gravis, a condition characterized by extreme muscular weakness, the body erroneously produces antibodies against its own motor end plate ACh receptors.

## Key Terms

acetylcholine (ACh)
acetylcholinesterase (AchE)
adrenal medulla
adrenergic fibers
agonists
alpha receptors
amyotropic lateral sclerosis (ALS)
angina pectoris
beta receptors
botulism
cholinergic fibers
collateral ganglion
curare
dystonias
end-plate potential
fight-or-flight response
final common pathway
Lou Gehrig's disease
motor end plate
motor neurons
muscle fiber
myasthenia gravis
neuromuscular junction
nicotinic receptors
noradrenalin
norepinephrine
organophosphates
parasympathetic nervous system
parasympathetic tone
postganglionic fiber
preganglionic fiber
sympathetic ganglion chain
sympathetic nervous system
sympathetic tone
sympathetic trunk
terminal button
terminal ganglia
tonic activity
varicosities

## Review Exercises

*Answers are in the appendix.*

**True/False**

_____ 1. As the axon approaches a muscle, it divides into many terminal branches and loses its myelin sheath.

_____ 2. Nerve and muscle cells occasionally come into direct contact at a neuromuscular junction.

_____ 3. A chemical messenger is used to carry the signal between the neuron terminal and the muscle fiber.

_____ 4. Propagation of an action potential to the axon terminal triggers the closing of voltage-gated calcium channels in the terminal button.

_____ 5. The released ACh diffuses across the cleft and binds with specific receptor sites.

_____ 6. The magnitude of an EPSP is much larger than an EPP.

_____ 7. The muscle's electrical response is turned on by an enzyme present in the motor end plate membrane, which inactivates ACh.

_____ 8. Motor neurons can only stimulate skeletal muscles.

_____ 9. The cells bodies of the somatic nervous system are located within the lateral horn of the spinal cord.

_____ 10. The somatic system is considered to be under voluntary control.

_____ 11. In the autonomic nervous system there is a single neuron from the origin in the CNS to the effector organ.

_____ 12. Sympathetic nerve fibers originate in the thoracic and lumbar regions of the spinal cord.

_____ 13. Short postganglionic fibers terminate on the effector organs.

_____ 14. Sympathetic preganglionic fibers arise from the cranial and sacral areas of the CNS.

_____ 15. Parasympathetic postganglionic fibers release acetylcholine.

_____ 16. Most visceral organs are innervated by both sympathetic and parasympathetic nerve fibers.

_____ 17. Both systems increase the activity of some organs and reduce the activity of others.

_____ 18. The parasympathetic system promotes responses that prepare the body for strenuous physical activity in the face of emergency or stressful situations.

**Fill in the Blank**

19. The skeletal muscle is innervated by ____________________.

20. Motor neurons are considered to be the __________________________ since the only way any other part of the nervous system can influence skeletal muscle activity is by acting on these motor neurons.

21. As the axon approaches a muscle, it divides into many terminal branches and loses its myelin sheath. Each of these axon terminals forms a special junction called _____________________________.

22. A single muscle cell is called a(n) _________________.

23. The axon terminal is enlarged into a knoblike structure called a(n) __________________.

24. The specialized portion of the muscle cell membrane immediately under the terminal button is known as the ____________________.

25. The venom of black widow spiders exerts its deadly effect by causing an explosive release of _______ from the storage vesicles.

26. __________________ exerts its lethal blow by blocking the release of ACh from the terminal button in response to an action potential in the motor neuron.

27. __________________ are a group of chemicals that modify neuromuscular junction activity.

28. One disease known as ___________________ is a condition characterized by extreme muscular weakness.

29. The ________________________ innervates the effector organ. The ________________________ synapses with the cell body of the second neuron.

30. Most sympathetic postganglionic fibers are called _____________________ because they release noradrenaline.

31. The terminal branches of autonomic fibers contain numerous swellings called _______ ____________.

32. About 20 percent of the adrenal medullary hormone output is norepinephrine, and the remaining 80 percent is the closely related substance ____________________.

33. __________________ respond to acetylcholine released from both sympathetic and parasympathetic preganglionic fibers.

34. __________________ are found on effector cell membranes.

**Matching**

*Match the distinguishing feature to the proper division of the autonomic nervous system.*

a. sympathetic system
b. parasympathetic system

_____ 35. Short cholinergic preganglionic fibers.
_____ 36. Origin of preganglionic fibers is the brain and sacral region of spinal cord.
_____ 37. Short cholinergic postganglionic fibers.
_____ 38. Long adrenergic postganglionic fibers.
_____ 39. Long cholinergic postganglionic fibers.
_____ 40. Long cholinergic preganglionic fibers.
_____ 41. Origin of postganglionic fibers is the terminal ganglia.
_____ 42. Nicotinic receptors for neurotransmitters.
_____ 43. This system dominates in emergency "fight or flight" situations.
_____ 44. Muscarinic receptors for neurotransmitters.

**Multiple Choice**

45. Atropine blocks the effect of:
a. acetylcholine at muscarinic receptors
b. acetylcholine at nicotinic receptors
c. acetylcholine at beta receptors
d. norepinephrine at alpha receptors
e. norepinephrine at beta receptors

46. Curare:
a. causes an explosive release of acetylcholine at the neuromuscular junction
b. blocks the effects of acetylcholine at muscarinic receptors
c. reversibly binds with acetylcholine receptor sites
d. irreversibly inhibits acetylcholinesterase
e. blocks the effect of norepinephrine at both beta receptors

47. Neostigmine:
a. causes an explosive release of norepinephrine at the neuromuscular junction
b. blocks the effect of acetylcholine at beta receptors
c. irreversibly inhibits acetylcholinesterase
d. prolongs the action of acetylcholine at the neuromuscular junction
e. blocks the release of acetylcholine at the neuromuscular junction

48. Botulinum toxin:
a. blocks the release of acetylcholine at the neuromuscular junction
b. prolongs the action of acetylcholine at the neuromuscular junction
c. blocks the effect of norepinephrine at the alpha receptors
d. reversibly binds with acetylcholine receptor sites
e. irreversibly inhibits acetylcholinersterase

49. Salbutamol:
a. activates beta adrenergic receptor sites
b. blocks the effect of norepinephrine at beta receptors
c. blocks the effect of acetylcholine at nicotinic receptors
d. activates the cholinergic receptors
e. irreversibly inhibits acetylcholinesterase

50. Mushroom poison:
a. blocks the effects of norepinephrine at muscarinic receptors
b. activates muscarinic receptors
c. activates nicotinic receptors
d. blocks the effect of norepinephrine at beta receptors
e. causes an explosive release of norepinephrine at the neuromuscular junction

51. Black widow spider venom:
a. blocks the effect of norepinephrine at beta receptors
b. activates muscarinic receptors
c. activates nicotinic receptors
d. causes an explosive release of acetylcholine at the neuromuscular junction
e. irreversibly inhibits acetylcholinesterase

52. Organophosphates:
a. blocks the effect of acetylcholine at alpha receptors
b. activates muscarinic receptors
c. activates nicotinic receptors
d. causes an explosive release of norepinephrine
e. irreversibly inhibits acetylcholinesterase

53. Military nerve gas:
a. irreversibly inhibits acetylcholinesterase
b. causes an explosive release of acetylcholine
c. blocks the release of acetylcholine at the neuromuscular junction
d. reversibly binds with acetylcholine receptor sites
e. blocks the release norepinephrine

54. Cocaine appears to:
a. block the parasympathetic innervation of the heart.
b. block the sympathetic innervation of the heart.
c. inhibit acetylcholinesterase.
d. bind with muscarine.
e. activate cholinergic acid.

55. The autonomic nervous system is:
a. Part of the somatic nervous system.
b. Considered to be the involuntary branch of the efferent nervous system.
c. Part of the efferent division of the peripheral nervous system.
d. Two of the above are correct.
e. All of the above are correct.

56. Parasympathetic postganglionic fibers:
a. Arise from the ganglion chain located along either side of the spinal cord.
b. Are cholinergic.
c. Secrete norepinephrine.
d. Both a and b are correct.
e. Both a and c are correct.

57. All of the following release acetylcholine except:
a. sympathetic preganglionic fibers
b. parasympathetic preganglionic fibers
c. sympathetic postganglionic fibers
d. parasympathetic postganglionic fibers
e. alpha motor neurons

58. The sympathetic nervous system:
a. is always excitatory
b. inervates only tissues concerned with protecting the body against challenges from outside
c. dominates in fight or flight situations
d. is part of the somatic nervous system
e. is part of the afferent nervous system

59. The sympathetic nervous system:
a. Is part of the somatic nervous system.
b. Has cholinergic preganglionic and adrenergic postganglionic fibers.
c. Originates in the thoracic and lumbar regions of the spinal cord.
d. Both b and c are correct.
e. A, b and c are correct.

60. Which of the following does not characterize the sympathetic nervous system?
a. It promotes responses that prepare the body for strenuous physical activity.
b. It is part of the autonomic nervous system.
c. It has norepinephrine as its postganglionic neurotransmitter.
d. It is always excitatory (that is, it increases the activity in every tissue it innervates).
e. It is part of the efferent division of the peripheral nervous system.

61. The parasympathetic nervous system:
a. Has long preganglionic fibers that end on terminal ganglia.
b. Dominates in quite, relaxed situations.
c. Releases postganglionic neurotransmitters that binds with muscarinic receptors.
d. Bind with norepinephrine released from sympathetic postgnglionic fibers.
e. More than one of the above are correct.

62. Nicotinic receptors:
a. Bind with acetylcholine released from parasympathetic postganglionic fibers.
b. Respond to acetylcholine released from both sympathetic and parasympathetic fibers.
c. Are found primarily in the heart.
d. Bind with norepiinephrine released from sympathetic postganglionic fibers.
e. More than one of the above are correct.

63. The chemical transmitter substance at the neuromuscular junction is:
a. Acetylcholine.
b. The same as the transmitter substance at parasympathetic postganglionic nerve endings.
c. Inactivated by organophosphates.
d. More than one of the above are correct.
e. None of the above are correct.

64. Acetylcholinesterase:
a. Is stored in vesicles in the terminal button.
b. When combined with receptor sites on the motor-end plate bring about an end-plate potential.
c. Is inhibited by organophosphates.
d. More than one of the above are correct.
e. None of the above are correct.

65. The motor end-plate:
a. Contains receptor sites that are capable of binding curare.
b. Contains acetylcholinesterase.
c. Experiences an increase in permeability to cations when combined with acetylcholine.
d. Both b and c are correct.
e. A, b and c are correct.

66. The neuromuscular junction:
    a. Is the junction between a motor neuron and a skeletal muscle fiber.
    b. Transmits an action potential between nerve cells.
    c. May produce either an EPSP or an IPSP on the motor end-plate.
    d. Both a and b are correct.
    e. A, b and c are correct.

67. Acetylcholinesterase:
    a. Is released from the terminal button.
    b. Destroys acetylcholine.
    c. Is blocked by curare.
    d. Both a and b are correct.
    e. A, b and c are correct.

68. Acetylcholine:
    a. Is released from the vesicles when an action potential is propagated to the terminal button of a motor neuron.
    b. Increases the permeability of the motor end-plate to sodium and potassium when combined at receptor sites on the motor end-plate.
    c. Is the chemical transmitter substance at the neuromuscular junction.
    d. Two of the above are correct.
    e. All of the above are correct.

69. Which of the following chemicals paralyzes skeletal muscle by binding to the acetylcholine receptor?
    a. black widow spider venom
    b. curare
    c. organophosphates
    d. DDT
    e. local anesthetics

70. Curare:
    a. Strongly binds to acetylcholine receptor sites.
    b. Inhibits acetylcholinesterase.
    c. Is found inpesticides and military nerve gases.
    d. Two of the above are correct.
    e. All of the above are correct.

**Modified Multiple Choice**

*Indicate which part of the autonomic nervous system is being described by writing the letter in the blank using the following code:*

A = sympathetic nervous system
B = parasympathetic nervous system
C = both sympathetic and parasympathetic nervous system
D = neither sympathetic nor parasympathetic nervous systems

_____ 71. Preganglionic fibers secrete acetylcholine
_____ 72. Preganglionic fibers secrete norepinephrine
_____ 73. Postganglionic fibers secrete acetylcholine
_____ 74. Postganglionic fibers secrete norepinephrine
_____ 75. Dominates in fight or flight situations
_____ 76. Dominates in relaxed situations
_____ 77. Has a long preganglionic fiber and a short postganglionic fiber
_____ 78. Has a short preganglionic fiber and a long postganglionic fiber
_____ 79. Originates in the cranial and sacral regions of the CNS
_____ 80. Originates in the thoracic and lumbar regions of the CNS

_____ 81. Innervates smooth muscle, cardiac muscle, and exocrine glands
_____ 82. Inervates skeletal muscle

*Indicate which type of neuron is associated with the characteristic by writing in the appropriate letter in the blank using the following code:*

A = all three types of neurons
B = both afferent and efferent
C = afferent neurons
D = efferent neurons
E = interneurons

_____ 83. Has a receptor at its peripheral ending
_____ 84. Autonomic nerves are this type of neuron
_____ 85. Lie primarily within the peripheral nervous system
_____ 86. Lie entirely within the central nervous system
_____ 87. Carry information from the central nervous system
_____ 88. Carry information to the central nervous system
_____ 89. Responsible for thoughts and other higher mental functions
_____ 90. Alpha motor neurons are this type of neuron
_____ 91. Terminate on effector organs

**Multiple Choice**

92. All of the following activities are correctly paired with the division of the autonomic nervous system except:
a. increased heart rate—sympathetic
b. decreased salivation—sympathetic
c. constricts skeletal muscles—parasympathetic
d. stimulates contraction of the bladder wall—parasympathetic
e. arises rostrally and sacrally—parasympathetic

93. The cell body of the postganglionic fiber of a parasympathetic pathway leading to a seat gland is located in the:
a. sweat gland itself
b. vertebral ganglion
c. spinal cord
d. brainstem
e. dorsal root ganglion

94. A comparison of the components of the autonomic visceral reflex arc with the somatic reflex arc reveals that they differ in the number of:
a. receptors
b. afferent neurons
c. interneurons
d. efferent neurons
e. effectors

95. Reflexes help to control:
    I. heart rate
    II. blood pressure
    III. digestive activities

    a. I only
    b. II only
    c. III only
    d. I and II only
    e. I, II, and III

96. If you were to electrically stimulate the parasympathetic nervous system, which of the following would occur?
    a. an erection of the penis
    b. dilation of the pupils
    c. an increased release of glucose by the liver
    d. constriction of the abdominal blood vessels
    e. constriction of the peripheral blood vessels

97. Which of the following does not occur following adrenergic stimulation?
    a. dilation of the pulmonary bronchi
    b. dilation of the pupils
    c. increased heart rate
    d. increased basal metabolism
    e. peristalsis of the GI tract

98. The parasympathetic and sympathetic nervous systems oppose each other in their effects on the:
    a. liver
    b. heart
    c. skeletal muscles
    d. sweat glands
    e. adrenal glands

99. Stimulation of skeletal muscles by motor neurons is essential for muscles to:
    a. maintain their strength
    b. induce contraction
    c. maintain their size
    d. both a and b
    e. a, b and c

100. One disease known to involve the neuromuscular junction is:
    a. myasthenia gravis
    b. polio
    c. Lou Gehrig's disease
    d. both a and b
    e. a, b and c

## *Points to Ponder*

1. Why do you think veterinarians find it difficult to diagnose toxic levels of organophosphate pesticides in animals?

2. When the opossum "plays possum," what is happening physiologically?

3. What is the relationship between biofeedback and the autonomic nervous system?

4. How does the central nervous system cooperate with the autonomic nervous system to maintain body temperature?

5. Which divisions of the autonomic nervous system prepare a physiology student, such as yourself, for intense muscular activity? Explain.

6. What is meant by the term "dual intervention"?

7. After being in space for long periods of time, why do space travelers have problems walking?

8. If you get bitten by a Black Widow spider, what problems might occur in your body?

9. Why is the "fight-or-flight" response necessary for the survival of a physiology student such as your self?

10. What happens in your body if you take the illegal drug cocaine?

## Clinical Perspectives

1. How does epinephrine help a person suffering from an asthma attack?

2. How does epinephrine help a person who has just been stung by a wasp and is hyper-sensitive to bee stings?

3. A patient is accidentally given curare. How would you keep this person alive? Would the same treatment work for an accidental dose of organophosphates?

4. An old antihypertensive drug known as propranolol was widely used because it blocked beta 1 receptors. However, it was discontinued because of its side affects on the respiratory system. What would these affects be?

5. When you get an eye examination, the optometrist or ophthalmologist puts drops of atropine in your eyes. Why? Explain the physiologically and pharmacology behind the action of the eye drops.

## PhysioEdge Activities

**Related to Text:**
No tutorials or media exercises directly related to text.

**Related to Figures:**
*Figure 7.1.* For an interaction related to this figure, see Media Exercise 7.1: Autonomic Nervous System.

*Figure 7.3.* For an interaction related to this figure, see Media Exercise 7.1: Autonomic Nervous System.
*Figure 7.6.* For an animation of this figure, click the Nerve-to-Muscle Communication tab in the Skeletal Muscle Contraction tutorial.

## Media Resources

**PhysioEdge CD-ROM**
For a visual review of concepts in this chapter, check out:

Tutorial: Skeletal Muscle Contraction, Nerve-to-Muscle Communication tab
Media Exercise 7.1: Autonomic Nervous System
Media Exercise 7.2: Neuromuscular Junction

**Book Companion Website**
The website for this book contains helpful study aids, and ideas for further reading and research. Log on to:
**http://www.brookscole.com/sherwoodhp6**

Select Chapter 7 from the drop-down menu, or click on one of the many resource areas.

For Suggested Readings, consult **InfoTrac® College Edition**, your online research library, at:
http://infotrac.thomsonlearning.com

8

# Muscle Physiology

## Chapter Overview

Muscle cells, and subsequently muscle fibers, are specialized to contract. Again the specialization is the extensive development of a basic fundamental function of most cells. Movement of the body, manipulation of objects, and movement of the contents of the body are all important in maintaining homeostasis. Muscles are classified by their appearance, their function, and the nerves that innervate them. The contractile elements are actin and myosin. Individual muscle fibers either contract or they do not contract. Muscles are under the control and coordination of the CNS. These contraction specialists respond to an electrical signal and convert chemical energy into mechanical energy. Muscles perform a wide variety of work in a broad array of responses. Again, we are looking at a body system working to maintain homeostasis, which is essential for the survival of all the cells. Considering all three types, muscles play a very large and important role in maintaining homeostasis.

## Chapter Outline

### INTRODUCTION

- Almost all living cells possess rudimentary intracellular machinery for producing such movement as redistributing various components of the cell during cell division.
- Muscle cells are the contraction specialists of the body.
- Controlled contraction of muscles allows: (1) purposeful movement of the whole body, or parts of the body, in relation to the environment; (2) manipulation of external objects; (3) propulsion of contents through various hollow internal organs; and (4) emptying the contents of certain organs to the external environment.
- Muscles can be classified in several different ways according to the common characteristics: striated or unstriated; voluntary (innervated by somatic nervous system) or involuntary (innervated by autonomic nervous system).

### STRUCTURE OF SKELETAL MUSCLE

*Skeletal muscle fibers are striated due to a highly organized internal arrangement.*

- Skeletal muscles are stimulated to contract via release of acetylcholine (ACh) at neuromuscular junctions between motor neuron terminals and muscle cells.
- A single skeletal muscle cell is known as a muscle fiber.
- A skeletal muscle consists of a number of muscle fibers lying parallel to each other.
- During embryonic development, the huge skeletal muscle fibers are formed by the fusion of many smaller cells; thus one striking feature is the presence of multiple nuclei in a single muscle cell.
- Another feature is the abundance of mitochondria.
- The most predominant structural feature of a skeletal muscle fiber is the presence of numerous myofibrils.
- Each myofibril consists of a regular arrangement of the proteins myosin (thick filaments) and actin (thin filaments).
- The bands of all the myofibrils lined up parallel to each other collectively lead to the striated appearance of a skeletal muscle fiber.
- An A band consists of a stacked set of thick filaments along with the portions of the thin filaments that overlap on both ends of the thick filaments.
- The thick filaments are found only in the A band and extend its entire width.
- The lighter area within the middle of the A band, where the thin filaments do not reach, is known as the H zone.
- Only the central portions of the thick filaments are found in this region.
- The I band consists of the remaining portion of the thin filaments that do not project into the A band.
- The I band contains only thin filaments but not the entire length of these filaments.
- In the middle of each I band is a dense, vertical Z line.
- The area between two Z lines is called a sarcomere, which is the functional unit of skeletal muscle.

*Myosin forms the thick filaments.*

- Each thick filament is composed of several hundred myosin molecules packed together in a specific arrangement.
- A myosin molecule is a protein consisting of two identical subunits shaped somewhat like a golf club.
- The globular heads form the cross bridges between thick and thin filaments.
- Each cross bridge has two important sites crucial to the contractile process: an actin binding site and a myosin ATPase site.
- Thin filaments are composed of three proteins: actin, tropomyosin, and troponin.

*Actin is the main structural component of the thin filaments.*

- Actin molecules are spherical in shape.
- The backbone of a thin filament is formed by actin molecules joined into two strands and twisted together.
- Tropomyosin molecules are threadlike proteins that lie end to end alongside the groove of the actin spiral.
- Tropomyosin covers the actin sites that bind with the cross bridges, thus blocking the interaction that leads to muscle contraction.
- Tropomyosin is stabilized in this blocking position by troponin molecules, which fasten down the ends of each tropomyosin molecule.
- When calcium binds to troponin, the shape of this protein is changed in such a way that tropomyosin is allowed to slide away from its blocking position.
- With tropomyosin out of the way, actin and myosin can bind and interact at the cross bridges, resulting in muscle contraction.

## Molecular Basis of Skeletal Muscle Contraction

*During contraction, cycles of cross-bridge binding and bending pull the thin filaments inward.*

- The thin filaments on each side of a sarcomere slide inward toward the A band's center during contraction.
- As they slide inward, the thin filaments pull the Z lines to which they are attached closer together, so the sarcomere shortens.
- This is known as the sliding-filament mechanism of muscle contraction.
- When myosin and actin make contact at a cross bridge, the conformation of the bridge is altered so that it bends inward.
- This so-called power stroke of a cross bridge pulls the thin filament to which it is attached inward.
- Complete shortening is accomplished by repeated cycles of cross-bridge binding and bending.
- Some cross bridges are "holding on" to the thin filaments, while others "let go" to bind with new actin.

*Calcium is the link between excitation and contraction.*

- Skeletal muscles are stimulated to contract by release of acetylcholine at neuromuscular junctions.
- At each junction of an A band and an I band, the surface membrane dips into the muscle fiber to form a transverse tubule (T tubule), which runs perpendicularly from the surface of the muscle cell membrane into the central portions of the muscle fiber.
- The T tubule provides a means of rapidly transmitting the surface electrical activity into the central portions of the fiber.
- The sarcoplasmic reticulum is a modified endoplasmic reticulum that consists of a fine network of interconnected tubules surrounding each myofibril.
- Separate segments of sarcoplasmic reticulum are wrapped around each A and I band.
- The ends of each segment expand to form saclike regions, the lateral sacs, which are separated from the adjacent T tubules by a slight gap.
- The sarcoplasmic reticulum lateral sacs store calcium.
- Spread of an action potential down a T tubule triggers the release of calcium from the sarcoplasmic reticulum into the cytosol.
- Foot proteins extend from the sarcoplasmic reticulum and span the gap between the lateral sac and the T tubule.
- Foot proteins serve as calcium-release channels, known as ryanodine receptors.
- The T tubule membrane also bears receptors at the points where it contacts the foot proteins protruding from the sarcoplasmic reticulum.
- These are known as dihydropyridine receptors.
- When an action potential is propagated down the T tubule, the dihydropyridine receptors trigger the opening of the ryanodine receptors.
- Consequently, calcium is released from the lateral sacs.
- This released calcium, by slightly repositioning the troponin and tropomyosin molecules, exposes the binding sites on the actin molecules so that they can link with the myosin cross bridges at their complementary binding sites.
- The ATPase site is an enzymatic site that can bind the energy carrier ATP, and split it into ADP and inorganic phosphate, yielding energy in the process.

- In skeletal muscle, magnesium must be attached to ATP before myosin ATPase can split the ATP.
- The generated energy is stored within the cross bridge to produce a high-energy form of myosin.
- When the muscle fiber is excited, calcium pulls the troponin-tropomyosin complex out of its blocking position so that the energized myosin cross bridge can bind with an actin molecule.
- The energy stored within the myosin cross bridge is released to cause the cross bridge bending responsible for the power stroke that pulls the thin actin filament inward.
- The contractile process is turned off when the calcium is returned to the lateral sacs upon cessation of local electrical activity.
- When acetylcholinesterase removes ACh from the neuromuscular junction, the muscle-fiber action potential ceases.
- Removal of cytosolic calcium allows the troponin-tropomyosin complex to slip back into its blocking position.
- Relaxation has occurred.

*Contractile activity far outlasts the electrical activity that initiated it.*

- The time delay between stimulation and the onset of contraction is known as the latent period.
- The time from the onset of contraction until peak tension is developed—the contraction time.
- The time from peak tension until relaxation is complete is called the relaxation time.

## Skeletal Muscle Mechanics

*Whole muscles are groups of muscle fibers bundled together and attached to bones.*

- Groups of muscle fibers are organized into whole muscles.
- Each muscle is covered by a sheath of connective tissue that penetrates from the surface into the muscle to envelop each individual fiber and divide the muscle into columns or bundles.
- The connective tissue further extends beyond the ends of the muscle to form tough, collagenous tendons that attach the muscle to bones.

*Contractions of whole muscles can be of varying strength.*

- A single action potential in a muscle fiber produces a brief, weak contraction known as a twitch.
- Muscle fibers are arranged into whole muscles where they can function cooperatively to produce contractions of variable grades of strength stronger than a twitch.
- Two primary factors can be adjusted to accomplish gradation of whole-muscle tension: (1) the number of muscle fibers contracting within a muscle and (2) the tension developed by each contracting fiber.

*The number of fibers contracting within a muscle depends on the extent of motor unit recruitment.*

- Larger muscles consisting of more muscle fibers are obviously capable of generating more tension than are smaller muscles with fewer fibers.
- One motor neuron innervates a number of muscle fibers, but each muscle fiber is supplied by only one motor neuron.
- This functional unit—one motor neuron plus all of the muscle fibers it innervates—is called a motor unit.
- For a weak contraction of the whole muscle, only one or a few of its motor units are activated.
- For stronger and stronger contractions, more and more motor units are recruited, or stimulated to contract, a phenomenon known as motor unit recruitment.
- To delay or prevent fatigue during a sustained contraction involving only a portion of a muscle's motor units, as is necessary in muscles supporting the weight of the body against the force of gravity, asynchronous recruitment of motor units takes place.
- The body alternates motor unit activity to give motor units that have been active an opportunity to rest while others take over.

*The frequency of stimulation can influence the tension developed by each muscle fiber.*

- Various factors influence the extent to which tension can be developed including: (1) the frequency of stimulation, (2) the length of the fiber at the onset of contraction, (3) the extent of fatigue, and (4) the thickness of the fiber.
- Even though a single action potential in a muscle fiber produces only a twitch, greater tension can be achieved by repetitive stimulation of the fiber.
- If the muscle fiber is stimulated a second time before it has completely relaxed from the first twitch, a second action potential occurs that causes a second contractile response.
- The two twitches resulting from the two action potentials add together, or sum, to produce greater tension than produced by a single action potential.
- This is known as twitch summation.
- If the muscle fiber is stimulated so rapidly that it does not have a chance to relax between stimuli, a smooth, sustained contraction of maximal strength known as tetanus occurs.

*Twitch summation results from a sustained elevation of cytosolic calcium.*
- Twitch summation depends on how long sufficient calcium is available.
- Additional factors not directly under nervous control also influence the tension developed during contraction.

*There is an optimal muscle length at which maximal tension can be developed.*
- For every muscle there is an optimal length at which maximal force can be achieved upon a subsequent tetanic contraction.
- In the body the muscles are so positioned that their relaxed length is approximately their optimal length.

*Muscle tension is transmitted to bone as the contractile component tightens the series-elastic component.*
- Tension is produced internally within the sarcomeres, considered the contractile component of the muscle.
- The noncontractile tissues are referred to as the series-elastic component.
- Typically, a muscle is attached to at least two different bones across a joint by means of tendons that extend from each end of the muscle.

*The two primary types of contraction are isotonic and isometric.*
- Tension is produced internally within the sarcomeres, the contractile component of the muscle.
- Tension generated by the contractile components must be transmitted to the bone via connective tissue and tendons.
- These noncontractile tissues are referred to as the series-elastic component.
- The end of the muscle attached to the more stationary part of the skeletal is called the origin, and the end attached to the skeletal part that moves is referred to as the insertion.
- The load is the weight of an object being lifted.
- In an isotonic contraction, muscle tension remains constant as the muscle changes length.
- In an isometric contraction, the muscle is prevented from shortening, so tension develops at constant muscle length.
- With concentric contractions the muscle shortens, whereas with eccentric contractions the muscle lengthens because it is being stretched by an external force while contracting.

*The velocity of shortening is related to the load.*
- The greater the load, the lower the velocity at which the muscle fiber shortens during an isotonic tetanic contraction.

*Although muscles can accomplish work, much of the energy is converted to heat.*
- Work is defined as force times distance.
- The amount of work depends on how much an object weighs and how far it is moved.
- All energy consumed by the muscle during contraction is converted to heat.
- Of the energy consumed by the muscle during contraction, 25 percent is realized as external work whereas the remaining 75 percent is converted to heat.
- This heat is not really wasted because it is used in maintaining body temperature.

*Interactive units of skeletal muscles, bones, and joints form lever systems.*
- Skeletal muscles are attached to bones across joints, forming lever systems.
- Skeletal muscles typically work at a mechanical disadvantage in that they must exert a considerably greater force than the actual load to be moved.
- The lever arrangement enables muscles to move loads faster over greater distances than would otherwise be possible.

## Skeletal Muscle Metabolism and Fiber Types

*Muscle fibers have alternate pathways for forming ATP.*
- Three different steps in the contraction-relaxation process require ATP: (1) the splitting of ATP by myosin ATPase; (2) the binding of a fresh molecule of ATP to myosin; and (3) the active transport of calcium back into the sarcoplasmic reticulum during relaxation which depends on energy derived from the breakdown of ATP.
- Only limited stores of ATP are immediately available in muscle, but three pathways supply additional ATP as needed during muscle contraction: (1) transfer of a high-energy phosphate from creatine phosphate to ADP, (2) oxidative phosphorylation, and (3) glycolysis.
- Creatine phosphate is the first energy storehouse tapped at the onset of contractile activity.
- The energy released from the hydrolysis of creatine phosphate, along with the phosphate, can be donated directly to ADP to form ATP.
- Oxidation phosphorylation takes place within the muscle mitochondria if sufficient oxygen is present.

- During light exercise to moderate exercise, muscle cells are able to form sufficient amounts of ATP through oxidative phosphorylation to keep pace with the modest energy demands of the contractile machinery for prolonged periods of time.
- Activity that can be supported in this way is known as endurance type exercise or aerobic exercise.
- The oxygen required for oxidative phosphorylation is primarily delivered by the blood.
- Some types of muscle fibers have an abundance of myoglobin, which increases the rate of oxygen transfer from the blood into muscle fibers.
- When oxygen delivery or oxidative phosphorylation cannot keep up with the demand for ATP formation as the intensity of exercise increases, the muscle fibers rely increasingly on glycolysis to generate ATP.
- Glycolysis alone has two advantages over the oxidative-phosphorylation pathway: (1) glycolysis can form ATP in the absence of oxygen, and (2) it can proceed more rapidly than oxidative phosphorylation because it requires fewer steps.
- Even though anaerobic glycolysis provides a means of performing intense exercise, using this pathway has two consequences: (1) large amounts of nutrient fuel must be processed, and (2) the end product of anaerobic glycolysis, pyruvic acid, is converted to lactic acid when it cannot be further processed by the oxidative-phosphorylation pathway.
- High-intensity or anaerobic exercise can be sustained for only a short duration in contrast to the body's prolonged ability to sustain aerobic activities.

*Fatigue may be of muscle or central origin.*

- There are three different types of fatigue: muscle fatigue, neuromuscular fatigue, and central fatigue.
- Muscle fatigue occurs when an exercising muscle can no longer respond to stimulation with the same degree of contractile activity.
- The primary implicated factors of muscle fatigue include: (1) accumulation of lactic acid and (2) depletion of energy reserves.
- With neuromuscular fatigue, the active motor neurons are not able to synthesize acetylcholine rapidly enough to sustain chemical transmissions of action potentials from the motor neurons to the muscles.
- Central fatigue, also known as psychological fatigue, occurs when the CNS no longer adequately activates the motor neurons supplying the working muscles.

*Increased oxygen consumption is necessary to recover from exercise.*

- The best known reason of elevated oxygen uptake during recovery is repayment of an oxygen debt that was incurred during exercise.
- During exercise, the creatine phosphate stores of active muscles are reduced, lactic acid may accumulate, and glycogen stores may be tapped.
- During the recovery period, fresh supplies of ATP are formed by oxidative phosphorylation using the newly acquired oxygen.
- This ATP is used to resynthesize creatine phosphate to restore its reserves.
- Any accumulated lactic acid is converted back into pyruvic acid, part of which is used in oxidative phosphorylation for ATP production.
- The remainder of the pyruvic acid is converted back into glucose by the liver.
- The glucose is used to replenish glycogen stores drained from the muscles and liver during exercise.
- These biochemical transformations involving pyruvic acid require oxygen.
- Part of the extra oxygen uptake during recovery is not directly related to the repayment of energy stores but instead is the result of a general metabolic disturbance following exercise.

*There are three types of skeletal muscle fibers based on differences in ATP hydrolysis and synthesis.*

- Based on their biochemical capacities, there are three major types of muscle fibers: (1) slow-oxidative (type I) fibers; (2) fast-oxidative (type IIa) fibers; and (3) fast-glycolytic (type IIb) fibers.
- Fast fibers have higher myosin ATPase activity than slow fibers.
- Fibers also differ in their ATP-synthesizing ability.
- Oxidative types of muscle fibers are more resistant to fatigue than are glycolytic fibers.
- Oxidative fibers also have a high myoglobin content and are referred to as red fibers.
- The glycolytic fibers contain very little myoglobin and therefore are pale in color, so they are sometimes called white fibers.
- In humans, most of the muscles contain a mixture of all three fiber types.

*Muscle fibers adapt considerably in response to the demands placed on them.*

- Two types of changes can be induced in muscle fibers: changes in their ATP-synthesizing capacity and changes in their diameter.
- Regular endurance exercise induces metabolic changes within the oxidative fibers, which are the ones primarily recruited during aerobic exercises.

- Muscles so adapted are better able to endure prolonged activity without fatiguing, but they do not change in size.
- The actual size of muscles can be increased by regular bouts of anaerobic, short duration, high-intensity resistance training, such as weight lifting.
- The resulting muscle enlargement comes primarily from an increase in diameter of the fast-glycolytic fibers that are called into play during such powerful contractions.
- The resultant bulging muscles are better adapted to activities that require intense strength for brief periods, but endurance has not been improved.
- The two types of fast fibers (type IIa and type IIb) are interconvertible.
- Slow and fast fibers are not interconvertible.
- Adaptive changes that take place in skeletal muscle gradually reverse to their original state over a period of months if the regular exercise program that induced these changes is discontinued.
- With age a person experiences a slow, continuous reduction in skeletal muscle mass and an accompanying loss of strength.

## Control of Motor Movement

*Multiple neural inputs influence motor unit output.*

- Three levels of input control motor neuron output: (1) input from afferent neurons, usually through intervening interneurons, at the level of the spinal cord—that is, spinal reflexes; (2) input from the primary motor cortex, the corticospinal (or pyramidal) motor system; and (3) input from the multineuronal (or extrapyramidal) motor system.
- The corticospinal system primarily mediates performances of fine, discrete, voluntary movements of the hands and fingers.
- The multineuronal system is primarily concerned with regulation of overall body posture involving involuntary movements of large muscle groups of the trunks and limbs.

*Muscle receptors provide afferent information needed to control skeletal muscle activity.*

- The CNS must know the starting position of your body to appropriately program muscle activity.
- The CNS must be constantly apprised of the progression of movement it has initiated so that it can make adjustments as needed.
- Your brain receives this information, which is known as proprioceptive input, from receptors in your eyes, joints, vestibular apparatus, and skin, as well as from the muscles themselves.
- Muscle length is monitored by muscle spindles, whereas changes in muscle tension are detected by Golgi tendon organs.
- Muscle spindles consist of collections of specialized muscle fibers known as intrafusal fibers, which lie within spindle-shaped connective tissue capsules parallel to the "ordinary" extrafusal fibers.
- The efferent neuron that innervates a muscle spindle's intrafusal fibers is known as a gamma motor neuron, whereas the motor neurons that supply the ordinary extrafusal fibers are designated as alpha motor neurons.
- The primary (annulospiral) endings are wrapped around the central portions of the intrafusal fibers; they detect changes in the length of the fibers during stretching as well as the speed with which it occurs.
- The secondary (flower-spray) endings, which are clustered at the end segments of many of the intrafusal fibers, are sensitive only to changes in length.
- Whenever the whole muscle is passively stretched, the intrafusal fibers are likewise stretched, increasing the rate of firing in the afferent nerve fibers.
- The afferent neuron directly synapses on the alpha motor neuron that innervates the extrafusal fibers of the same muscle, resulting in contraction of that muscle.
- This stretch reflex serves as a local negative-feedback mechanism to resist any passive changes in muscle length so that optimal resting length can be maintained.
- The primary purpose of the stretch reflex is to resist the tendency for the passive stretch of extensor muscles caused by gravitational forces when a person's standing upright.
- Golgi tendon organs are located in the tendons of the muscle, where they are able to respond to changes in the muscle's externally applied tension rather than to changes in its length.
- It is essential that motor control systems be apprised of the tension actually achieved so that adjustments can be made if necessary.
- The Golgi tendon organs consist of endings of afferent fibers entwined within bundles of connective tissue fibers that make up the tendon.
- When the extrafusal muscle fibers contract, the entwined Golgi organ afferent receptor endings are stretched, causing the afferent fibers to fire; the frequency of firing is directly related to the tension developed.
- Input from the activated Golgi tendon organs counterbalances excitatory inputs to the alpha motor neurons.

- This inhibitory response halts further contraction and brings about sudden reflex relaxation, thus helping prevent damage to muscle or tendons from excessive, tension-developing muscle contractions.

## Smooth and Cardiac Muscle

*Smooth and cardiac muscle share some basic properties with skeletal muscle.*

- Smooth muscle and cardiac muscle share some basic properties with skeletal muscle, but each also displays unique characteristics.

*Smooth muscle cells are small and unstriated.*

- The majority of smooth muscle cells are found in the walls of hollow organs and tubes.
- Smooth muscle cells are spindle-shaped, have a single nucleus, and are considerably smaller than skeletal muscle cells.
- Also, a single smooth muscle cell does not extend the full length of a muscle.
- Three types of filaments are found in a smooth muscle cell: (1) thick myosin filaments, (2) thin actin filaments, and (3) unique filaments of intermediate size.
- Lacking sarcomeres, smooth muscle does not have Z lines, but dense bodies containing the same protein constituent found in Z lines are present.
- Dense bodies are positioned throughout the smooth muscle cell as well as being attached to the internal surface of the plasma membrane.
- The actin filaments are anchored to the dense bodies.
- More actin is present in smooth muscle cells than skeletal muscle cells.

*Smooth muscle cells are turned on by calcium-dependent phosphorylation of myosin.*

- Smooth muscle myosin is able to interact with actin only when the myosin is phosphorylated.
- Smooth muscle calcium binds with calmodulin.
- This calcium-calmodulin complex binds to and activates myosin kinase, which in turn phosphorylates myosin.
- Phosphorylated myosin then binds with actin so that cross-bridge cycling can begin.
- When calcium is removed the muscle relaxes.

*Multiunit smooth muscle is neurogenic.*

- Smooth muscle is grouped into two categories—multiunit and single-unit smooth muscle based on differences in how the muscle fibers become excited.
- Contractile activity in both skeletal muscle and multiunit smooth muscle is neurogenic.
- Multiunit smooth muscle is supplied by the involuntary autonomic nervous system.
- Multiunit smooth muscle is found: (1) in the walls of large blood vessels, (2) in large airways to the lungs, (3) in the muscle of the eye that adjusts the lens, (4) in the iris of the eye, and (5) at the base of hair follicles.

*Single-unit smooth muscle cells form functional syncytia.*

- Most smooth muscle is of the single-unit variety.
- Single-unit is alternately called visceral smooth muscle because it is found in the walls of the hollow organs or viscera (for example, the digestive, reproductive, and urinary tracts and small blood vessels).
- The muscle fibers in single-unit smooth muscle are electrically linked by gap junctions.
- When an action potential occurs anywhere within a sheet of single-unit smooth muscle, it is quickly propagated throughout the entire group of interconnected cells, which then contract as a single coordinated unit, known as a functional syncytium.

*Single-unit smooth muscle is myogenic.*

- Single-unit smooth muscle is self-excitable rather than requiring nervous stimulation for contraction.
- There are two major types of spontaneous depolarizations displayed by self-excitable cells: pacemaker activity and slow-wave potentials.
- In pacemaker activity the membrane potential gradually depolarizes on its own.
- When the membrane has depolarized to threshold, an action potential is initiated.
- Self-generated action potentials are cyclically produced.
- Slow-wave potentials are gradually alternating hyperpolarizing and depolarizing swings in the potential, caused by automatic cyclical changes in the rate at which sodium ions are actively transported across the membrane.
- If threshold is reached, a burst of action potentials occurs at the peak of a depolarizing swing.
- Such nerve-independent contractile activity initiated by the muscle itself is called myogenic activity.

*Gradation of single-unit smooth muscle contraction differs from that of skeletal muscle.*

- In smooth muscle, the gap junctions assure than an entire smooth muscle mass contracts as a single unit, making it impossible to vary the number of muscle fibers contracting.
- The portion of cross bridges activated and the tension subsequently developed in single-unit

smooth muscle can be graded by varying the cytosolic calcium concentration.
- Many single-unit smooth muscle cells have sufficient levels of cytosolic calcium to maintain a low level of tension, or tone.
- Smooth muscle is typically innervated by both branches of the autonomic nervous system.
- In single-unit smooth muscle, this nerve supply does not initiate contraction, but it can modify the rate and strength of contraction, either enhancing or retarding the inherent contractile activity of a given organ.
- Each terminal branch of a postganglionic autonomic fiber travels across the surface of one or more smooth muscle cells, releasing transmitter from the vesicles within its multiple varicosities as an action potential passes along the terminal.
- The transmitter diffuses to the many receptor sites specific for it on the cells underlying the terminal.
- A given smooth muscle cell can be influenced by more than one type of neurotransmitter, and each autonomic terminal can influence more than one smooth muscle cell.

*Smooth muscle can still develop tension yet inherently relaxes when stretched.*
- The ability of a considerably stretched smooth muscle fiber to still develop tension is important, because the smooth muscle fibers within the wall of a hollow organ are progressively stretched as the volume of the organ's contents increases.
- When smooth muscle is suddenly stretched, the muscle quickly adjusts to this new length, and inherently relaxes to the tension level prior to the stretch.

*Smooth muscle is slow and economical.*
- A smooth muscle contractile response proceeds at a more leisurely pace than does a skeletal muscle twitch.
- Because of the low rate of cross bridge cycling, cross bridges are maintained in the attached state for a longer period of time during each cycle; that is, the cross bridges "latch onto" the thin filaments for a longer time each cycle.
- This so-called latch phenomenon enables smooth muscle to maintain tension with comparatively less ATP consumption.

*Cardiac muscle blends features of both skeletal and smooth muscle.*
- Cardiac muscle, found only in the heart, shares structural and functional characteristics with both skeletal and single-unit smooth muscle.
- Cardiac muscle is striated.
- Cardiac thin filaments contain troponin and tropomyosin.
- Cardiac muscle has a clear-cut length-tension relationship.
- Cardiac muscle cells have an abundance of mitochondria and myoglobin and possess T tubules.
- As in smooth muscle, calcium enters the cytosol from both the ECF and the sarcoplasmic reticulum during cardiac excitation.
- Calcium entry from the ECF through voltage-gated dihydropyridine receptors in the T tubule membrane triggers the release of calcium intracellular from the sarcoplasmic reticulum.
- Like single-unit smooth muscle, the heart displays pacemaker activity.
- Cardiac cells are interconnected by gap junctions.
- The heart is innervated by the autonomic nervous system, which along with certain hormones and local factors, can modify the rate and strength of contraction.
- Unique to cardiac muscle, the cardiac fibers are joined together in a branching network.

## Key Terms

A-band
actin
aerobic exercise
alpha motor neurons
atrophies
calmodulin
cardiac muscle
central fatigue
contraction time
creatine phosphate
cross bridges
denervation atrophy
dense bodies
dihydropydine receptors
disuse atrophy
endurance exercise
excitation-contraction coupling
extrafusal fibers
fatigue
flacid paralysis
foot proteins
force
fulcrum
glycolysis
Golgi tendon organs
hyperplasia
hypertrophy
H-zone
I-band
intrafusal fibers
isometric contraction
isotonic contraction
latch phenomenon

latent period
lateral sacs
lever
load arm
m line
motor unit
motor unit recruitment
multiunit smooth muscle
muscle fatigue
muscle fiber
muscle spindle
muscular dystrophy
myofibrils
myogenic activity
myoglobin
myosin
myosin kinase
neuromusular fatigue
optimal length
oxidative phosphorylation
oxygen debt
pacemaker potential
paraplegia
patella tendon reflex
power arm
power stroke
pyruvic acid
quadriplegia
regulatory proteins
relaxation time
rigor mortis
ryanodine receptors
sarcomere
sarcoplasmic reticulum
series elastic component
skeletal muscle
slow wave potential
smooth muscle
stretch reflex
summation
tendons
tension
tetanus
thick filaments
thin filaments
transverse tubule
tropomyosin
troponin
twitch
twitch summation
utrophin
white fibers
work
Z-line (disc)

## *Review Exercises*

*Answers are in the appendix.*

**True/False**

_____ 1. The corticospinal system primarily mediates performance of fine, discrete, voluntary movements of the hands and fingers.

_____ 2. Muscle spindles consist of specialized muscle fibers known as extrafusal fibers.

_____ 3. The efferent neuron that innervates a muscle spindle's intrafusal fibers is known as a gamma motor neuron.

_____ 4. The stretch reflex serves as a local positive-feedback mechanism to resist any passive changes in muscle length.

_____ 5. Golgi tendon organs respond to changes in muscle length.

_____ 6. Hemiplegia is the paralysis of one side of the body.

_____ 7. Ordinary muscle fibers are known as extrafusal fibers.

_____ 8. Smooth muscle cells are spindle-shaped with multiple nuclei.

_____ 9. In smooth muscle, myosin kinase phosphorylates myosin.

_____ 10. Smooth muscle calcium binds with calmodulin, an intracellular protein similar to troponin.

_____ 11. Contractile activity initiated by the muscle itself is called myogenic activity.

_____ 12. Cardiac cells are interconnected by gap junctions.

_____ 13. The portion of cross bridges activated and the tension subsequently developed in a single-unit smooth muscle can be graded by varying the cytosolic magnesium concentration.

_____ 14. The splitting of ATP by actin ATPase provides the energy for the power stroke of the cross bridge.

_____ 15. Creatine phosphate is the first energy storehouse tapped at the onset of contractile activity.

_____ 16. Hemoglobin increases the rate of oxygen transfer from the blood into the muscle fiber.

_____ 17. The end product of anaerobic glycolysis is lactic acid.

_____ 18. Psychological fatigue occurs when a person stops exercising even though the muscles are still able to perform.

_____ 19. Glycolytic fibers are more resistant to fatigue.

_____ 20. A single skeletal muscle cell is known as a myofibril.

_____ 21. Thick filaments are special assemblies of the protein myosin.

_____ 22. Each myofibril consists of a regular arrangement of the proteins myosin and actin.

_____ 23. Tropomyosin molecules lie end to end alongside the groove of the myosin spiral.

_____ 24. The globular heads of the myosin molecule form the cross bridges between the thick and thin filaments.

_____ 25. The area between two Z lines is called a sarcomere.

_____ 26. When magnesium binds to troponin, the shape of this protein is changed in such a way that tropomyosin is allowed to slide away from its blocking position.

_____ 27. One motor neuron plus all of the muscle fibers it innervates is called a motor unit.

_____ 28. A single action potential in a muscle fiber produces a brief, weak contraction known as a twitch.

_____ 29. The epithelial tissue further extends beyond the ends of the muscle to form a tough, collagenous tendon.

_____ 30. The asynchronous recruitment of motor units produces muscle spasms.

_____ 31. The thickness of the fiber influences the extent to which tension can be developed.

_____ 32. The phenomenon known as summation is when more and more motor units are stimulated for stronger contractions.

_____ 33. A sustained contraction of maximal strength is known as tetanus.

_____ 34. The end of the muscle attached to the more stationary part of the skeleton is known as the insertion.

_____ 35. In eccentric contractions the muscle shortens.

_____ 36. At each junction of an A band and an I band, the surface membrane dips into the muscle fiber to form a transverse tubule.

_____ 37. The sarcoplasmic tubule is a modified endoplasmic reticulum.

_____ 38. T tubules store calcium.

_____ 39. In skeletal muscle, magnesium must be attached to ATP before myosin ATPase can split the ATP.

_____ 40. The time delay between stimulation and the onset of contraction is known as the contraction time.

_____ 41. The contractile process is turned off when calcium is returned to the latent sacs upon cessation of local electrical activity.

_____ 42. Removal of cytosolic potassium allows the troponin-tropomyosin complex to slip back into its blocking position.

**Fill in the Blank**

43. ______________________________ refers to the series of events linking muscle excitation to muscle contraction.

44. At each junction of an A band and an I band, the surface membrane dips into the muscle fiber to form a(n) __________________, which runs perpendicularly from the surface of the muscle cell membrane into the central portions of the muscle fiber.

45. The ______________________ is a modified endoplasmic reticulum that consists of a fine network of interconnected tubules surrounding each myofibril.

46. The ends of each sarcoplasmic reticulum segments expand to form saclike regions, the ___________________, which lie in close proximity to the adjacent T tubules.

47. A myosin cross bridge has two special sites, a(n) ______________ site and a(n) ______ site.

48. The necessity for ATP in the separation of myosin and actin is amply demonstrated by the phenomenon of __________________.

49. The time delay between stimulation and the onset of contraction is known as the ________________.

50. _________________ is an intracellular protein found in most cells that is structurally similar to troponin.

51. ____________________________ exhibits properties and pathway between skeletal muscle and single unit smooth muscle.

52. ______________ means they depend on their nerve supply to initiate contraction. Examples are skeletal muscle and multiunit smooth muscle.

53. A group of interconnected muscle cells that function electrically and mechanically as a unit is known as a(n)_________________________.

54. _______________________ are gradually alternating hyperpolarizing and depolarizing swings in potential caused by automatic cyclical changes in the rate at which sodium ions are actively transported across the membrane.

55. _______________________ enables smooth muscle to maintain tension with comparatively less ATP consumption, because each cross-bridge cycle uses up one molecule of ATP.

56. A single skeletal muscle cell is known as _______________.

57. The most predominant structural feature of a skeletal muscle fiber is the presence of numerous ____________.

58. A(n) ___________ consists of a stacked set of thick filaments along with the portions of the thin filaments that overlap on both ends of the thick filaments.

59. The area between two Z lines is called a(n) ________________.

60. A(n) __________ molecule is a protein consisting of two identical subunits, each shaped somewhat like a golf club. __________ molecules are spherical in shape.

61. ______________ molecules are threadlike proteins that lie end to end alongside the groove of the actin spiral.

62. ______________ and ___________ are often referred to as regulatory proteins.

63. ______________ is paralysis of one side of the body.

64. ______________ is paralysis of all four limbs.

65. ______________ is paralysis of the legs.

66. The efferent neuron that innervates a muscle spindle's intrafusal fibers is known as a(n) ______________________, whereas the motor neurons that supply the ordinary extrafusal fibers are designated as ______________________.

67. ______________________ are located in the tendons of the muscle, where they are able to respond to changes in the muscle's externally applied tension rather than to changes in its length.

68. One motor neuron plus all of the muscle fibers it innervates is called a(n) ____________.

69. Cross-bridge activity produces __________ within the sarcomeres.

70. The noncontractile tissues are referred to as the ____________________ component of the muscle.

71. The end of the muscle attached to the more stationary part of the skeleton is called the _________, whereas the end attached to the skeletal part that moves is referred to as the ___________.

72. In a(n) __________ contraction, muscle tension remains constant as the muscle changes length.

73. There are actually two types of isotonic contractions, ________________ and ___________________.

74. ____________ muscles guard the exit of urine and feces from the body by isotonically contracting.

75. _______ is defined as force times distance.

76. A(n) _______ is a rigid structure capable of moving around a pivot point known as a(n) ____________.

77. ______________________ is the first energy storehouse tapped at the onset of contractile activity.

78. ______________________ takes place within the muscle mitochondria if sufficient oxygen is present.

79. ___________ increases the rate of oxygen transfer from the blood into muscle fibers.

80. __________ atrophy occurs when a muscle is not used for a long period of time even though the nerve supply is intact.

81. ______________________ is a hereditary pathological condition characterized by progressive degeneration of contractile elements, which are ultimately replaced by fibrous tissue.

**Matching**
*Match the characteristic to the type of fiber.*

a. fast-glycolytic type iib
b. fast-oxidative type iia
c. slow-oxidative type i

_____ 82. intermediate intensity of contraction
_____ 83. few capillaries
_____ 84. high resistance to fatigue
_____ 85. few mitochondria
_____ 86. small fiber diameter
_____ 87. low oxidative phosphorylation capacity
_____ 88. slow speed of contraction
_____ 89. white fibers
_____ 90. intermediate amount of enzymes for anaerobic glycolysis
_____ 91. intermediate glycogen content
_____ 92. low myoglobin content
_____ 93. low myosin-ATPase activity

**Multiple Choice**

94. Which of the following consists of a stacked set of thick filaments along with the portions of the thin filaments that overlap on both ends of the thick filaments?
a. Z-line
b. M-line
c. A-band
d. I-band
e. H-zone

95. Which of the following are threadlike proteins that lie end to end alongside the groove of the actin spiral?
a. tropomyison
b. troponin
c. myosin
d. tubulin
e. myoglobin

96. Which of the following ions must be attached to ATP before myosin ATPase can split the ATP in skeletal muscle?
a. magnesium
b. calcium
c. potassium
d. sodium
e. calmodulin

97. Which of the following represents the region where only the central portions of the thick filaments are found?
a. Z-line
b. M-line
c. A-band
d. I-band
e. H-zone

98. Which of the following pairs of cross-bridge sites is crucial to the contractile process?
a. troponin binding site and myosin ATPase site
b. tropomyosin binding site and actin ATPase site
c. myosin binding site and myosin ATPase site
d. actin binding site and myosin ATPase site
e. myoglobin binding site and troponin ATPase site

99. Which of the following ions binds to calmodulin in smooth muscles?
a. magnesium
b. calcium
c. potassium
d. sodium
e. troponin

100. Which of the following types of muscle has no gap junctions?
a. multiunit smooth
b. skeletal
c. single-unit smooth
d. cardiac
e. all of the above

101. Which of the following types of muscle has cross bridges turned on by calcium?
a. multiunit smooth
b. skeletal
c. single-unit smooth
d. cardiac
e. all of the above

102. Which of the following is innervated by the somatic nervous system?
a. multiunit smooth
b. skeletal
c. single-unit smooth
d. cardiac
e. all of the above

103. Which of the following is the functional unit of a skeletal muscle?
a. golgi organ
b. thick filaments
c. sarcomere
d. sliding filaments
e. transverse-contractile coupler

104. Which of the following receptor combinations is responsible for releasing calcium from the lateral sacs?
a. tropomyosine and troponin receptors
b. myosin and ryanodine receptors
c. dihydropyridine and ryanodine receptors
d. dihydropyridine and ACh receptors
e. ryanodine and AChase receptors

105. Which type of fatigue occurs due to an accumulation of lactic acid?
a. muscle fatigue
b. neuromuscular fatigue
c. central fatigue
d. all of the above
e. none of the above

106. Which of the following skeletal muscle structures has the same protein constituents as the dense bodies of smooth muscle?
 a. M-ine
 b. A-band
 c. I-band
 d. H-zone
 e. Z-zone

**Modified Multiple Choice**

*Indicate which type(s) of muscle is associated with the property in question by writing the appropriate letter in the blank using the code below.*

A = skeletal muscle only
B = single-unit smooth muscle only
C = cardiac muscle only
D = skeletal muscle and cardiac muscle
E = skeletal muscle and single-unit smooth muscle
F = single-unit smooth muscle and cardiac muscle
G = skeletal, single-unit smooth, and cardiac muscle

_____ 107. Contains actin, myosin, troponin, and tropomyosin.
_____ 108. Contains gap junctions.
_____ 109. Innervated by alpha motor neurons.
_____ 110. Is self excitable.
_____ 111. Maintains a constant membrane potential unless stimulated.
_____ 112. Innervated by the autonomic nervous system.
_____ 113. Is attached to bones.
_____ 114. Considered to be involuntary.
_____ 115. Thick and thin filaments are highly organized into a banding pattern.
_____ 116. Found in the heart.
_____ 117. Can exist over a variety of lengths with little change in tension.
_____ 118. Is striated.
_____ 119. Is found in the walls of hollow organs such as the digestive tract, bladder, uterus, blood vessels.
_____ 120. Behaves as a functional syncytium.
_____ 121. Is under voluntary control.
_____ 122. Has a clear-cut length tension relationship.
_____ 123. Basis of contraction is cross bridge interaction between actin and myosin.
_____ 124. Contraction is triggered as calcium physically pulls troponin and tropomyosin from its blocking position over actin's binding sites for cross bridges.
_____ 125. Myosin must be phosphorylated before it can bind with actin.
_____ 126. Contains T tubules.
_____ 127. Displays pacemaker potentials and slow wave potentials.
_____ 128. Is neurogenic.

*Indicate which bands are being described in each statement by writing the appropriate letter(s) in the blank using the answer code below. Note that more than one answer may apply.*

A = A-band
B = I-band
C = H-zone

_____ 129. composed of thin filaments only
_____ 130. composed of thick filaments only
_____ 131. composed of both thick and thin filaments

_____ 132. shortens during muscular contraction
_____ 133. remains the same size during muscular contraction

## Points to Ponder

1. Can you think of an example where involuntary muscles are stimulated to contract voluntarily?

2. How can an average-sized woman beat a contestant for Mr. Universe in arm wrestling?

3. How does weight-lifting shape up as a healthy form of exercise?

4. Why do you think the average life expectancy for NFL players is only 57 years?

5. Why does skeletal muscle constitutes about 40-50 percent of an adult male's body weight and 30-40 percent of an adult female's body weight?

6. What makes one muscle larger than another muscle?

7. What causes your foot to "fall asleep"?

8. What causes rigor mortis to occur after death?

9. While running on a hot day, you develop muscle cramps. Which of the following will help the muscle cramps: Drinking a lot of water, massaging the muscles, stretching the cramped muscles, or eating a banana as a source of potassium?

10. Thirty-five percent of each cardiac muscle cell is made up of mitochondria compared with 2 percent in skeletal muscle cells. What does this reflect physiologically?

## Clinical Perspectives

1. Explain why heat and/or cold packs are used to treat the so-called "pulled muscle."

2. How does rigor mortis help the medical examiner determine the time of death?

3. Why does Duchenne muscular dystrophy only occur in males?

4. Rapid stretching of skeletal muscles produces very forceful muscle contractions. This can result in painful muscle spasms. How can these muscle spasms be prevented?

5. How does the disease polio cause flaccid paralysis, reduced muscle tone, depressed stretch reflexes, and atrophy?

6. Parkinson's disease causes a resting tremor to occur. This "shaking of the limbs" tends to disappear during voluntary movements and then reappears when the limbs are at rest again. Why does this occur?

7. The enzyme that transfers phosphate between creatine and ATP is called creatine kinase. Skeletal muscle and heart muscle have two different forms of this enzyme (isozymes). What diseases can these isozymes detect in muscle tissue?

8. How do calcium channel blocking drugs work in the walls of blood vessels to help reduce blood pressure in those patients who have hypertension?

9. In some patients, calcium channel blocking drugs can cause fatigue and muscle pain. Explain why this occurs physiologically.

10. From 12-48 hours after strenuous exercise, skeletal muscles often become sore. Such delayed onset muscle soreness is accompanied by stiffness, tenderness, and swelling. Why do these symptoms occur?

## *Experiments of the Day*

1. Go to the meat department of your favorite grocery store and explain the appearance of the meats in terms of muscle structure.

2. Go to the vitamin supplement department of your favorite pharmacy and find supplements that claim to increase muscle mass. Are these the same as metabolic steroids? Are they good for you? Explain.

3. The next time you eat a steak, try to identify the anatomical parts of the steak.

## *PhysioEdge Activities*

**Related to Text:**
Media Exercise 8.1: Muscle Types.
Media Exercise 8.2: Skeletal Muscle Mechanics.
Media Exercise 8.3: Skeletal Muscle Structure.
Media Exercise 8.4: Smooth Muscle Structure.

**Related to Figures:**
*Figure 8.2.* For an interaction related to this figure, see the Anatomy Review tab in the Skeletal Muscle Contraction tutorial and Media Exercise 8.2: Skeletal Muscle Mechanics.
*Figure 8.3.* For an interaction related to this figure, see Media Exercise 8.1: Muscle Types.
*Figure 8.10.* For interactions related to this figure, see the Anatomy Review tab in the Skeletal Muscle Contraction tutorial and Media Exercise 8.2: Skeletal Muscle Mechanics.
*Figure 8.12.* For an animation of this figure, click the Contraction/Relaxation tab in the Skeletal Muscle Contraction tutorial.
*Figure 8.13.* For an animation of this figure, click the Contraction/Relaxation tab in the Skeletal Muscle Contraction tutorial.
*Figure 8.25.* For an animation of this figure, click the Muscle Spindle tab (p.4) in the Skeletal Muscle Contraction tutorial.
*Figure 8.26.* For an animation of this figure, click the Control of Motor Movement tab in the Skeletal Muscle Contraction tutorial.
*Figure 8.27.* For an interaction related to this figure, see Media Exercise 8.4: Smooth Muscle.

## *Media Resources*

**PhysioEdge CD-ROM**
For a visual review of concepts in this chapter, check out:

Tutorial: Skeletal Muscle Contraction
Media Exercise 8.1: Muscle Types
Media Exercise 8.2: Skeletal Muscle Mechanics
Media Exercise 8.3: Skeletal Muscle Structure
Media Exercise 8.4: Smooth Muscle

**Book Companion Website**
The website for this book contains a wealth of helpful study aids, as well as many ideas for further reading and research. Log on to:

**http://www.brookscole.com/sherwoodhp6**

Select Chapter 8 from the drop-down menu, or click on one of the many resource areas.

For Suggested Readings, consult **InfoTrac® College Edition**, your online research library, at:

**http://infotrac.thomsonlearning.com**

# Cardiac Physiology

## Chapter Overview

The importance of the circulatory system, particularly the heart, in maintaining homeostasis has been known to some degree by most of us since our childhood days. This heart-homeostasis relationship is probably the best known aspect of homeostasis. From the bending and twisting of an embryonic blood vessel, an organ develops that will pump the blood. Complete with muscular walls and valves, the heart responds to the autorhythmicity of the sinoatrial node. Using specialized conductive cells, action potentials are spread throughout the organ and coordinated contractions occur. After a brief period of relaxation the cycle begins again.

The electrical and mechanical activities of the heart produce sounds, pressures, and electrical changes that can be detected on the surface of the body. Electrocardiography is an important tool for monitoring cardiac activities.

To meet the changing homeostatic needs of all the cells, the heart must be able to change its output. Cardiac output is controlled through changes in heart rate and stroke volume. While the heart is working for all the cells it must also nourish itself. This is one of the most frequent areas of complications in the circulatory system.

Even though the heart is emphasized in science and culture, in reality it is but a part of an organ system controlled by the nervous and endocrine systems. The remarkable properties of the heart are specializations of the general properties and functions of all cells. The circulatory system aids in maintaining homeostasis, which is essential for the survival of cells.

## Chapter Outline

INTRODUCTION

- The circulatory system consists of the heart, the blood vessels, and the blood.
- The pulmonary circulation consists of a closed loop of vessels carrying blood between the heart and lungs, whereas the systemic circulation consists of a circuit of vessels carrying blood between the heart and organ systems.

ANATOMY OF THE HEART

*The heart is located in the middle of the chest cavity.*

- The heart is situated at an angle under the sternum so that its base lies predominately to the right and the apex to the left of the sternum.
- The fact that the heart is positioned between two bony structures, the sternum and vertebrae, make it possible to manually drive blood out of the heart.

*The heart is a dual pump.*

- Even though anatomically the heart is a single organ, the right and left sides of the heart function as two separate pumps.
- The upper chambers, the atria, receive blood returning to the heart and transfer it to the lower chambers, the ventricles, which pump the blood from the heart.
- The vessels that return blood from the tissues to the atria are veins, and those that carry blood away from the ventricles to the tissues are arteries.
- The right side of the heart pumps blood into the pulmonary circulation.
- The right ventricle pumps blood out through the pulmonary artery to the lungs.
- The left side of the heart pumps blood into the systemic circulation.
- The large artery carrying blood away from the left ventricle is the aorta.
- Both sides of the heart simultaneously pump equal amounts of blood.
- The volume of oxygen-poor blood being pumped to the lungs by the right side of the heart soon becomes the same volume of oxygen-rich blood being delivered to the tissues by the left side of the heart.
- The pulmonary circulation is a low-pressure, low-resistance system, whereas the systemic circulation is a high-pressure, high-resistance system.

*Pressure operated valves ensure that the blood flows in the right direction through the heart.*

- Blood flows through the heart in one fixed direction from veins to atria to ventricles to arteries.
- Two of the heart valves, the right and left atrioventricular (AV) valves, are positioned between the atrium and the ventricle on the right and left sides, respectively.
- The two remaining heart valves, the aortic and pulmonary valves, are located at the junction where the major arteries leave the ventricles.
- They are known as semilunar valves because they are composed of three cusps, each resembling a shallow half-moon shaped pocket.
- The heart forms from a single tube that bends upon itself and twists on its axis during embryonic development.

*The heart walls are composed primarily of spirally arranged cardiac muscle fibers.*

- The heart wall consists of three distinct layers: (l) The endocardium is a thin inner layer of endothelium that lines the entire circulatory system. (2) The myocardium, the middle layer composed of cardiac muscle, constitutes the bulk of the heart wall. (3) The epicardium is a thin external membrane covering the heart.
- The myocardium consists of interlacing bundles of cardiac muscle fibers arranged spirally around the circumference of the heart.
- The spiral arrangement is due to the heart's complex twisting during development.
- When the ventricular muscle contracts and shortens, the diameter of the ventricular chambers is reduced while the apex is simultaneously pulled upward toward the top of the heart in a rotating manner.

*Cardiac muscle fibers are interconnected by intercalated discs and form functional synctia.*

- The individual cardiac muscle cells are interconnected to form branching fibers, with adjacent cells joined end to end at specialized structures known as intercalated discs.
- Within an intercalated disc, there are two types of membrane junctions: desmosomes and gap junctions.
- Cardiac muscle is capable of generating action potentials without nervous stimulation.
- There are no gap junctions between atrial and ventricular contractile cells, and furthermore, these muscle masses are separated by the electrically nonconductive fibrous skeleton that surrounds the valves.
- An important specialized conducting system is present to facilitate and coordinate the transmission of electrical excitation from the atria to the ventricles.
- Either all of the cardiac muscle fibers contract or none of them do.
- Gradation of cardiac contraction is accomplished by varying the strength of contraction of all the cardiac muscle cells.
- Cardiac muscle cells contain an abundance of mitochondria and myoglobin.
- No new cardiac-muscle cells are produced after infancy.

*The heart is enclosed by the pericardial sac.*

- The heart is enclosed in the double-walled, membranous pericardial sac.
- The sac is lined by a membrane that secretes a thin pericardial fluid, which provides lubrication to prevent friction between the pericardial layers as they glide over each other.

## ELECTRICAL ACTIVITY OF THE HEART

*Cardiac autorhythmic cells display pacemaker activity.*

- The heart contracts, or beats, rhythmically as a result of action potentials that it generates by itself, a property known as autorhythmicity.
- There are two types of cardiac muscle cells: (2) Ninety-nine percent of the cardiac muscle cells are contractile cells. (2) The autorhythmic cells do not contract, but instead are specialized for initiating and conducting the action potentials responsible for contraction.
- The cardiac autorhythmic cells display pacemaker activity.
- The membrane potential's slow drift to threshold is caused by a cyclical decrease in passive outward flux of potassium superimposed on a slow, unchanging inward leak of sodium.
- The cardiac cells capable of autorhythmicity are found in the following specific locations: (1) the sinoatrial node, (2) the atrioventricular node, (3) the bundle of His, and (4) the Purkinje fibers.
- Various autorhythmic cells differ in the rates at which they are normally capable of generating action potentials.

*The sinoatrial node is the normal pacemaker of the heart.*

- The SA node, which normally exhibits the fastest rate of autorhythmicity, is known as the pacemaker of the heart.
- The non-SA nodal autorhythmic tissues are latent pacemakers.

*The spread of cardiac excitation is coordinated to ensure efficient pumping.*

- For efficient cardiac function, the spread of excitation should satisfy three criteria: (1) Atrial excitation and contraction should be complete before the onset of ventricular contraction. (2) Excitation of cardiac muscle fibers should be coordinated to ensure that each heart chamber contracts as a unit to accomplish efficient pumping. (3) The pair of atria and the pair of ventricles should be functionally coordinated so that both members of the pair contract simultaneously.
- Simultaneous contraction permits synchronized pumping of blood into the pulmonary and systemic circulation.
- The normal spread of cardiac excitation is carefully orchestrated to ensure that these criteria are met and the heart functions efficiently.
- An action potential originating in the SA node first spreads throughout both atria, primarily from cell to cell via gap junctions.
- The interatrial pathway extends from the SA node within the right atrium to the left atrium.
- This pathway ensures that both atria become depolarized to contract more or less simultaneously.
- The internodal pathway extends from the SA node to the AV node.
- The internodal conduction pathway directs the spread of an action potential originating at the SA node to the AV node to ensure sequential contraction of the ventricles following atrial contraction.
- The action potential is conducted relatively slowly through the AV node.
- The slowness is advantageous because it allows time for complete ventricular filling to occur.
- The impulse rapidly travels down the bundle of His and throughout the ventricular myocardium via the Purkinje fibers.
- The ventricular conduction system is more highly organized and more important than the interatrial and internodal conduction pathways.
- The rapid conduction of the action potential down the bundle of His and its swift, diffuse distribution throughout the Purkinje network lead to almost simultaneous activation of the ventricular myocardial cells in both ventricular chambers, which ensures a single, smooth, coordinated contraction that can efficiently eject blood into both the systemic and pulmonary circulations at the same time.

*The action potential of contractile cardiac muscle cells shows a characteristic plateau.*

- Unlike autorhythmic cells, the membrane of contractile cells remains essentially at rest, at about -90mV, until excited by electrical activity propagated from the pacemaker.
- Once the membrane is excited, an action potential is generated by a complicated interplay of permeability changes and membrane potential changes.
- During the rising phase of the action potential, the membrane potential rapidly becomes reversed to a positive value of +30mV as a result of an explosive sodium influx.
- The membrane potential is maintained at this positive level for several hundred milliseconds, producing a plateau phase of the action potential.
- Two voltage-dependent permeability changes are responsible for maintaining this plateau: activation of slow calcium channels and a marked decrease in potassium permeability.
- The rapid falling phase of the action potential results from inactivation of the calcium channels and activation of potassium channels.
- The mechanism by which an action potential in a cardiac muscle fiber brings about contraction of that fiber is quite similar to the excitation-contraction coupling process of skeletal muscle and smooth muscle.
- Cardiac muscle's T tubular dihydropyridine receptors serve as calcium channels that are opened in response to local membrane depolarization.

*Calcium entry from the ECF induces a much larger calcium release from the sarcoplasmic reticulum.*

- Calcium diffuses into the cytosol across the T tubule membrane during cardiac action potential.
- Entering calcium triggers release of calcium from the sarcoplasmic reticulum.
- Calcium also diffuses into the cytosol across the plasma membrane from the ECF during a cardiac action potential.
- This extra supply of calcium is not only the major factor responsible for the prolongation of the cardiac action potential but is also responsible for the subsequent lengthening of the period of cardiac contraction.
- In cardiac muscle the extent of cross-bridge activity varies with the amount of cytosolic calcium.

*A long refractory period prevents tetanus of cardiac muscle.*

- During the refractory period, which occurs immediately after the initiation of an action potential, an excitable membrane's responsiveness is totally abolished.

- Cardiac muscle has a long refractory period.
- Consequently, cardiac muscle cannot be restimulated until contraction is almost over, making summation of contractions and tetanus of cardiac muscle impossible.

*The ECG is a record of the overall spread of electrical activity through the heart.*

- An ECG is a recording of that portion of the electrical activity induced in the body fluids by the cardiac impulse that reaches the surface of the body, not a direct recording of the actual electrical activity of the heart.
- The ECG is a complex recording, representing the overall spread of activity throughout the heart during depolarization and repolarization.
- The ECG is not a recording of a single action potential in a single cell.
- The recording represents comparisons in voltage detected by electrodes at two different points on the body surface, not the actual potential.
- To provide standard comparisons, ECG records routinely consist of twelve conventional electrode systems or leads.

*Different parts of the ECG record can be correlated to specific cardiac events.*

- A normal ECG exhibits three distinct waveforms: the P wave, the QRS complex, and the T wave.
- The P wave represents atrial depolarization.
- The QRS complex represents ventricular depolarization.
- The T wave represents ventricular repolarization.
- The electrical activity associated with atrial repolarization normally occurs simultaneously with ventricular depolarization and is marked by the QRS complex.
- The P wave is much smaller than the QRS complex because the atria have a much smaller muscle mass than the ventricles.
- There are three times when no current is flowing in the heart musculature and the ECG remains at baseline: (1) during the AV nodal delay, (2) the ST segment, and (3) the TP interval.

*The ECG can be useful in diagnosing abnormal heart rates, arrhythmias, and damage of heart muscle.*

- Because electrical activity triggers mechanical activity, abnormal electrical patterns are usually accompanied by abnormal contractile activity of the heart.
- The principle deviations from normal that can be ascertained through electrocardiography are as follows: tachycardia, bradycardia, arrhythmia, atrial flutter, atrial fibrillation, ventricular fibrillation, and heart block.
- Abnormal ECG waves are also important in the recognition and assessment of cardiac myopathies, such as myocardial ischemia, necrosis, coma, and acute myocardial infarction.

## MECHANICAL EVENTS OF THE CARDIAC CYCLE

*The heart alternately contracts to empty and relaxes to fill.*

- The cardiac cycle consists of alternate periods of systole and diastole.
- The atria and ventricles go through separate cycles of systole and diastole.
- Because of the continuous inflow of blood from the venous system into the atrium, atrial pressure slightly exceeds ventricular pressure even though both chambers are relaxed.
- The AV valve opens and blood flows directly from the atrium into the ventricle throughout ventricular diastole.
- Ventricular volume slowly continues to rise even before atrial contraction takes place.
- Late in ventricular diastole the SA node fires.
- The impulse spreads throughout the atria (P wave).
- Atrial depolarization brings about atrial contraction.
- Throughout atrial contraction, atrial pressure still slightly exceeds ventricular pressure, so the AV valve remains open.
- Ventricular diastole ends at the onset of ventricular contraction.
- The volume of blood in the ventricle at the end of diastole is known as the end-diastolic volume (EDV).
- Following atrial excitation, the impulse passes through the AV node and specialized conducting system to excite the ventricle.
- Simultaneously, atrial contraction is occurring.
- By the time ventricular activation is complete, atrial contraction is already accomplished.
- The QRS complex represents this ventricular excitation, which induces ventricular contraction.
- As ventricular contraction begins, ventricular pressure immediately exceeds atrial pressure.
- The backward pressure differential forces the AV valve closed.
- There is a brief period of time between closure of the AV valve and opening of the aortic valve when the ventricle remains a closed chamber.
- This interval is termed the period of isovolumetric ventricular contraction.
- When ventricular pressure exceeds aortic pressure, the aortic valve is forced open and ejection of blood begins.

- Ventricular systole includes both the period or isovolumetric contraction and the ventricular ejection phase.
- About half of the blood contained within the ventricle at the end of diastole is pumped out during the subsequent systole.
- The amount of blood remaining in the ventricle at the end of ventricular systole is known as end-systolic volume (ESV).
- The amount of blood pumped out of each ventricle with each contraction is known as the stroke volume, the difference between the volume of blood in the ventricle before contraction and the volume after contraction (EDV-ESV).
- The T wave signifies ventricular repolarization occurring at the end of ventricular systole.
- As the ventricle starts to relax, the aortic valve closes.
- Closure of the aortic valve produces a disturbance or notch on the aortic pressure curve known as the dicrotic notch.
- All valves are once again closed for a brief period of time, known as isovolumetric ventricular relaxation.
- When the ventricular pressure falls below the atrial pressure, the AV valve opens and ventricular filling occurs once again.

*Two heart sounds associated with valve closures.*

- Two major heart sounds normally can be heard during the cardiac cycle when listening with a stethoscope.
- The first sound is associated with the closure of the semilunar valves.
- Closure of the AV node occurs at the onset of ventricular contraction.
- Closure of the semilunar valves occurs at the onset of ventricular relaxation.

*Turbulent blood flow produces heart murmurs.*

- Abnormal heart sounds, or murmurs, are usually associated with cardiac disease.
- Blood normally flows in a laminar fashion and does not produce any sound.
- When blood flow becomes turbulent, however, a sound can be heard and is due to vibrations created in the surrounding structures by the turbulent flow.
- The most common cause of turbulence is valve malfunction, either a stenotic (a stiff, narrowed valve) or an insufficient valve (one that cannot close completely).
- Most often, both valvular stenosis and insufficiency are caused by rheumatic fever.
- The valve involved and the type of defect can usually by detected by the location and timing of the murmur.
- A murmur occurring between the first and second heart sounds signifies a systolic murmur.
- A diastolic murmur occurs between the second and first heart sounds.
- The sound of the murmur characterizes it as either a stenotic (whistling) murmur or an insufficient (swishy) murmur.

## Cardiac Output and Its Control

*Cardiac output depends on the heart rate and the stroke volume.*

- Cardiac output is the volume of blood pumped by each ventricle per minute.
- The two determinants of cardiac output are heart rate and stroke volume.

*Heart rate is determined primarily by autonomic influences on the SA node.*

- The heart is innervated by both divisions of the autonomic nervous system, which can modify the rate of contraction, even though nervous stimulation is not required to initiate contraction.
- The parasympathetic nerve to the heart, the vagus nerve, primarily supplies the atrium, especially the SA and AV nodes.
- There is no significant parasympathetic innervation to the ventricles.
- The cardiac sympathetic nerves also supply the atria, including the SA and AV nodes, and richly innervate the ventricles as well.
- The parasympathetic nervous system's influence on the SA node is to decrease the heart rate.
- Parasympathetic influence on the AV node decreases the node's excitability, prolonging the AV nodal delay.
- Parasympathetic stimulation of the atrial contractile cells shortens the action potential, that is, the plateau phase is reduced and atrial contraction is weakened.
- The sympathetic nervous system speeds up the heart rate through its effect on the pacemaker tissue.
- Sympathetic stimulation of the AV node reduces the AV nodal delay.
- Similarly, sympathetic stimulation speeds up the spread of the action potential throughout the specialized conducting pathway.
- Sympathetic stimulation increases contractile strength so that the heart beats more forcefully and squeezes out more blood.

*Stroke volume is determined by the extent of venous return and by sympathetic activity.*

- Two types of controls influence stroke volume: (1) intrinsic control related to the extent of venous return, and (2) extrinsic control related to the extent of sympathetic stimulation of the heart.

*Increased end-diastolic volume results in increased stroke volume.*

- The direct correlation between end-diastolic volume and stroke volume constitutes the intrinsic control of stroke volume, which refers to the heart's inherent ability to vary the stroke volume.
- For cardiac muscle, the resting cardiac muscle fiber length is less than optimal length.
- An increase in cardiac muscle fiber length, by moving closer to the optimal length, increases the contractile tension of the heart on the following systole.
- The main determinant of cardiac muscle fiber length is the degree of diastolic filling.
- This intrinsic relationship between end-diastolic volume and stroke volume is known as the Frank-Starling law of the heart.
- Stated simply, the law says that the heart normally pumps all the blood returned to it.

*Sympathetic stimulation increases the contractility of the heart.*

- Stroke volume is also subject to extrinsic control.
- Sympathetic stimulation and epinephrine enhance the heart's contractility.
- Sympathetic stimulation increase stroke volume by enhancing venous return.
- Sympathetic stimulation constricts the veins, which squeezes more blood forward from the veins to the heart, increasing the end-diastolic volume.
- The strength of cardiac contraction and the stroke volume can thus be graded by: (1) varying the initial length of the muscle fibers, and (2) varying the extent of sympathetic stimulation.

*High blood pressure increases the workload of the heart.*

- When the ventricles contract, they must generate sufficient pressure to exceed the blood pressure in the major arteries in order to force open the semilunar valves.
- The arterial blood pressure is referred to as the afterload because it is the workload imposed on the heart.
- If the arterial blood pressure is chronically elevated, the ventricle has to generate more pressure to eject blood.
- A chronically elevated afterload is one of the two major factors that cause heart failure.

*In heart failures, contractility of the heart is decreased.*

- Heart failure refers to the inability of the cardiac output to keep pace with the body's demands for supplies and removal of wastes.
- Heart failure may occur for a variety of reasons, but the two most common are: (1) damage to the heart muscle, and (2) prolonged pumping against a chronically increased afterload.
- The prime defect in heart failure is a decrease in cardiac contractility.
- A failing heart will pump out a smaller stroke volume than a normal healthy heart.
- Two major compensatory measures help restore the stroke volume to normal; sympathetic activity to the heart is reflexly increased and when cardiac output is reduced the kidneys, in a compensatory attempt to improve their reduced blood flow, retain extra salt and water in the body during urine formation to expand the blood volume.

## NOURISHING THE HEART MUSCLE

*The heart receives most of its own blood supply through the coronary circulation during diastole.*

- The heart muscle must receive blood through blood vessels, specifically by means of the coronary circulation.
- The coronary arteries branch from the aorta just beyond the aortic valve, and the coronary veins empty into the right atrium.
- Most coronary arterial flow occurs during diastole.
- Coronary blood flow is adjusted primarily in response to changes in the heart's oxygen requirements.
- The link that coordinates coronary blood flow with myocardial oxygen needs is adenosine.
- Increased formation and release of adenosine from the cardiac cells occurs: (1) when there is cardiac oxygen deficit, or (2) when cardiac activity is increased and the heart accordingly requires more oxygen and is using more ATP as an energy source

*Atherosclerotic coronary artery disease can deprive the heart of essential oxygen.*

- Coronary artery disease can cause myocardial ischemia by the three following mechanisms: (1) profound vascular spasm of the coronary arteries, (2) the formation of atherosclerotic plaques, and (3) thromboembolism.

- Vascular spasm is an abnormal spastic constriction that transiently narrows the coronary vessels.
- Atherosclerosis is a progressive, degenerative arterial disease that leads to occlusion of affected vessels.
- Potential complications of coronary atherosclerosis are angina pectoris and thromboembolism.
- The amount of good cholesterol versus bad cholesterol in the blood is linked to atherosclerosis.
- There are two sources of cholesterol for the body: (1) dietary intake of cholesterol, and (2) manufacture of cholesterol by many organs within the body.
- There are three major lipoproteins, named for their density of protein as compared to lipid: (1) high-density lipoproteins (HDL), (2) low-density lipoproteins (LDL), and (3) very-low-density lipoproteins (VLDL).
- Cholesterol carried in LDL complexes has been termed bad cholesterol, because cholesterol is transported to the cells, including those lining the blood-vessel walls, by means of LDL.
- In contrast, cholesterol carried in HDL complexes has been dubbed good cholesterol, because HDL removes cholesterol from the cells and transports it to the liver for partial elimination from the body.
- It is difficult to significantly reduce cholesterol levels in the blood by decreasing cholesterol intake.
- The liver has a primary role in determining total blood cholesterol levels, and the interplay between LDL and HDL determines the traffic flow of cholesterol between the liver and the individual cells of the body.
- Evidence suggests that the propensity toward developing atherosclerosis substantially increases with elevated levels of LDL.
- Elevated levels of HDL are associated with a low incidence of atherosclerotic heart disease.
- The higher the HDL-cholesterol concentration in relationship to the total blood cholesterol level, the lower the risk of atherosclerosis.
- Some other factors known to influence atherosclerotic risk are: (1) cigarettes smoking which lowers HDL, (2) higher HDL levels in individuals who exercise regularly, (3) estrogen in premenopausal women causes higher HDL levels, and (4) vitamins E and C have been shown to slow plaque deposition.
- The ingestion of polyunsaturated fatty acids, the predominant fatty acids of most plants, tends to reduce blood cholesterol levels by enhancing the elimination of both cholesterol and cholesterol-derived bile salts in the feces.

## Key Terms

acute myocardial infarction
after load
angina pectoris
apolipoprotein A-I, A-Ii
arrhythmia
arteries
atheromas
atherosclerosis
atrial fibrillation
atrial flutter
atrioventricular (AV) node
atrioventricular (AV) valves
atrium
autorhythmicity
AV nodal delay
bicuspid valve
blood
blood vessels
bradycardia
bundle of his (atrioventricular bundle)
cardiac myopathies
cardiac output (CO)
cardiopulmonary resuscitation (CPR)
chordae tendineae
complete heart block
congestive heart failure
contractility
coronary artery disease (CAD)
C-reactive protein
diastole
diastolic murmur
electrical defibrillation
electrocardiogram
end-diastolic volume
endocardium
endothelium
end-systolic volume
epicardium
extrasystole (premature beat)
fibrillation
fibrous skeleton
first heart sound
foam cells
frank-starling law of the heart
functional murmurs
gingivitis
heart
heart attack
heart block
high density lipoproteins
homocysteine
intercalated discs
internodal pathway
isovolumetric ventricular contraction
latent pacemaker
lead
lipoprotein (a)
low density lipoproteins
mitral valve
myocardial infarction
myocardial ischemia
myocardium
necrosis

nitric oxide
P wave
pacemaker
pacemaker activity
pacemaker potential
papillary muscle
pericardial fluid
pericardial sac
pericarditis
plaque
preload
pulmonary circulation
Purkinje fibers
QRS complex
regurgitation
rheumatic fever
second heart sound
semilunar valves
septum
sinoatrial (SA) node
stenotic valve
sternum
stroke volume
systemic circulation
systole
systolic murmur
T wave
tachycardia
thoracic cavity
thromboembolism
thrombus
tricuspid valve
ventricular fibrillation

## Review Exercises

*Answers are in the appendix*

### True/False

_____ 1. Ninety-nine percent of the cardiac muscle cells are autorhythmic cells.

_____ 2. The AV node, which normally exhibits the fastest rate of autorhythmicity is known as the pacemaker of the heart.

_____ 3. An action potential originating in the SA node first spreads throughout both atria primarily from cell to cell, via gap junctions.

_____ 4. The internodal conduction pathway directs the spread of an action potential originating at the SA node to the AV node to assure sequential contraction of the ventricles following atrial contraction.

_____ 5. The ventricular conduction system is more highly organized and more important than the interatrial and internodal conduction pathways.

_____ 6. Cardiac muscle has a short refractory period.

_____ 7. A normal ECG exhibits three distinct wave forms: the P wave, the QRS complex, and the T wave.

_____ 8. The QRS complex represents atrial depolarization.

_____ 9. The two determinants of cardiac output are heart rate and stroke volume.

_____ 10. The parasympathetic nervous system's influence on the SA node is to increase the heart rate.

_____ 11. The parasympathetic nervous system stimulation of the AV node reduces the AV nodal delay.

_____ 12. Sympathetic stimulation increases stroke volume by enhancing venous return.

_____ 13. The prime defect in heart failure is a decrease in cardiac contractility.

_____ 14. Cholesterol carried in HDL complexes has been termed bad cholesterol.

_____ 15. It is difficult to significantly reduce cholesterol levels in the body by decreasing cholesterol intake.

_____ 16. The liver has a primary role in determining total blood cholesterol levels.

_____ 17. Elevated levels of HDL are associated with a low incidence of atherosclerotic heart disease.

_____ 18. The cardiac cycle consists of alternate periods of systole and diastole.

_____ 19. Early in ventricular diastole the SA node reaches threshold and fires.

_____ 20. Atrial depolarization brings about atrial contraction.

_____ 21. Ventricular systole includes both the period of isovolumetric contraction and the ventricular ejection phase.

_____ 22. The first heart sound is associated with closure of the AV valve.

_____ 23. Laminar flow produces a faint sound.

_____ 24. The sound of the murmur characterizes it as either a stenotic murmur or an insufficient murmur.

_____ 25. There are no gap junctions between the atrial and ventricular contractile cells.

_____ 26. Either all the cardiac muscle fibers contract or none of them do.

_____ 27. Cardiac muscle cells contain no mitochondria and no myoglobin.

_____ 28. Few new cardiac muscle cells are produced after infancy.

_____ 29. Blood serves as the transport medium.

_____ 30. Pulmonary circulation consists of a circuit of vessels carrying blood between the heart and organ systems.

_____ 31. Anatomically the heart is a single organ though it has a right and left side, they still function as a single pump.

_____ 32. The left side of the heart pumps blood into the pulmonary circulation.

_____ 33. Both sides of the heart simultaneously pump equal amounts of blood.

_____ 34. The systemic circulation is a low-pressure, low-resistance system.

_____ 35. Blood flows through the heart in one fixed direction from veins to atria to ventricles to arteries.

_____ 36. Cardiac muscle is capable of generating action potentials without any nervous stimulation.

**Fill in the Blank**

37. ____________________ is an abnormal, spastic constriction that transiently narrows the coronary vessels.

38. ____________________ is a progressive, degenerative arterial disease that leads to occlusion of affected vessels.

39. An abnormal clot attached to a vessel wall is known as a(n) ____________________. A freely floating clot is known as ____________________.

40. ____________________ exists when small terminal branches from adjacent blood vessels nourish the same area.

41. There are three major lipoproteins: (1) ________________, which contain the most protein and least cholesterol, (2) ________________, which contain less protein and more cholesterol, and (3) ________________, which contain the least protein and most lipid, but the lipid carried is neutral fat, not cholesterol.

42. Random, uncoordinated excitation and contraction of the cardiac cells is known as ____________.

43. The __________________ pathway extends from the SA node within the right atrium to the left atrium.

44. The __________________ represents ventricular repolarization.

45. A rapid heart rate of more than 100 beats per minute is known as __________________.

46. A slow heart rate of fewer than 60 beats per minute is known as __________________.

47. __________________ fibrillation is characterized by rapid, irregular, uncoordinated depolarization of the atria with no definite P waves.

48. The heart has a broad base at the top and tapers to a pointed tip, known as the ________________, at the bottom.

49. CPR is short for __________________.

50. __________________ pump blood from the heart.

51. The vessels that return blood from the tissues to the atria are __________________.

52. The large artery carrying blood away from the left ventricle is the __________________.

53. The right AV valve is also called the __________________.

54. The __________________ is a thin external membrane covering the heart.

55. The non-SA nodal autorhythmic tissues are __________________.

56. __________________ is the volume of blood pumped by each ventricle per minute.

57. The difference between the cardiac output at rest and the maximum volume of blood the heart is capable of pumping per minute is known as the __________________.

58. The direct correlation between end-diastolic volume and stroke volume constitutes the __________________ of stroke volume, which refers to the heart's inherent ability to vary the stroke volume.

59. The intrinsic relationship between end-diastolic volume and stroke volume is known as the __________________.

60. __________________ refers to the inability of the cardiac output to keep pace with the body's demands for supplies and removal of wastes.

61. The volume of blood in the ventricle at the end of diastole is known as the __________________.

62. When all the valves are closed and no blood can enter or leave the ventricle during this time, this interval is termed the period of __________________.

63. The amount of blood remaining in the ventricle at the end of systole when ejection is complete is known as the __________________.

64. Abnormal heart sounds are called ____________________.

65. An abnormal heart sound occurring between the first and second heart sounds signifies a(n) ____________________.

**Matching**
*Match the division of the autonomic nervous system to the effect it exerts on the heart or structures that influence the heart.*

a. sympathetic
b. parasympathetic

_____ 66. has no effect on the adrenal medulla
_____ 67. increases contractility of the atrial muscle and strengths contraction
_____ 68. decreases excitability; increases AV nodal delay
_____ 69. has no effect on the ventricular conduction pathway
_____ 70. increases rate of depolarization to threshold of SA node; increases heart rate
_____ 71. promotes adrenomedullary secretion of epinephrine
_____ 72. increases contractility of the ventricular muscle; strengthens contraction
_____ 73. increases venous return
_____ 74. decreases contractility of atrial muscle
_____ 75. increases excitability; decreases AV nodal delay

**Multiple Choice**

76. This cardiac disorder is most often caused by a streptococcus bacterial infection.
    a. Myocardial infarction
    b. Myocardial ischemia
    c. Atherosclerosis
    d. Congestive heart failure
    e. Valvular stenosis

77. This cardiac complication can be caused by prolonged pumping against a chronically increased afterload.
    a. Myocardial infarction
    b. Myocardial ischemia
    c. Atherosclerosis
    d. Congestive heart failure
    e. Valvular stenosis

78. This chest pain is associated with myocardial ischemia.
    a. Myocardial nociceptosis
    b. Angina pectoris
    c. Atheroma
    d. Valvular stenosis
    e. Heartburn

79. This is a noncancerous tumor of smooth-muscle cells within the blood-vessel walls.
    a. Myocardial infarction
    b. Myocardial ischemia
    c. Atheroma
    d. Fibroblastoma
    e. Glioma

80. This cardiac disorder can be caused by profound vascular spasm of the coronary arteries.
    a. Myocardial infarction
    b. Myocardial ischemia
    c. Atherosclerosis
    d. Congestive heart failure
    e. Valvular stenosis

81. This is a progressive, degenerative arterial disease that leads to occlusion of affected vessels.
    a. Myocardial infarction
    b. Myocardial ischemia
    c. Atherosclerosis
    d. Congestive heart failure
    e. Valvular stenosis

82. This heart disorder can be caused by thromboembolism.
    a. Myocardial infarction
    b. Myocardial ischemia
    c. Atherosclerosis
    d. Congestive heart failure
    e. Valvular stenosis

83. This disease has a close relationship with cholesterol.
    a. myocardial infarction
    b. myocardial ischemia
    c. atherosclerosis
    d. congestive heart failure
    e. valvular stenosis

84. This disease is characterized by abnormally large amounts of blood being dammed up in the venous system.
    a. Myocardial infarction
    b. Myocardial ischemia
    c. Atherosclerosis
    d. Congestive heart failure
    e. Valvular stenosis

85. This condition is commonly known as a heart attack.
    a. Myocardial infarction
    b. Myocardial ischemia
    c. Atherosclerosis
    d. Congestive heart failure
    e. Valvular stenosis

86. Which of the following compounds reduces the force of cardiac contraction by blocking calcium influx during an action potential?
    a. digitalis
    b. nicotine
    c. verapamil
    d. caffeine
    e. all of the above

**Modified Multiple Choice**

*Indicate the proper order of the events during the cardiac cycle by placing the numbers in the blank preceding the events in sequence. The first and last events are already so indicated as a guide.*

__1__ 87. AV vale open; aortic valve closed; ventricular filling occurring
_____ 88. blood ejected from the ventricle
_____ 89. isovolumetric ventricular relaxation
_____ 90. atrial contraction
__12_ 91. AV valves open; ventricular filling occurs again; one cardiac cycle is complete
_____ 92. aortic valve opens
_____ 93. SA node discharges
_____ 94. ventricular filling complete
_____ 95. ventricular relaxation begins
_____ 96. aortic valve closes
_____ 97. isovolumetric ventricular contraction
_____ 98. ventricular contraction begins AV valves close

*Indicate which valve abnormality is being described using the answer code below.*

A = valvular stenosis
B = valvular insufficiency

_____ 99. produces a "gurgling" murmur
_____ 100. produces a "whistling" murmur
_____ 101. valve does not close completely
_____ 102. valve does not open completely

*Indicate which ion is involved in each event being described by using the following answer code.*

A = $K^+$
B = $Na^+$
C = $Ca^{2+}$

103. Inactivity of _____ channels brings about the slow drift of membrane potential to threshold in the cardiac autorhythmic cells.
104. Explosive increase in membrane permeability to _____ brings about the rapidly rising phase of the action potential in contractile cardiac cells.
105. Slow inward diffusion of _____ is largely responsible for the plateau portion of the cardiac action potential.
106. The rapid falling phase of the cardiac action potential is brought about primarily by the outward diffusion of _____.
107. Changes in cytosolic _____ concentration bring about changes in the strength of cardiac contraction.
108. Parasympathetic stimulation increases the permeability of the SA node to _____, whereas sympathetic stimulation decreases the permeability to this same ion.

*Complete the following discussion by circling the correct italicized phrase within the parentheses.*

> If the venous return increases, at the end of diastole the ventricular volume will be *(109. increased, decreased, the same as before)*. Therefore, the length of the cardiac muscle cells will be *(110. increased, decreased, the same as before)*. Consequently, during the next contraction the tension developed by the heart will be *(111. greater than before, less than before, the same as before)*. The amount of blood pumped out as a result of this contraction will be *(112. more than, less than, the same as)* the amount pumped out by the contraction prior to the increase in venous return. Therefore, as venous return to the heart increases, the stroke volume ejected by the heart *(113. increases, decreases, remains unchanged)*.

*Use the following answer code to indicate which factor involved in the initiation and spread of cardiac excitation is being identified.*

A = SA node
B = AV node
C = His and Purkinje system
D = gap junctions

_____ 114. has the fastest rate of pacemaker activity
_____ 115. allows the impulse to spread from cell to cell
_____ 116. delays conduction of the impulse
_____ 117. only point of electrical contact between the atria and ventricles
_____ 118. normal pacemaker of the heart
_____ 119. rapidly conducts the impulse down the ventricular septum and throughout much of the ventricular musculature

## Points to Ponder

1. Congenital heart defects are routinely corrected through the miracles of modern surgery. What happens to the number of people having such disorders as a result of these surgical techniques? What is the solution to this problem?

2. After carefully studying the sections of this chapter involving cholesterol and lipids, why do you suppose a good physician checks the triglycerides in the blood?

3. Speculate as to how much more work the heart of a physically inactive person has to perform than that of a physically active person.

4. Do adults and babies have the same heart rate? Blood pressure? Explain your answers.

5. How do the AV valves ensure a one-way flow of blood?

6. How does aerobic exercise help the heart?

7. How would you differentiate between heart flutter and fibrillation?

8. How does a defibrillator work physiologically?

9. What risk factors for heart disease can be modified? Which can't be modified?

## *Clinical Perspectives*

1. With respect to the heart, why do people generally sleep in a prone position?

2. What happens to the heart in these two situations?
   A. You sense you are about to sneeze so you close your mouth and hold your nose to muffle the sneeze. Then you sneeze.
   B. You are extremely constipated. In an effort to have a bowel movement you strain very hard.

3. Why do you suppose the sympathetic nervous system has an effect on the adrenal medulla while the parasympathetic has no effect at all?

4. Why is the nitroglycerin tablet placed under the tongue in cardiac patients?

5. Why does a physician or nurse practitioner tap a patient's chest wall during a physical examination?

6. Why is it that individuals who have atrial fibrillation do not appear to have a higher mortality rate than those who have normal functioning atria?

7. What is meant by the phrase "splitting the heart sounds?"

8. How would you explain to a patient the relationship between plasma lipids and heart disease?

9. If a patient has premature ventricular contractions, how would this affect stroke volume?

## *PhysioEdge Activities*

**Related to Text:**
Media Exercise 9.2: The Heart: A Dual Pump.
Media Exercise 9.1: The Electrocardiogram.
Media Exercise 9.3: The Heart: Cardiac Output.

**From Figures:**
*Figure 9.1*. For an interaction related to this figure, see Media Exercise 9.2.
*Figure 9.4*. For an animation of this figure, click the Cardiac Anatomy tab in the Cardiovascular Physiology tutorial. For an interaction related to this figure, see Media Exercise 9.2: The Heart: A Dual Pump.
*Figure 9.8*. For an animation of this figure, click the Cardiac Electrical Activity tab in the Cardiovascular Physiology tutorial. For an interaction related to this figure, see Media Exercise 9.2: The Heart: A Dual Pump.
*Figure 9.15*. For an interaction related to this figure, see Media Exercise 9.1: The Electrocardiogram.
*Figure 9.17*. For an animation of this figure, click the ECG and the Cardiac Cycle tab in the Cardiovascular Physiology tutorial.
*Figure 9.23*. For an interaction related to this figure, see Media Exercise 9.3: The Heart: Cardiac Output.

## *Media Resources*

**PhysioEdge CD-ROM**
For a visual review of concepts in this chapter, check out:

Tutorial: Cardiovascular Physiology
Media Exercise 9.1: The Electrocardiogram
Media Exercise 9.2: The Heart: A Dual Pump
Media Exercise 9.3: The Heart: Cardiac Output

**Book Companion Website**
The website for this book contains a wealth of helpful study aids, as well as many ideas for further reading and research. Log on to:

**http://www.brookscole.com/sherwoodhp6**

Select Chapter 9 from the drop-down menu, or click on one of the many resource areas, including **Case Histories**, which introduce clinical aspects of human physiology. For this chapter check out: Case Histories 7: Why Am I So Tired?; 9: Endocarditis; 10: Blue Baby; and 11: Congestive Heart Failure.

For Suggested Readings, consult **InfoTrac® College Edition**, your online research library, at:

**http://infotrac.thomsonlearning.com**

# The Blood Vessels and Blood Pressure

## Chapter Overview

The circulatory system contributes to homeostasis by serving as the body's transport system. This aids in the performance of many vital functions. The circulatory contribution to all these functions is flow. Flow depends on pressure and resistance. The circulatory system includes arteries, arterioles, capillaries, venules, and veins. It is the flow of blood through these vessels that is so important to homeostasis. For this system to function, pressure and resistance are essential. The heart and the blood vessels provide the pressure. The pressure is monitored by baroreceptors, which provide the CNS with the information necessary to control the blood flow. Arteries and veins play a major role in the gross transport of blood. The cardiac output is distributed throughout the body in a highly variable pattern. In this capacity it is the arterioles and capillaries that play the major role. Through changes in resistance the arterioles provide the mechanism for the distributional changes. The distribution of cardiac output is adjusted to maintain homeostasis. Consider all the substances released into or removed from the blood. Many of these molecules have typical cells as their destinations, but most are being transported to specific tissues. The capillaries are the sites of exchange. Arterioles, which respond to intrinsic and extrinsic controls, provide the proper pressure for the exchanges to occur. The circulatory system, like the other body systems, functions to maintain homeostasis. Again, all levels of organization are performing to maintain homeostasis for the survival of cells.

## Chapter Outline

INTRODUCTION

*To maintain homeostasis, reconditioning organs receive blood flow in excess of their own needs.*

- The majority of the body cells are not in direct contact with the external environment, yet these cells must make exchanges with the environment, such as picking up oxygen and nutrients and eliminating wastes.
- All blood pumped by the right side of the heart passes through the lungs for oxygen pickup and carbon dioxide removal.
- The blood pumped by the left side of the heart is parceled out in various proportions to the systemic organs through a parallel arrangement of vessels that branch from the aorta.
- Blood is constantly "reconditioned" so that its composition remains relatively constant despite an ongoing drain of supplies to support metabolic activities and the continual addition of wastes from the tissues.
- The organs that recondition the blood (digestive tract, kidneys, and skin) normally receive substantially more blood than is necessary to meet their basic metabolic needs so that they can perform homeostatic adjustments on the blood.
- Reconditioning organs can withstand temporary reductions in blood flow.
- In contrast, the brain can least tolerate a disruption in its blood supply.

*Blood flow through vessels depends on the pressure gradient and vascular resistance.*

- Arteries carry blood from the heart.
- When a small artery reaches the organ it is supplying, it branches into numerous arterioles.
- Arterioles branch further within the organs into capillaries.
- Capillaries rejoin to form small venules, which further merge to form small veins.
- The small veins unite to form larger veins.
- The arterioles, capillaries, and venules are collectively referred to as the microcirculation.
- The flow rate of blood through a vessel is directly proportional to the pressure gradient and inversely proportional to vascular resistance.
- The pressure gradient—the difference in pressure between the beginning and end of a vessel—is the main driving force for flow through the vessel.
- The greater the pressure gradient forcing blood through a vessel, the greater the rate of flow through that vessel.
- Resistance is a measure of the hindrance to blood flow through a vessel caused by friction between the moving fluid and the stationary vascular walls.

- Resistance to blood flow depends on three factors: (1) viscosity of the blood; (2) vessel length, and (3) vessel radius.

*The vascular tree consists of arteries, arterioles, capillaries, venules, and veins.*

- The systemic and pulmonary circulation each consist of a closed system of vessels called arteries, arterioles, capillaries, venules, and veins.

## ARTERIES

*Arteries serve as rapid-transit passageways to the tissues and as a pressure reservoir.*

- Arteries are specialized to serve as rapid-transit passageways for blood from the heart to the tissues and to act as a pressure reservoir to provide the driving force for blood when the heart is relaxing.
- Capillary flow does not fluctuate between cardiac systole and diastole; blood flow is continuous through the capillaries supplying the tissues.
- The driving force for the continued flow of blood to the tissues during cardiac relaxation is provided by the elastic properties of the arterial wall.
- The arteries' elasticity enables them to expand to temporarily hold an excess volume of ejected blood, storing some of the pressure energy imparted by the cardiac contraction.
- When the heart relaxes and ceases pumping blood into the arteries, the stretched arterial walls passively recoil and push the excess blood contained in the arteries into the vessels downstream.

*Arterial pressure fluctuates in relation to ventricular systole and diastole.*

- The maximum pressure exerted in the arteries when the blood is ejected into them during systole, the systolic pressure, averages 120 mm Hg.
- The minimum pressure within the arteries when blood is draining off into the remainder of the vessels during diastole, the diastolic pressure, averages 80 mm Hg.

*Blood pressure can be measured indirectly by using a sphygmomanometer.*

- It is convenient and reasonably accurate to measure the pressure indirectly through the use of a sphygmomanometer, an externally applied inflatable cuff attached to a pressure gauge.
- During the determination of blood pressure, a stethoscope is placed over the brachial artery.
- At the onset of a blood pressure determination, the cuff is inflated to a pressure greater than systolic blood pressure so that the brachial artery collapses.
- The pressure in the cuff is gradually reduced.
- The highest cuff pressure at which the first sound can be heard is indicative of the systolic blood pressure.
- The highest cuff pressure at which the last sound can be detected is indicative of the diastolic pressure.
- The pressure difference between systolic and diastolic pressures is known as the pulse pressure.

*Mean arterial pressure is the main driving force for blood flow.*

- The mean arterial pressure is the average pressure responsible for driving blood forward into the tissues throughout the cardiac cycle.
- The mean arterial pressure equals the diastolic pressure plus one-third the pulse pressure.
- The mean arterial pressure averages 93 mmHg.
- Arterial pressure is essentially the same throughout the arterial tree.

## ARTERIOLES

*Arterioles are the major resistance vessels.*

- The arterioles are the major resistance vessels in the vascular tree, even though the capillaries have smaller radii than the arterioles.
- The radii of arterioles supplying individual organs can be adjusted independently to determine the distribution of cardiac output and to regulate arterial blood pressure.
- Arteriolar walls have a thick layer of smooth muscle that is richly innervated by sympathetic nerve fibers.
- The smooth muscle layer runs circularly around the arteriole so when it contracts, the vessel's circumference becomes smaller, thus increasing resistance and decreasing the flow through that vessel.
- Local (intrinsic) controls and extrinsic controls influence the level of contractile activity in arteriolar smooth muscle.

*Local control of arteriolar radius is important in determining the distribution of cardiac output.*

- Blood is delivered to all tissues at the same mean arterial pressure.
- The distribution of cardiac output can be varied by differentially adjusting arteriolar resistance in the various vascular beds.
- Local controls are changes within a tissue that alter the radii of the vessels and hence adjust blood flow through the tissue by directly affecting the smooth muscle of the tissue's arterioles.

*Local metabolic influences on arteriolar radius help match blood flow with the "organs" needs.*

- Local influences may be either chemical or physical in nature.
- Local arteriolar vasodilation increasing blood flow to a particular area is called active hyperemia.
- Local chemical changes produce dilation without involving nerves or hormones.
- The following local chemical factors produce relaxation of arteriolar smooth muscles: decreased oxygen, increased carbon dioxide, increased acid, increased potassium, increased osmolarity, adenosine release, and prostaglandin release.
- The single layer of specialized epithelial cells that line the lumen of all blood vessels–the endothelial cells–release nitric oxide.
- Nitric oxide causes relaxation of arteriolar smooth muscle.
- By dilating the arterioles of the penis, nitric oxide is the direct mediator of penile erection.
- Macrophages produce nitric oxide, which they use as "chemical warfare" against bacteria and cancer cells.
- Nitric oxide interferes with platelet function and blood clotting at sites of vessel damage.
- Nitric oxide serves as a neurotransmitter in the brain and elsewhere.
- Nitric oxide helps regulate peristalsis.
- Nitric oxide may play a role in relaxation of skeletal muscle.
- Nitric oxide plays a role in the changes underlying memory.
- The endothelial cells release other important chemicals, such as endothelin, which bring about vasoconstriction by causing arteriolar smooth muscle contraction.

*Local histamine release pathologically dilates arterioles.*

- Histamine is another local chemical mediator that influences arteriolar smooth muscle, but it is not released in response to local metabolic changes and is not derived from the endothelial cells.
- Histamine is synthesized and stored within special connective tissue cells in many tissues and in certain types of circulating white blood cells.
- When tissues are injured or during allergic reactions, histamine is released in the damaged region.
- Histamine is the major cause of vasodilation in an injured area.

*Local physical influences on arteriolar radius include temperature changes, shear stress, and stretch.*

- Local physical influences on arterioles include heat or cold and myogenic responses to stretch.
- Myogenic responses appear to be important in reactive hyperemia and pressure autoregulation.
- When the blood supply to a region is completely occluded, arterioles in the region dilate due to: (1) myogenic relaxation, which occurs in response to the diminished stretch accompanying no blood flow, and (2) changes in local chemical composition.
- After the occlusion is removed, blood flow to the previously deprived tissue is transiently much higher than normal because the arterioles are widely dilated.
- This is called reactive hyperemia.
- When mean arterial pressure falls, the driving force is reduced, so blood flow to tissues decreases.
- Widespread arteriolar dilation reduces the mean arterial pressure, which aggravates the problem.
- In the presence of sustained elevations in mean arterial pressure, local chemical and myogenic influences triggered by the initial increased flow of blood bring about an increase in arteriolar tone and resistance.
- This greater degree of vasoconstriction reduces tissue blood flow toward normal despite the elevated blood pressure.
- Pressure autoregulation is the term applied to these local arteriolar mechanisms that are aimed at keeping tissue blood flow fairly constant.

*Extrinsic sympathetic control of arteriolar radius is important in the regulating of blood pressure.*

- Extrinsic control of arteriolar radius includes both neural and hormonal influences, with the effects of the sympathetic nervous system being the most important.
- Sympathetic nerve fibers supply arteriolar smooth muscle everywhere except in the brain.
- A certain level of ongoing sympathetic activity contributes to vascular tone.
- Increased sympathetic activity produces generalized arteriolar vasoconstriction, whereas decreased sympathetic activity leads to generalized arteriolar vasodilation.
- The extent to which each organ actually receives blood flow is determined by local arteriolar adjustments that override the sympathetic constrictor effect.

*The medullary cardiovascular control center and several hormones regulate blood pressure.*

- The main region of the brain responsible for adjusting sympathetic output to the arterioles is the cardiovascular control center in the medulla of the brain stem.

- Several hormones also extrinsically influence arteriolar radius including epinephrine, norepinephrine, vasopressin, and angiotensin II.

## CAPILLARIES

*Capillaries are ideally suited to serve as sites of exchange.*

- Capillaries, the sites of exchange of materials between the blood and tissues, branch extensively to bring blood within the reach of every cell.
- Exchange of materials across capillary walls is accomplished primarily by the process of diffusion.
- Diffusing molecules have only a short distance to travel between the blood and surrounding cells because of the thin capillary wall and small capillary diameter, coupled with the close proximity of each and every cell to a capillary.
- Because capillaries are distributed in such incredible numbers, a tremendous total surface area is available for exchange.
- Diffusion is enhanced because blood flows more slowly in the capillaries than elsewhere in the circulatory system.

*Water-filled pores permit passage of small, water-soluble substances.*

- Diffusion across capillary walls also depends on the walls' permeability to the materials being exchanged.
- In most capillaries, narrow, water-filled clefts, or pores, are present at the junctions between the cells.
- The size of the capillary pore varies from organ to organ.
- In response to appropriate signals, the endothelial cells can readjust themselves to vary the size of the pores.
- Vesicular transport also plays a limited role in the passage of materials across the capillary wall.

*Many capillaries are not open under resting conditions.*

- Precapillary sphincters act as stopcocks to control blood flow through a particular capillary.
- Capillaries themselves have no smooth muscle, so they cannot actively participate in the regulation of their own blood flow.

*Interstitial fluid is a passive intermediary between the blood and cells.*

- Interstitial fluid acts as the go-between between the blood and the tissues.
- Exchanges across the capillary wall between the plasma and interstitial fluid are largely passive.

*Diffusion across the capillary wall is important in solute exchange.*

- Exchanges are not made directly between blood and the tissue cells.
- Cells exchange materials directly with the interstitial fluid, the type and extent of exchange being governed by the properties of the cellular plasma membranes.
- Passive diffusion down concentration gradients is the primary mechanism for exchange of individual solutes.

*Bulk flow across the capillary wall is important in extracellular fluid distribution.*

- Bulk flow is the process whereby a volume of protein-free plasma filters out of the capillary, mixes with the surrounding interstitial fluid, and is subsequently reabsorbed.
- Bulk flow occurs because of differences in the hydrostatic and colloid osmotic pressures between the plasma and interstitial fluid.
- Bulk flow does not play an important role in the exchange of individual solutes between blood and tissues, because the quantity of solutes moved across the capillary wall by bulk flow is extremely small compared to the much larger transfer of solutes by diffusion.
- Bulk flow plays an extremely important role in regulating the distribution of ECF between plasma and interstitial fluid.

*The lymphatic system is an accessory route by which interstitial fluid can be returned to the blood.*

- Even under normal circumstances, slightly more fluid is filtered out of the capillaries into the interstitial fluid than is reabsorbed from the interstitial fluid back into the plasma.
- The extra fluid filtered out as a result of this filtration-reabsorption imbalance is picked up by the lymphatic system.
- Small, blind-ended terminal lymph vessels permeate almost every tissue of the body.
- Once interstitial fluid enters a lymphatic vessel, it is called lymph.
- Lymphatics converge to form larger and larger lymph vessels, which eventually empty into the venous system.
- The most important functions of the lymphatic system are as follows: (1) the return of excess filtered fluid to the blood, (2) the defense against disease provided by phagocytic cells in the lymph nodes, (3) the transport of absorbed fat, and (4) the return of filtered protein to the blood.

*Edema occurs when too much interstitial fluid accumulates.*

- Swelling of the tissues because of excess interstitial fluid is known as edema.
- The cause of edema can be grouped into four general categories: (1) a reduced concentration of plasma proteins; (2) an increased permeability of the capillary walls; (3) an increased venous pressure; and (4) the blockage of lymph vessels.
- Whatever the cause of edema, an important consequence is a reduction in exchange of materials between the blood and cells.

## VEINS

*Veins serve as a blood reservoir as well as passageways back to the heart.*

- Because the total cross-sectional area of the venous system gradually decreases as smaller veins converge into progressively fewer but larger vessels, the velocity of blood flow increases as the blood approaches the heart.
- Systemic veins serve as a blood reservoir.
- When the stored blood is needed, such as during exercise, extrinsic factors drive the extra blood from the veins to the heart.
- A delicate balance exists between the capacity of the veins, the extent of venous return, and the cardiac output.

*Venous return is enhanced by a number of extrinsic factors.*

- Changes in venous capacity directly influence the magnitude of venous return, which in turn is an important determinant of effective circulating blood volume.
- Sympathetic stimulation produces venous vasoconstriction, which modestly elevates venous pressure; this in turn, increases the pressure gradient to drive more blood from the veins into the right atrium.
- Venous vasoconstriction enhances venous return by decreasing venous capacity.
- Many of the large veins in the extremities lie between skeletal muscles so when the muscles contract, the veins are compressed.
- External venous compression decreases venous capacity and increases venous pressure, in effect squeezing fluid contained in the veins forward toward the heart.
- This pumping action, known as the skeletal muscle pump, is one way by which extra blood stored in the veins is returned to the heart during exercise.
- Increased muscular activity pushes more blood out of the veins and into the heart.
- Blood can only be driven forward because the large veins are equipped with one-way valves.
- These valves permit blood to move forward toward the heart but prevent it from moving back toward the tissues.
- Varicose veins occur when the venous valves become incompetent and can no longer support the column of blood above them.
- The most serious consequence of varicose veins is the possibility of abnormal clot formation in the sluggish, pooled blood.
- As a result of respiratory activity, the pressure within the chest cavity averages 5 mm Hg less than atmospheric pressure.
- The pressure difference between the lower veins and the chest veins squeezes blood from the lower veins to the chest veins, promoting venous return.
- This mechanism of facilitating venous return is known as the respiratory pump.
- During ventricular contraction, the AV valves are drawn downward, enlarging the atrial cavities, which drop the atrial pressure to below 0 mm Hg, thus enhancing venous return.
- Thus, the heart functions as a "suction pump" to facilitate cardiac filling.

## BLOOD PRESSURE

*Blood pressure is regulated by controlling cardiac output, total peripheral resistance, and blood volume.*

- Mean arterial blood pressure is the main driving force for propelling blood to the tissues.
- The mean arterial blood pressure must be high enough to ensure sufficient driving pressure, but not be so high that it creates extra work for the heart and increases the risk of vascular damage and possible rupture of small blood vessels.
- Let's review all the factors that have an effect on mean arterial blood pressure: (1) Mean arterial pressure depends on cardiac output and total peripheral resistance. (2) Cardiac output depends on heart rate and stroke volume. (3) Heart rate depends on the relative balance of parasympathetic and sympathetic activity. (4) Stroke volume increases in response to sympathetic activity. (5) Stroke volume increases as venous return increases. (6) Venous return is enhanced by sympathetically induced venous vasoconstriction, the skeletal pump, the respiratory pump, and cardiac suction. (7) The effective circulating blood volume influences how much blood returns to the heart. (8) The blood volume depends on the magnitude of passive bulk-flow fluid shifts between the plasma and interstitial fluid and the salt and water balance, which are hormonally controlled by the renin-angiotensin-aldosterone system and vasopressin, respectively. (9) Total peripheral resistance

depends on the radius of all arterioles as well as blood viscosity. (10) The major factor determining blood viscosity is the number of red blood cells. (11) Arteriolar radius is influenced by intrinsic metabolic controls that cause local arteriolar vasodilation and sympathetic activity, an extrinsic mechanism that causes arteriolar vasoconstriction. (12) Arteriolar radius is also extrinsically controlled by the hormones vasopressin and angiotensin II, which are potent vasoconstrictors.

- Mean arterial pressure is constantly monitored by baroreceptors within the circulatory system.
- Short-term adjustments are accomplished by alterations in cardiac output and total peripheral resistance, mediated by means of autonomic nervous system influences on the heart, veins, and arterioles.
- Long-term control involves adjusting total blood volume by restoring normal salt and water balance through mechanisms that regulate urine output and thirst.

*The baroreceptor reflex is an important short-term mechanism for regulating blood pressure.*

- Any change in mean blood pressure triggers an autonomically mediated baroreceptor reflex that influences the heart and blood vessels to adjust cardiac output and total peripheral resistance in an attempt to restore blood pressure to normal.
- The carotid sinus and aortic arch baroreceptors are sensitive to changes in both mean arterial pressure and pulse pressure.
- The integrating center that receives the afferent impulses about the status of arterial pressure is the cardiovascular control center.
- The cardiovascular control center alters the ratio between sympathetic and parasympathetic activity to the effector organs.
- If arterial pressure becomes elevated above normal, the carotid sinus and aortic arch baroreceptors increase the rate of firing in their respective afferent neurons.
- The cardiovascular control center responds by decreasing sympathetic and increasing parasympathetic activity.
- These efferent signals decrease heart rate, decrease stroke volume, and produce arteriolar and vasodilation.
- When blood pressure falls below normal, baroreceptor activity decreases, inducing the cardiovascular center to increase sympathetic cardiac and vasoconstrictor nerve activity while decreasing its parasympathetic output.
- This efferent activity leads to an increase in heart rate and stroke volume coupled with arteriolar and venous vasoconstriction.

*Other reflexes and responses influence blood pressure.*

- The sole function of the baroreceptor reflex is blood pressure regulation. Other reflexes and responses that influence the cardiovascular system include the following: (1) left atrial volume receptors and hypothalamic osmoreceptors, (2) chemoreceptors located in the carotid and aortic arteries, (3) cardiovascular responses associated with behaviors and emotions, (4) cardiovascular responses associated with exercise, (5) hypothalamic control over cutaneousarterioles, (6) vasoactive substances released by endothelial cells, and (7) numerous neurotransmitters from various regions of the brain.

*Hypertension is a serious national public health problem, but its causes are largely unknown.*

- The causes of secondary hypertension fall into four categories: (1) cardiovascular hypertension, (2) renal hypertension, (3) endocrine hypertension, and (4) neurogenic hypertension.
- The underlying cause is known in primary hypertension.
- There is a strong genetic tendency to develop primary hypertension, which can be hastened or worsened by contributing factors such as obesity, stress, smoking, and excessive ingestion of salt.
- Whatever the underlying defect, once initiated, hypertension appears to be self-perpetuating.
- Constant exposure to elevated blood pressure predisposes vessel walls to the development of atherosclerosis, which further elevates blood pressure.
- The baroreceptors do not respond to bring the blood pressure back to normal during hypertension because they adapt or are "reset" to operate at a higher level.
- Complications of hypertension include congestive heart failure, strokes, or heart attacks.

*Orthostatic hypotension results from transient inadequate sympathetic activity.*

- Orthostatic hypotension occurs either when there is a disproportion between vascular capacity and blood volume or when the heart is too weak to impart sufficient driving pressure to the blood.
- Two common situations in which hypotension occur transiently are orthostatic hypotension and emotional fainting.
- Both are due to inadequate sympathetic activity.

*Circulatory shock can become irreversible.*

- When blood pressure falls so low that adequate blood flow to the tissues can no longer be maintained, the condition known as circulatory shock occurs.

- Circulatory shock is categorized into four main types: (1) hypovolemic shock, (2) cardiogenic shock, (3) vasogenic shock, and (4) neurogenic shock.
- Following severe loss of blood, the resultant reduction in circulating blood volume leads to a decrease in venous return and a subsequent fall in cardiac output and arterial pressure.
- The baroreceptor reflex to the fall in blood pressure brings about increased sympathetic and decreased parasympathetic activity to the heart.
- As a result of increased sympathetic activity to the veins, generalized venous vasoconstriction occurs.
- Sympathetic stimulation of the heart increases the heart's contractility, as does increasing the stroke volume.
- Sympathetically induced generalized arteriolar vasoconstriction leads to an increase in total peripheral resistance.
- The increase in cardiac output and total peripheral resistance bring about a compensatory increase in arterial pressure.
- The original fall in arterial pressure results in fluid shifts from the interstitial fluid into the capillaries to expand the plasma volume.
- The ECF fluid shift is enhanced by plasma-protein synthesis by the liver following the hemorrhage.
- Urinary output is reduced, thereby conserving water.
- The reduced plasma volume also triggers increased secretion of the hormone vasopressin and activation of the salt-and-water-conserving renin-angiotensin-aldosterone hormonal pathway.
- Increased thirst is also stimulated by a fall in plasma volume.
- Lost red blood cells are replaced through increased red blood cell production, triggered by a reduction in oxygen delivery to the kidneys.
- Compensatory mechanisms are often insufficient in the face of substantial fluid loss.
- Fluid volume must be replaced from the outside through drinking, transfusion, or a combination of both.
- A point may be reached at which blood pressure continues to drop rapidly because of tissue damage, despite vigorous therapy.
- This condition is termed irreversible shock, in contrast to reversible shock, which can be corrected by compensatory mechanisms and effective therapy..

## *Key Terms*

active hyperemia
arteries
arterioles
baroreceptors
baroreceptor reflex
blood pressure
bulk flow
capacitance vessels
capillaries
capillary blood pressure
cardiovascular control center
carotid sinus
circulatory shock
compliance
diastolic pressure
distensibility
edema
endothelial cells
endothelin
endothelial-derived relaxing factor (EDRF)
effective circulating volume
essential or idiopathic hypertension
flow rate
hypertension
hypotension
initial lymphatics
interstitial fluid-colloid osmotic pressure
interstitial fluid hydrostatic pressure
Korotkoff sounds
lymph
lymph nodes
lymphatic system
mean arterial blood pressure
metarteriole
microcirculation
myocardial toxic factor
nitric oxide
orthostatic (postural) hypotension
plasma-colloid osmotic pressure
pores
Poiseuille's law
precapillary sphincters
pressure autoregulation
pressure gradient
pulse pressure
reabsorption
reactive hyperemia
resistance
respiratory pump
secondary hypertension
skeletal muscle pump
sphygmomanometer
systolic pressure
total peripheral resistance
ultrafiltration
varicose veins
vascular tone
vasodilation
veins
venous capacity
venous return
venules
viscosity

## *Review Exercises*

*Answers are in the appendix.*

**True/False**

_____ 1. Blood flows more slowly in capillaries than elsewhere in the circulatory system.

_____ 2. Diffusion across capillary walls also depends on the wall's permeability to the materials being exchanged.

_____ 3. The size of the capillary pores is consistent throughout the body.

_____ 4. Capillaries have no smooth muscle, so they cannot actively participate in the regulation of their own blood flow.

_____ 5. Solute exchanges are made directly between blood and the tissue cells.

_____ 6. Bulk flow occurs because of differences in the hydrostatic and colloid osmotic pressures between the plasma and interstitial fluid.

_____ 7. The inability to return filtered protein is a cause of edema.

_____ 8. All blood pumped by the left side of the heart passes through the lungs for oxygen pickup and carbon dioxide removal.

_____ 9. Reconditioning organs can withstand temporary reductions in blood flow.

_____ 10. The greater the pressure gradient forcing blood through a vessel, the lesser the rate of flow through that vessel.

_____ 11. Capillary flow fluctuates between cardiac systole and diastole.

_____ 12. Systolic pressure averages 120 mm Hg and diastolic pressure averages 80 mm Hg.

_____ 13. To determine blood pressure, a stethoscope is placed over the brachial artery.

_____ 14. Arterial pressure is essentially the same throughout the arterial tree.

_____ 15. Orthostatic hypotension caused by emotional stress can also cause dizziness or fainting.

_____ 16. Anaphylactic shock, which may accompany massive infections, is due to vasodilator substances released from the infective agents.

_____ 17. Mean arterial blood pressure is the main driving force for propelling blood to the tissues.

_____ 18. Long-term adjustments are accomplished by alterations in cardiac output and total peripheral resistance, mediated by means of autonomic nervous system influences on the heart, veins, and arterioles.

_____ 19. The cardiovascular control center alters the ratio between sympathetic and parasympathetic activity to the effector organs.

_____ 20. The blood flow to the brain is increased during exercise.

_____ 21. Baroreceptors respond and help bring the blood pressure back to normal during hypertension.

_____ 22. Arterioles are the major resistance vessels in the vascular tree.

_____ 23. Local (intrinsic) controls are changes within a tissue that alter the radii of the vessels and hence adjust blood flow through the tissue by directly affecting the smooth muscle of the tissue's arterioles.

_____ 24. Local influences can only be chemical, not physical in nature.

_____ 25. Nitric oxide causes relaxation of arteriolar smooth muscle.

_____ 26. Histamine is a local chemical mediator that influences arteriolar smooth muscle.

_____ 27. Sympathetic nerve fibers supply arteriolar smooth muscle everywhere including the brain.

_____ 28. Extrinsic control of arteriolar radius includes both neural and hormonal influences.

_____ 29. Systemic veins also serve as a blood reservoir.

_____ 30. Venous vasoconstriction enhances venous return by increasing venous capacity.

_____ 31. Increased muscular activity pushes more blood out of the veins and into the heart.

_____ 32. Blood can be driven forward or backward through the veins.

_____ 33. Abnormal clot formation in the sluggish, pooled blood is a serious risk to a person with varicose veins.

_____ 34. The pressure within the chest cavity averages 11 mm Hg less than atmospheric pressure.

**Fill in the Blank**

35. ____________________ is when the blood pressure is above 140/90 mm Hg.

36. ____________________ in its extreme form is circulatory shock.

37. The causes of secondary hypertension are: (1) ____________________, (2) ______________, (3) ______________, and (4) _____________ hypertension.

38. ________________________________ is a transient hypotensive condition resulting from insufficient compensatory responses to the gravitational shifts in blood that occur when a person moves from a horizontal to a vertical position.

39. ___________ shock is due to a weakened heart's failure to pump blood adequately.

40. ___________ shock is induced by a fall in blood volume, through hemorrhage or loss of fluids like diarrhea and sweating.

41. ___________ shock is when a point is reached where the blood pressure continues to drop rapidly because of tissue damage, despite vigorous therapy.

42. Veins are often referred to as ___________________________.

43. _________________ refers to the volume of blood that the veins can accommodate.

44. _________________ refers to the volume of blood entering each atrium per minute from the veins.

45. _________________ occur when the venous valves become incompetent and can no longer support the column of blood above them.

46. Mean arterial pressure is constantly monitored by ________________ within the circulatory system.

47. ________________ and __________________________ are mechanoreceptors sensitive to changes in both mean arterial pressure and pulse pressure.

48. ___________ carry blood from the heart.

49. Arterioles, capillaries, and venules are collectively referred to as the ______________ __________________.

50. _______________ of blood through a vessel is directly proportional to the pressure gradient and inversely proportional to vascular resistance.

51. The maximum pressure exerted in the arteries when blood is ejected into them during systole averages ________________.

52. The pressure difference between systolic and diastolic is known as the ______________________.

53. ____________________________ is the average pressure responsible for driving blood forward into the tissues throughout the cardiac cycle.

54. When a small artery reaches the organ it is supplying, it branches into numerous ____________.

55. Exchange of materials across capillary walls is accomplished primarily by the process of ____________.

56. Capillaries typically branch either directly from an arteriole or from a thoroughfare channel known as a(n) ______________________.

57. ________________ is a volume of protein-free plasma that actually filters out of the capillary, mixes with the surrounding interstitial fluid, and is subsequently reabsorbed.

58. ____________________ is the fluid or hydrostatic pressure exerted on the inside of the capillary walls by the blood.

59. The extra fluid filtered out as a result of this ultrafiltration reabsorption imbalance is picked up by the __________________.

60. Swelling of the tissues because of excess interstitial fluid is known as __________.

61. ____________________ is the term applied to such narrowing of a vessel.

62. Arteriolar smooth muscle normally displays a state of partial constriction known as __________________.

63. _______________________________________________ causes local arteriolar vasodilation by inducing relaxation of arteriolar smooth muscle in the vicinity.

64. __________________ is a single layer of specialized epithelial cells that line the lumen of all blood vessels.

65. The main region of the brain responsible for adjusting sympathetic output to the arterioles is the ______________________________.

**Matching**

*Match the cardiovascular variable to the change as a result of exercise.*

a. increases
b. decreases
c. unchanged

_____ 66. Blood flow to skin
_____ 67. Cardiac output
_____ 68. Venous return
_____ 69. Blood flow to digestive track
_____ 70. Blood flow to brain
_____ 71. Stroke volume
_____ 72. Blood flow to bone
_____ 73. Blood flow to heart
_____ 74. Blood flow to kidney
_____ 75. Heart rate
_____ 76. Median arterial blood pressure
_____ 77. Blood flow to active skeletal muscles
_____ 78. Total peripheral resistance
_____ 79. Stored blood

**Multiple Choice**

80. Which of the following is caused by atherosclerosis?
    a. neurogenic hypertension
    b. endocrine hypertension
    c. cardiovascular hypertension
    d. pheochromocytoma
    e. hypotension

81. Which disorder is associated with an increased production of aldosterone by the adrenal cortex?
    a. neurogenic hypertension
    b. pheochromocytoma
    c. cardiovascular hypertension
    d. Conn's syndrome
    e. primary hypertension

82. Which disorder may be characterized by a defect in the baroreceptors?
    a. neuorgenic hypertension
    b. cardiovascular hypertension
    c. Conn's syndrome
    d. hypotension
    e. none of the above

83. Which of the following does not hasten the development of primary hypertension?
    a. obesity
    b. smoking
    c. stress
    d. excessive ingestion of salt
    e. none of the above

84. Which of the following is characterized by excessive secretions of epinephrine?
   a. varicose veins
   b. renal hypertension
   c. pheochromocytoma
   d. Conn's syndrome
   e. cardiovascular syndrome

85. Which disorder is brought about by massive hemorrhage?
   a. varicose veins
   b. hypovolemic shock
   c. vasogenic shock
   d. Conn's syndrome
   e. endocrine hypertension

86. Which of the following is due to a weakened heart's failure to pump blood adequately?
   a. hypovolemic shock
   b. pheochromocytoma
   c. cardiogenic shock
   d. Conn's syndrome
   e. vasogenic shock

87. Which condition is characterized by an inability to eliminate the normal salt load?
   a. Conn's syndrome
   b. hypovolemic shock
   c. cardiovascular hypertension
   d. neurogenic hypertension
   e. renal hypertension

88. Which disorder is caused by a loss of sympathetic vascular tone?
   a. cardiogenic shock
   b. neurogenic shock
   c. hypovolemic shock
   d. vasogenic shock
   e. none of the above

89. Which of the following disorders has an unknown cause?
   a. renal hypertension
   b. neurogenic shock
   c. Conn's syndrome
   d. pheochromocytoma
   e. primary hypertension

**Modified Multiple Choice**

*Indicate which of the vessels performs the function listed by writing the appropriate letter in the blank using the answer code below.*

A = arteries
B = arterioles
C = capillaries
D = veins
E = lymphatics

_____ 90. site of exchange of nutrients and waste products between the blood and tissues.
_____ 91. serve as low resistance passageways from the heart to the tissues.
_____ 92. serve as blood reservoir to accommodate variations in blood volume.
_____ 93. major resistance vessels.
_____ 94. portion of the circulatory system through which the velocity of blood flow is the slowest.
_____ 95. serve as low-resistance passageways from the tissues to the heart.
_____ 96. act as a pressure reservoir to drive blood forward through the vasculature during diastole.
_____ 97. changes in the radius of this vessel type regulate the distribution of the cardiac output to various areas of the body.
_____ 98. vessels that pick up fluid that is filtered but not reabsorbed.

*If the blood pressure is recorded as 118/76, indicate the correct value of the pressure in question.*

A = 118 mm Hg
B = 42 mm Hg
C = 97 mm Hg
D = 76 mm Hg
E = 90mm Hg

_____ 99. What is the systolic pressure?
_____ 100. What is the diastolic pressure?
_____ 101. What is the pulse pressure?
_____ 102. What is the mean pressure?

*The calculations below are based on the following pressures:*

Blood capillary pressure at arteriolar end of tissue capillaries = 35 mm Hg.
Blood capillary pressure at venule end of tissue capillaries = 15 mm Hg.
Blood-colloid osmotic pressure = 20 mm Hg.
Interstitial-fluid hydrostatic pressure = 1 mm Hg.
Interstitial-fluid colloid osmotic pressure = 0 mm Hg.

103. What would the ultrafiltration pressure be?
a. 14 mm Hg
b. 16 mm Hg
c. 9 mm Hg
d. 10 mm Hg
e. 5 mm Hg

104. What would the reabsorption pressure be?
a. 21 mm Hg
b. 15 mm Hg
c. 6 mm Hg
d. 14 mm Hg
e. 20 mm Hg

105. Would edema occur in this situation?
   a. yes
   b. no

*Indicate the relative comparison of each of the paired items by writing the appropriate letter in the blank using the answer code below.*

A = A is greater than B
B = B is greater than A
C = A and B are equal

106. _____ A. Blood flow through an arteriole upon increased sympathetic activity.
B. Blood flow through an arteriole upon decreased sympathetic activity.

107. _____ A. Blood flow through a vein upon increased sympathetic activity.
B. Blood flow through a vein upon decreased sympathetic activity.

108. _____ A. Velocity of blood flow through the veins.
B. Velocity of blood flow through the capillaries.

109. _____ A. Local arteriole radius in the presence of local decreased oxygen concentration and increased carbon dioxide concentration.
B. Local arteriole radius with normal local concentration of both oxygen and carbon dioxide.

110. _____ A. Circulation through the skin during exercise.
B. Circulation through the skin at rest.

111. _____ A. Circulation to the brain at rest.
B. Circulation to the brain during an examination.

112. _____ A. Net ultrafiltration pressure at the arteriolar end of the capillary.
B. Net reabsorption pressure at the venous end of the capillary.

## Points to Ponder

1. Having learned about active hyperemia can you think of an example of passive hyperemia?
2. How do the palace guards in London avoid fainting?
3. Why is the cardiovascular center in the medulla as opposed to some cortical, hypothalamic, or thalamic area?
4. Since there is blood in the blood vessels, why do they need an additional external blood supply?
5. With respect to blood vessels, what causes dark circles under your eyes?
6. Based on your knowledge of capillaries, why do your cheeks turn red on a very cold day?
7. What changes occur in the circulatory system during exercise?
8. How does lifting weights cause a fall in cardiac output and blood pressure?
9. One of the consequences of the Frank-Starling law is that the outputs of the right and left ventricles are matched. Why is this important and how is this matching accomplished?

10. Why is hypotension a result of hypovolemia?

## Clinical Perspectives

1. How does hypertension affect the risk of coronary heart disease?

2. How would you explain to your friend the fact that blood pressure is higher at one time of the day than during other times?

3. Since you are taking a physiology course, how would you explain to your friend the cause of elephantiasis and how it is related to edema?

4. During prolonged exercise on a very hot day, a substantial amount of water will be lost from the body through sweating. This causes hyperthermia. Why can't the body be immediately rehydrated by drinking pure water?

5. How would you explain to your friend how the class of drugs known as ASE inhibitors work?

6. Cocaine inhibits the re-uptake of norepinephrine in the adrenergic axons. Based on this fact, why is myocardial ischemia a common side effect of cocaine abuse?

7. How does nitric oxide contribute to hypertension?

8. How would you explain to your friend why he becomes dizzy when he stands up too fast?

9. Preeclampsia is a toxemia of late pregnancy characterized by high blood pressure, proteinuria, and edema. Based on your knowledge of blood vessels and blood pressure, why does preeclampsia occur?

10. People with congestive heart failure are often treated with the drug digitalis. Digitalis works by inhibiting the sodium-potassium pump and causing the intracellular concentration of calcium to increase. What effect is this drug going to have on the heart and blood pressure?

## Experiments of the Day

1. On a classmate's arm try to locate the venous valves.

2. The next time you go to the grocery store or you local pharmacy, if there is a blood pressure machine there, take your blood pressure.

## PhysioEdge Activities

**Related to Text:**
Media Exercise 10.1: Blood Flow and Total Cross-Sectional Area.
Media Exercise 10.2: Arteriolar Resistance and Flow in Capillaries.
Media Exercise 10.3: Circulation.

**Related to Figures:**
*Figure 10.2.* For an interaction related to this figure, see Media Exercise 10.1: Blood Flow and Total Cross-Sectional Area.
*Figure 10.3.* For an interaction related to this figure, see Media Exercise 10.1: Blood Flow and Total Cross-Sectional Area.
*Figure 10.6.* For an animation of this figure, click the Arteries tab in the Cardiovascular Physiology tutorial.

*Figure 10.16.* For an interaction related to this figure, see Media Exercise 10.2: Arteriolar Resistance and Flow in Capillaries.
*Figure 10.34.* For an interaction related to this figure, see Media Exercise 10.3: Circulation.

## Media Resources

**PhysioEdge CD-ROM**
For a visual review of concepts in this chapter, check out:

Tutorial: Cardiovascular Physiology – Arteries tab
Media Exercise 10.1: Blood Flow and Total Cross-Sectional Area
Media Exercise 10.2: Arteriolar Resistance and Flow in capillaries
Media Exercise: 10.3: Circulation

**Book Companion Website**
The website for this book contains a wealth of helpful study aids, as well as many ideas for further reading and research. Log on to:

**http://www.brookscole.com/sherwoodhp6**

Select Chapter 10 from the drop-down menu, or click on one of the many resources areas, including **Case Histories**, which introduce clinical aspects of human physiology. For this chapter check out Case History 8: Toxic Shock Syndrome.

For Suggested Readings, consult **InfoTrac® College Edition**, your online research library, at:

**http://infotrac.thomsonlearning.com**

# The Blood

## Chapter Overview

Blood plays a vital role in homeostasis through the transport of nutrients, dissolved gases, wastes and substances such as hormones. While the blood and the circulatory system are of no greater importance than many other systems or organs, they are one of the weaker links in maintaining homeostasis. There are two kidneys, things can be relearned, and the liver can regenerate, but breakdowns in the blood and circulatory system can result in greater consequences.

There are two general components in blood besides water: plasma and the cellular elements. All vascular transportation relies on the fluid portion of the blood. In addition, the plasma maintains an osmotic gradient that is important in the distribution of extracellular fluid. The plasma also has a buffer system that functions against changes in pH and has the necessary constituents to begin the clotting process.

Erythrocytes contain hemoglobin, which plays a vital role in oxygen and carbon dioxide transport, and the buffering capacity of the blood. Anemia results any time the oxygen carrying capacity is reduced. There are many causes of anemia.

Leucocytes defend the body against pathogens, identify and destroy cancer cells and clean up cellular debris. Platelets are the major elements in maintaining hemostasis. All the cellular components of blood are produced in the bone marrow even though lymphocytes simply divide in the normal healthy body.

## Chapter Outline

INTRODUCTION

- Blood consists of three types of specialized cellular elements--erythrocytes, leukocytes, and platelets--suspended in the complex liquid plasma.
- The hematocrit represents the percentage of total blood volume occupied by erythrocytes.
- Plasma accounts for the remaining volume.
- The white blood cells and platelets represent less than 1 percent of the blood volume.

PLASMA

*Plasma water is a transport medium for many inorganic and organic substances.*

- Plasma is composed of 90 percent water.
- Because water has a high capacity to hold heat, plasma is able to absorb and distribute much of the heat generated metabolically within the tissues.
- Heat energy not needed to maintain body temperature is eliminated to the environment.
- The most plentiful organic constituents by weight are the plasma proteins, which compose 6 percent to 8 percent of plasma's total weight.
- Inorganic constituents account for approximately 1 percent of plasma weight.
- The most abundant electrolytes in the plasma are sodium and chloride ions.
- The remaining small percentage of plasma is occupied by nutrients, waste products, dissolved gases and hormones.
- Most of these substances are merely being transported in the plasma.

*Many functions of plasma are carried out by plasma proteins.*

- The plasma proteins are the one group of plasma constituents not present just for the ride.
- Because they are the largest of the plasma constituents, plasma proteins usually do not exit through the narrow pores in the capillary walls.
- Plasma proteins exist in a colloidal dispersion.
- There are three groups of plasma proteins: albumins, globulins, and fibrinogen.
- Plasma proteins establish an osmotic gradient between blood and interstitial fluid.
- Plasma proteins are partially responsible for the plasma's capacity to buffer changes in pH.
- Plasma proteins contribute to blood viscosity.
- In a state of starvation, plasma proteins can be degraded to provide energy for cells.
- Albumins bind many substances for transport through the plasma.
- Albumins contribute most extensively to the colloid osmotic pressure.

- Specific alpha and beta globulins bind and transport a number of substances in the plasma, such as thyroid hormone, cholesterol, and iron.
- Alpha and beta globulins are involved in the process of blood clotting.
- Inactive precursor protein molecules, which are activated as needed by specific regulatory inputs, belong to the alpha-globulin group.
- The gamma globulins are the immunoglobulins.
- Fibrinogen is a key factory in the blood-clotting process.
- The plasma proteins are generally synthesized by the liver, with the exception of the gamma globulins, which are produced by lymphocytes.

## ERYTHROCYTES

*The structure of erythrocytes is well suited to their main function of oxygen transport in the blood.*

- Erythrocytes are flat biconcave discs.
- The biconcave shape provides a larger surface area for diffusion of oxygen.
- The thinness of the cell enables oxygen to diffuse rapidly between the exterior and the innermost regions of the cell.
- The flexibility of the erythrocytes membrane facilitates their transport function.
- Erythrocytes' most important feature which enables them to transport oxygen, is the hemoglobin they contain.
- Hemoglobin consists of: (1) the globin portion, a protein made up of four highly folded polypeptide chains, and (2) four iron-containing, nonprotein nitrogenous groups known as heme groups.
- Each hemoglobin molecule can pick up four oxygen passengers.
- Hemoglobin can also combine with carbon dioxide, the acidic hydrogen-ion portion of ionized carbonic acid, and carbon monoxide.
- Hemoglobin plays the key role in oxygen transport while contributing significantly to carbon dioxide transport and the buffering capacity of blood.
- Erythrocytes contain no nucleus, organelles, or ribosomes.
- Only a few crucial nonrenewable enzymes remain within the mature erythrocyte: (1) The glycolytic enzymes are necessary for generating the energy needed to fuel the active transport mechanisms involved in maintaining proper ionic concentrations within the cell; (2) Carbonic anhydrase is critical in carbon dioxide transport.

*The bone marrow continuously replaces worn-out erythrocytes.*

- Most old red blood cells meet their final demise in the spleen.
- The spleen has a limited ability to store healthy erythrocytes in its pulpy interior, serves as a reservoir site for platelets, and contains an abundance of lymphocytes.
- Old erythrocytes must be replaced by new cells, which are produced in the bone marrow.
- The bone marrow normally generates new red blood cells, a process known as erythropoiesis, keeping pace with the demolition of old cells.
- Red bone marrow that is capable of blood cell production is found in the sternum, vertebrae, ribs, base of the skull, and upper ends of the long limb bones.
- Red marrow not only produces red blood cells but is the ultimate source for leucocytes and platelets.
- Undifferentiated pluripotent stem cells continuously divide and differentiate to give rise to each of the types of blood cells.
- Regulatory factors act on the hemopoietic marrow to govern the number and type of cells generated and discharged into the blood.

*Erythropoiesis is controlled by erythropoietin from the kidneys.*

- Reduced oxygen delivery to the kidneys stimulates them to secrete the hormone erythropoietin into the blood.
- This hormone stimulates erythropoiesis by the bone marrow.
- The gene that directs erythropoietin synthesis has been identified.
- This hormone can now be produced in the laboratory.
- Testosterone increases the basal rate of erythropoiesis.
- Testosterone is undoubtedly responsible for the normally larger hematocrit in males compared to females.

*Anemia can be caused by a variety of disorders.*

- Anemia refers to a reduction below normal in the oxygen-carrying capacity of the blood and is characterized by a low hematocrit.
- Nutritional anemia is caused by a dietary deficiency of a factor needed for erythropoiesis.
- Iron-deficiency anemia occurs when insufficient iron is available for the synthesis of hemoglobin.
- A dietary deficiency of folic acid can produce anemia.
- Pernicious anemia is caused by an inability to absorb adequate amounts of vitamin B12 from the digestive tract.
- Aplastic anemia is caused by failure of the bone marrow to produce adequate numbers of red blood cells.

- Renal anemia is caused by inadequate erythropoietin secretion as a result of kidney disease.
- Hemorrhagic anemia is caused by the loss of substantial quantities of blood.
- Hemolytic anemia is caused by the rupture of excessive numbers of circulating erythrocytes.

*Polycythemia is an excess of circulating erythrocytes.*

- Polycythemia is characterized by an excess of circulating red blood cells and an elevated hematocrit.
- Primary polycythemia is caused by a tumorlike condition of the bone marrow in which the erythropoiesis proceeds at an excessive uncontrolled rate.
- Secondary polycythemia is an appropriate erythropoietin-induced adaptive mechanism to improve the blood's oxygen-carrying capacity.

## LEUKOCYTES

*Leukocytes primarily function as defense agents outside the blood.*

- Leukocytes are the mobile units of the body's immune defense system.
- The leukocytes and their derivatives: (1) defend against invasion by pathogens by phagocytizing the foreigners or causing their destruction by more subtle means; (2) identify and destroy cancer cells that arise within the body; and (3) function as a "cleanup crew" that removes the body's "litter" by phagocytizing debris resulting from dead or injured cells.

*There are five types of leukocytes.*

- The five types of leukocytes fall into two main categories: (1) polymorphonuclear granulocytes (neutrophils, eosinophils, and basophils) and (2) mononuclear agranulocytes (monocytes and lymphocytes).

*Leukocytes are produced at varying rates depending on the changing defense needs of the body.*

- Granulocytes and monocytes are produced only in the bone marrow.
- Lymphocytes are actually produced by already-existing lymphocytes residing in the lymphoid tissues.
- Neutrophils are phagocytic specialists.
- An increase in circulating eosinophils is associated with allergic conditions and internal parasite infestations such as worms.
- Eosinophils attach to the worm and secrete substances that kill it.
- Basophils synthesize and store histamine and heparin.
- Histamine release is important in allergic reactions, whereas heparin hastens the removal of fat particles from the blood following a fatty meal.
- Monocytes are destined to become professional phagocytes.
- Monocytes mature and enlarge to become the large tissue phagocytes known as macrophages.
- Lymphocytes provide immune defense against targets for which they are specifically programmed.
- There are two types of lymphocytes, B lymphocytes and T lymphocytes.
- B lymphocytes produce antibodies that circulate in the blood.
- An antibody binds with and marks for destruction specific kinds of foreign matter.
- T lymphocytes destroy their specific target cells, a process known as cell-mediated immune response.

## PLATELETS AND HEMOSTASIS

*Platelets are cell fragments derived from megakaryocytes.*

- Platelets are a third type of cellular element present in the blood.
- Platelets are small cell fragments that are shed off the outer edges of extraordinarily large bone marrow-bound cells known as megakaryocytes.
- The hormone thrombopoietin increases the number of megakaryocytes in the bone marrow and stimulates each megakaryocyte to produce more platelets.

*Hemostasis prevents blood loss from damaged small vessels.*

- Hemostasis is the arrest of bleeding from a broken blood vessel.
- Hemostasis involves three major steps: (1) vascular spasm, (2) formation of a platelet plug, and (3) blood coagulation.

*Vascular spasm reduces blood flow through an injured vessel.*

- A cut or torn blood vessel immediately constricts.
- The opposing endothelial surfaces of the vessel are pressed together by this initial vascular spasm.
- The surfaces become sticky and adhere to each other.

*Platelets aggregate to form a plug at a vessel tear or cut.*

- When the smooth endothelial lining is disrupted because of vessel injury, platelets attach to the exposed collagen.

- Platelets release adenosine diphosphate, which causes the surface of nearby circulating platelets to become sticky so that they adhere to the first layer of platelets.
- This aggregating process is reinforced by the formation of thromboxane A2 from a component of the platelet plasma membrane.
- Thromboxane A2 directly promotes platelet aggregation.
- Normal endothelium releases prostacyclin, which inhibits platelet aggregation.
- Thus, the platelet plug is limited to the defect and does not spread to normal vascular tissue.
- The aggregated platelet plug also performs three other important roles: (1) the actin-myosin protein complex within the plug contracts to compact and strengthen the plug, (2) the platelet plug releases several powerful vasoconstrictors, and (3) the platelet plug releases chemicals that enhance coagulation.

*Clot formation results from a triggered chain reaction involving plasma clotting factors.*
- Blood coagulation, or clotting, is the transformation of blood from a liquid into a solid gel.
- The conversion of fibrinogen into fibrin is catalyzed by the enzyme thrombin at the site of vessel injury.
- Fibrin molecules adhere to the damaged vessel surface, forming a loose, netlike meshwork that traps the cellular elements of the blood.
- Fibrin strands are only loosely interlaced.
- Chemical linkages rapidly form between adjacent strands.
- This cross-linkage process is catalyzed by a clotting factor known as factor XIII.
- Thrombin activates factor XIII.
- Thrombin exists in the plasma in the form of an inactive precursor called prothrombin.
- Factor X converts prothrombin into thrombin.
- Altogether, 12 plasma clotting factors participate in essential steps that lead to the final conversion of fibrinogen into a stabilized fibrin network.
- Once the first factor in the sequence is activated, it in turn activates the next factor, and so on, in a series of sequential reactions known as a cascade, until thrombin catalyzes the final conversion of fibrinogen into fibrin.
- The clotting cascade may be triggered by the intrinsic pathway or extrinsic pathway.
- The intrinsic pathway is set off when factor XII is activated by coming in contact with exposed collagen.
- The extrinsic pathway requires contact with tissue factors external to the blood.
- When a tissue is traumatized, it releases a protein complex known as tissue thromboplastin, which directly activates factor X, thereby bypassing all preceding steps of the intrinsic pathway.

*Fibrinolytic plasmin dissolves clots.*
- Fibroblasts form a scar at the vessel defect.
- The clot is slowly dissolved by a fibrinolytic enzyme called plasmin.
- Phagocytic white blood cells gradually remove the products of clot dissolution.

*Inappropriate clotting is responsible for thromboembolism.*
- An abnormal intravascular clot attached to a vessel wall is known as a thrombus, and free floating clots are called emboli.
- Several factors can cause thromboembolism: (1) roughened vessel surfaces associated with atherosclerosis, (2) imbalances in the clotting-anticlotting systems, (3) slow-moving blood, and (4) the release of tissue thromboplastin.

*Hemophilia is the primary condition responsible for excessive bleeding.*
- The most common cause of excessive bleeding is hemophilia, which is caused by a deficiency in one of the factors in the clotting cascade.
- Eighty percent of all hemophiliacs lack the genetic ability to synthesize factor VIII.
- Vitamin K deficiency can cause a bleeding tendency.

## Key Terms

adenosine diphosphate (ADP)
albumins
anemia
antibodies
aplastic anemia
B lymphocytes
basophils
bicarbonate ion
blood coagulation (clotting)
bone marrow
carbonic anhydrase
clot retraction
embolus
eosinophils
eosinophilia
erythrocytes (red blood cells)
erythropoiesis
erythropoietin
extrinsic pathway
factor X
fibrin

fibrinogen
globulins
glycolytic enzymes
granulocyte colony-stimulating factor
hematocrit (packed cell volume)
hemolysis
hemostasis
hemoglobin
hemophilia
heparin
immunity
infectious mononucleosis
intrinsic pathway
leukemia
leukocytes
lymphocytes
lymphoid tissue
macrophages
megakaryocytes
malaria
mast cells
monocytes
mononuclear granulocytes
neutrophils
neutrophilia
nitric oxide
nutritional anemia
pathogens
perfluocarbons
pernicious anemia
plasma
plasma proteins
plasmin
plasminogen
platelet plug
platelets (thrombocytes)
pluripotential stem cells
polycythemia
polymorphonuclear granulocytes
prothrombin
prostacyclin
red blood cell count
red bone marrow
reticulocyte
serum
spleen
T lymphocytes
thrombin
thrombopoietin
thrombus
thromboxane
tissue plasminogen activator (tPA)
tissue thromboplastin
vascular spasm
yellow bone marrow

## *Review Exercises*

*Answers are in the appendix.*

### True/False

_____ 1. Erythrocytes most important feature which enables them to transport oxygen is the hemoglobin they contain.

_____ 2. Hemoglobin can also combine with carbon monoxide.

_____ 3. Each of us has a total of 25 to 30 billion red blood cells.

_____ 4. Red marrow not only produces red blood cells but it is the ultimate source for leukocytes and platelets as well.

_____ 5. Estrogen, a sex hormone, increases the basal rate of erythropoiesis.

_____ 6. Nutritional anemia is caused by an inability to absorb adequate amounts of vitamin $B_{12}$ from the digestive tract.

_____ 7. Secondary polycythemia is caused by a tumorlike condition of the bone marrow in which erythropoiesis proceeds at an excessive, uncontrolled rate.

_____ 8. Platelets remain functional for an average of 20 days.

_____ 9. The conversion into fibrin is catalyzed by the enzyme thromboxane at the site of vessel injury.

_____ 10. The clotting cascade may be triggered by the intrinsic pathway or the extrinsic pathway.

_____ 11. Intrinsic pathway requires only four steps to bring about clotting.

_____ 12. Phagocytic white blood cells gradually remove the products of clot dissolution.

_____ 13. An abnormal intravascular clot attached to a vessel wall is called a thrombus.

_____ 14. Eighty percent of all hemophiliacs lack the genetic ability to synthesize factor VIII.

_____ 15. Vitamin A deficiency can cause a bleeding tendency.

_____ 16. Plasma is composed of 80 percent water.

_____ 17. Inorganic constituents account for approximately 1 percent of plasma weight.

_____ 18. The most abundant electrolytes in the plasma are sodium and potassium.

_____ 19. Plasma proteins usually do not exit through the narrow pores in the capillary walls.

_____ 20. Plasma proteins exist in a colloidal dispersion.

_____ 21. Plasma proteins contribute to blood viscosity.

_____ 22. The gamma globulins are the immunoglobulins.

_____ 23. Plasma proteins do not affect the plasma's capacity to buffer changes in pH.

_____ 24. Globulins are the most abundant of the plasma proteins.

_____ 25. Leukocytes lack hemoglobin.

_____ 26. Eosinophils have an affinity for the blue dye.

_____ 27. Monocytes have a large spherical nucleus that occupies most of the cell.

_____ 28. Granulocytes and monocytes are produced only in the bone marrow.

_____ 29. The total number of leukocytes normally ranges from five to ten million cells per milliliter of blood.

_____ 30. Basophils store histamine and heparin.

_____ 31. Heparin release is important in allergic reactions.

_____ 32. T lymphocytes produce antibodies.

**Fill in the Blank**

33. _____________ is the arrest of bleeding from a broken blood vessel.

34. ______________________ or ______________ is the transformation of blood from a liquid into a solid gel.

35. When a tissue is traumatized, it releases a protein complex known as _____________.

36. ____________ is the fluid squeezed from the clot during clot retraction.

37. The clot is slowly dissolved by a fibrinolytic enzyme called _____________.

38. Free-floating clots are called _____________.

39. The most common cause of excessive bleeding is _________________.

40. ______________ enzymes are necessary for generating the energy needed to fuel the active transport mechanisms involved in maintaining proper ionic concentrations within the cell.

41. Most old red blood cells meet their final demise in the __________.

42. The bone marrow normally generates new red blood cells, a process known as ________________.

43. ___________ anemia is caused by failure of the bone marrow to produce adequate numbers of red blood cells.

44. _________ refers to a reduction below normal in the oxygen carrying capacity of the blood and is characterized by a low hematocrit.

45. ____________ anemia is caused by the rupture of excessive numbers of circulating erythrocytes.

46. ________________________ is the best known example among various hereditary abnormalities of red blood cells.

47. The remaining small percentage of plasma is occupied by (1)___________, (2) ________________, (3) ________________, and (4) ______________.

48. __________________ are the one group of plasma constituents not present just for the ride.

49. Name the three groups of plasma proteins: (1) _____________, (2) ______________ and (3) ______________.

50. ________________ bind many substances for transport through the plasma.

51. Alpha, beta, and gamma are the three subclasses of __________________.

52. ________________ is a key factor in the blood-clotting process.

53. ________________ are the mobile units of the body's immune defense system.

54. Neutrophils, eosinophils, and basophils are categorized as __________________________________.

55. _________________ are the smallest of the leukocytes.

56. Leukocytes and their derivatives defend against invasion by __________________.

57. ________________ are phagocytic specialists.

58. ____________ are the least numerous and most poorly understood of the leukocytes.

59. T lymphocytes destroy their specific target cells, a process known as a(n) ____________ ______________________.

60. ____________ are destined to become professional phagocytes.

**Matching**
*Match the blood constituent to their function.*

| | | |
|---|---|---|
| _____ | 61. carries heat | a. platelets |
| _____ | 62. hemostasis | b. erythrocyte |
| _____ | 63. become macrophages | c. fibrinogen |
| _____ | 64. transport oxygen | d. basophils |
| _____ | 65. precursor for fibrin | e. neutrophils |

| | | |
|---|---|---|
| _____ 66. | cell-mediated immune response | f. eosinophils |
| _____ 67. | release histamine | g. B lymphocytes |
| _____ 68. | attacks parasitic worms | h. water |
| _____ 69. | antibody production | i. T lymphocytes |
| _____ 70. | phagocytizes bacteria | j. monocytes |

**Multiple Choice**

71. Which cellular element has the largest percentage of total blood volume?
    a. eosinophils
    b. erythrocytes
    c. basophils
    d. neutrophils
    e. platelets

72. Which type of anemia is caused by failure of the bone marrow to produce adequate numbers of red blood cells?
    a. nutritional anemia
    b. pernicious anemia
    c. hemolytic anemia
    d. aplastic anemia
    e. renal anemia

73. Which enzyme is critical in carbon dioxide transport?
    a. glycolytic enzyme
    b. carbonic anhydrase
    c. erythropoietin
    d. plasminogen

74. Which of the plasma proteins are antibodies?
    a. fibrinogen
    b. alpha globulins
    c. gamma globulins
    d. beta globulins
    e. albumins

75. Which organ secretes erythropoietin?
    a. bone marrow
    b. heart
    c. lungs
    d. spleen
    e. kidney

76. Which type of anemia is caused by inadequate erythropoietin production?
    a. nutritional anemia
    b. pernicious anemia
    c. hemolytic anemia
    d. aplastic anemia
    e. renal anemia

77. Which disorder is characterized by a tumorlike condition of the bone marrow?
    a. hemorrhagic anemia
    b. sickle-cell anemia
    c. secondary polycythemia
    d. leukemia
    e. primary polycythemia

78. Which substance causes platelets to become sticky?
    a. adenosine diphosphate
    b. plasminogen
    c. fibrinogen
    d. tissue thromboplastin
    e. prothrombin

79. Which factor do most hemophiliacs lack the ability to synthesize?
    a. factor X
    b. factor VII
    c. factor XII
    d. factor VIII
    e. factor XIII

80. Which substance slowly dissolves the clot?
    a. thromboplastin
    b. plasmin
    c. carbonic anhydrase
    d. fibrinogen
    e. plasminogen

81. Hemophilia results from:
    a. a deficiency of platelets
    b. inadequate hemoglobin production
    c. vitamin $B_{12}$ deficiency
    d. a genetic inability to produce one of the factors in the clotting cascade
    e. excess production of heparin

82. Which of the following plays a role in dissolving clots?
    a. tissue thromboplastin
    b. prostacyclin
    c. plasmin
    d. calcium
    e. exposed collagen

83. Which of the following describes the major function of tissue thromboplastin?
    a. It is released from traumatized tissue and triggers the extrinsic clotting pathway.
    b. It converts fibrinogen to fibrin.
    c. It forms the network of the clot.
    d. It activates Hageman factor.
    e. More than one of the above are correct.

84. Which of the following is not accomplished by thrombin?
    a. It stimulates the conversion of fibrinogen to fibrin.
    b. It activates tissue thromboplastin.
    c. It enhances factor III.
    d. It enhances platelet aggregation.
    e. It facilitates its own formation.

85. Which of the following is true with respect to the extrinsic clotting pathway?
    a. It is set off by factor XII.
    b. It clots blood in an injured blood vessel.
    c. It has more steps than the intrinsic pathway.
    d. More than one of the above are correct.
    e. None of the above are correct.

86. Platelets:
    a. are important in homeostasis
    b. convert prothrombin to thrombin
    c. form the meshwork upon which the erythrocytes become trapped to produce a clot
    d. are white blood cells
    e. are produced from red blood cells

87. Platelets:
    a. convert prothrombin to thrombin
    b. form the meshwork upon which the erythrocytes become entrapped
    c. adhere to collagen
    d. release fibrinogen when a platelet plug is formed
    e. produce antibodies

88. When small blood vessels are damaged, loss of blood is prevented by:
    a. Platelet aggregation.
    b. Vasoconstriction of these vessels.
    c. Formation of a platelet plug.
    d. Both a and b are correct.
    e. A, b and c are correct.

89. What forms the meshwork of a blood clot?
    a. red blood cells
    b. fibrinogen
    c. platelets
    d. thrombin
    e. Hageman factor

90. What does exposed collagen in a damaged vessel cause?
    a. Activates factor XII
    b. Initiates platelet aggregation
    c. Secretes ADP
    d. Both a and b are correct
    e. A, b and c are correct

91. What does prostacyclin do in the body?
    a. It activates the clotting cascade.
    b. It induces profound vasoconstriction of an injured blood vessel.
    c. It profoundly inhibits platelet aggregation.
    d. It is released by aggregated platelets.
    e. It dissolves the blood clot.

92. Which of the following types of blood cellular elements lack nuclei?
    a. Platelets
    b. Erythrocytes
    c. Leukocytes
    d. Both a and b are correct
    e. A, b and c are correct

93. Platelets:
    a. Are small cell fragments derived from large megakaryocytes in the bone marrow.
    b. Lack nuclei.
    c. Contain high concentrations of actin and myosin.
    d. Both a and b are correct.
    e. A, b and c are correct.

94. Neutrophilia frequently accompanies:
    a. Bacterial infections.
    b. Viral infections.
    c. Parasite infections.
    d. Allergic condictions.
    e. All of the above are correct.

95. Basophils:
    a. Contain granules that preferentially take up basic blue dyes.
    b. Leave the blood to become macrophages.
    c. Synthesize and store histamine and heparin.
    d. Both a and b are correct.
    e. A, b and c are correct.

96. Lymphocytes:
    a. Are polymorphonuclear granulocytes.
    b. Are the most abundant type of leukocyte.
    c. Can be produced in lymphoid organs as well as in the bone marrow.
    d. Both a and b are correct.
    e. A, b and c are correct.

97. Polymorphonuclear granulocytes include:
    a. Neutrophils.
    b. Eosinophils.
    c. Basophils.
    d. Both a and b are correct.
    e. A, b and c are correct.

98. Which of the following is not a function of white blood cells?
    a. Activation of factor XII.
    b. The production of antibodies.
    c. The destruction of cancer cells.
    d. The phagocytosis of foreign invaders.
    e. Phagocytosis of cellular debris.

99. Hemoglobin:
    a. Is found in the nuclei of red blood cells.
    b. Contains carbonic anhydrase.
    c. Can combine with oxygen, carbon dioxide, hydrogen ions, and carbon monoxide.
    d. Both a and b are correct.
    e. A, b and c are corect.

100. Iron:
    a. Can combine reversibly with oxygen.
    b. Deficiency can produce anemia.
    c. Is converted into bilirubin and secreted in the bile when an old red blood cell is destroyed.
    d. Both a and b are correct.
    e. A, b and c are correct.

## Points to Ponder

1. Why do you suppose that the clotting cascade is so long and complicated?
2. Over the years we have gained much knowledge about the composition of blood. Why then is it so difficult to make artificial blood?
3. What is the advantage in having short-lived erythrocytes with no nuclei, organelles, or ribosomes, as compared to longer-lived erythrocytes that are complete cells?
4. Does drinking coffee after binge drinking help a person "sober up." Explain your answer.
5. In your body, where is most of your blood located?
6. What does the term "blue blooded" mean?
7. Why doesn't blood clot in blood vessels?
8. When a blood panel is done, blood is taken from a vein and not an artery. Explain why.
9. How is hemoglobin recycled?
10. What is the difference between hematocrit and reticulocyte count?

## Clinical Perspectives

1. When a person cuts their wrists to commit suicide does it make a difference whether or not they completely sever the vessels?
2. Why is 20 percent of a cigarette smokers' hemoglobin nonfunctional?
3. In what type of clinical situation would you receive just blood platelets during a blood transfusion?
4. Your friend tells you she has iron deficiency anemia. She knows you are taking a physiology course and asks you to explain to her what this means. What would your answer be?
5. Your mother tells you that after her yearly physical the doctor recommends that she take one baby aspirin each day. She knows that you are taking a physiology course and asks you to explain why the doctor recommended aspirin therapy. How would you explain this therapy to her?
6. What are the reasons you give a patient tissue plasminogen activator (tPA) or streptokinase?
7. Your athlete friend asks you if she should try induced erythrocythemia (blood doping). What would you tell her?
8. In a blood panel, what does a differential white blood cell count tell the physician about homeostasis in your body?
9. How would you explain to your friend why blood is typed and cross-matched before a blood transfusion?
10. Your college roommate has infectious mononucleosis. She asks you why she is so tired all the time. What would you tell her?

## *PhysioEdge Activities*

**Related to Text:**
Media Exercise 11.1: Blood.
Media Exercise 11.2: Abnormal Hemostasis.
Media Exercise 11.3: Thrombocytes and Hemostasis.

**Related to Figures:**
*Figure 11.1.* For an interaction related to this figure, see Media Exercise 11.1: Blood.
*Figure 11.7.* For an interaction related to this figure, see Media Exercise 11.1: Blood.
*Figure 11.9.* For an interaction related to this figure, see Media Exercise 11.1: Blood.
*Figure 11.10.* For an interaction related to this figure, see Media Exercise 11.3: Thrombocytes and Hemostasis.
*Figure 11.11.* For an interaction related to this figure, see Media Exercise 11.3: Thrombocytes and Hemostasis.

## *Media Resources*

**PhysioEdge CD-ROM**
For a visual review of concepts in this chapter, check out:

Media Exercise 11.1: Blood
Media Exercise 11.2: Abnormal Hemostasis
Media Exercise 11.3: Thrombocytes and Hemostasis

**Book Companion Website**
The website for this book contains a wealth of helpful study aids, as well as many ideas for further reading and research. Log on to:

**http://www.brookscole.com/sherwoodhp6**

Select Chapter 11 from the drop-down menu, or click on one of the many resources areas, including **Case Histories**, which introduce clinical aspects of human physiology. For this chapter check out Case History 17: Pneumococcal Bacteremia.

For Suggested Readings, consult **InfoTrac® College Edition**, your online research library, at:

**http://infotrac.thomsonlearning.com**

# 12

# Body Defenses

## Chapter Overview

Humans are constantly coming into contact with external agents that can be harmful if they gain entry into the body. To help prevent disease, the immune system contributes to homeostasis by providing a sophisticated, highly effective defense system. The system is essential in defending against viruses and bacteria, removing worn out cell and tissue debris, and destroying mutant cells.

However, these benefits are not without cost. Inappropriate immune responses and tissue rejection are problems due to the efficacy of the immune system as a defense system.

The primary defensive cells are neutrophils, eosinophils, basophils, B lymphocytes, T lymphocytes, and macrophages.

The methods of defense include phagocytizing, lysing secreting chemicals, marking for destruction, producing antibodies and inactivating the invaders. Also involved are memory and the transfer of information by cloning. The immunity derived from this system is active, passive, or acquired. In most systems the cells function as part of a tissue or organ while in the immune system the defense is primarily from individual cells.

The skin and all surfaces that communicate with the external environment provide either physical and/or chemical defenses. Defense of the body begins at the environmental body interface and continues throughout the body. Thus, homeostasis is maintained for the survival of cells, which make up body systems, which maintain homeostasis.

## Chapter Outline

INTRODUCTION

*The immune defense system provides protection against foreign and abnormal cells and removes cellular debris.*

- Immunity refers to the body's ability to resist or eliminate potentially harmful foreign materials or abnormal cells.
- The following activities are attributed to the immune defense system: (l) defense against invading pathogens; (2) removal of worn out cells and tissue debris; (3) identification and destruction of abnormal or mutant cells that have originated in the body, (4) elicitation of inappropriate immune responses that lead either to allergies, or to autoimmune disease; and (5) rejection of tissue cells of foreign origin.

*Pathogenic bacteria and viruses are the major targets of the immune system.*

- Bacteria are non-nucleated, single-celled microorganisms self-equipped with all machinery essential for their own survival and reproduction.
- Pathogenic bacteria induce tissue damage and produce disease by releasing enzymes or toxins that injure or functionally disrupt cells and organs.
- Viruses are not self-sustaining cellular entities.
- Viruses consist only of nuclei acids enclosed in a protein coat.
- Viruses invade a host cell and take over the cellular biochemical facilities for their own purposes.

*Leukocytes are the effector cells of the immune system.*

- Neutrophils are highly mobile phagocytic specialists that engulf and destroy unwanted materials.
- Eosinophils secrete chemicals that destroy parasitic worms and are involved in allergic manifestations.
- Basophils release histamine and heparin and also are involved in allergic manifestations.
- The B lymphocytes secrete antibodies that indirectly lead to the destruction of foreign material.
- The T lymphocytes are responsible for cell-mediated immunity involving direct destruction of virus-invaded cells and mutant cells through nonphagocytic means.
- Monocytes are transformed into macrophages, which are large, tissue-bound phagocytic specialists.

*Immune responses may be either innate and nonspecific, or adaptive and specific.*

- Nonspecific immune responses are inherent defense responses that nonselectively defend against foreign or abnormal material of any type.
- Specific immune responses are selectively targeted against particular foreign material to which the body has previously been exposed.

## INNATE IMMUNITY

*Innate defenses include inflammation, interferon, natural killer cells, and the complement system.*

- Nonspecific defenses that come into play whether or not there has been prior experience with the offending agent include the following: inflammation, interferon, natural killer cells, and the complement system.

*Inflammation is a nonspecific response to foreign invasion or tissue damage.*

- The ultimate goal of inflammation is to bring to the invaded or injured area phagocytes and plasma proteins that can: (l) isolate, destroy, or inactivate the invaders; (2) remove debris; and (3) prepare for subsequent healing and repair.
- Upon bacterial invasion, macrophages already present in the area immediately begin phagocytizing the foreign microbes.
- Almost immediately upon microbial invasion arterioles within the area dilate, increasing blood flow to the site of injury.
- Histamine is released in the area of tissue damage by mast cells.
- Released histamine increases the capillaries' permeability.
- Plasma proteins that normally are prevented from leaving the blood escape into the inflamed tissue.
- As the leaked plasma proteins accumulate in the interstitial fluid, they exert a colloid osmotic pressure.
- The elevation in local osmotic pressure and the increased capillary blood pressure result in localized edema.
- Upon exposure to tissue thromboplastin in the injured tissue and to specific chemicals secreted by phagocytes on the scene, fibrinogen, the final factor in the clotting system, is converted into fibrin.
- Fibrin forms interstitial fluid clots in the spaces around the bacterial invaders and damaged cells.
- This walling off of the injured region from the surrounding tissues prevents or at least delays the spread of bacterial invaders and their toxic products.
- Within an hour after the injury, the area is teeming with leukocytes that have exited from the vessels.
- Neutrophils are the first to arrive.
- During the next eight to twelve hours, monocytes swell and mature into macrophages.
- Leukocyte emigration from the blood into the tissues involves the processes of margination, diapedesis, amoeboid movement and chemotaxis.
- Margination refers to the sticking of blood-borne leucocytes to the inner endothelial lining of capillaries.
- Cell adhesion molecules (CAM's) are important in the process of margination.
- Selectins, one type of CAM, cause leucocytes that are flowing by in the bloodstream to slow down.
- This slowing down allows the leucocytes enough time to check for local activating factors.
- The activating factors cause the leukocytes to adhere firmly to the endothelial lining by means of interaction with another type of CAM, the integrins.
- The adhered leucocytes start exiting by a mechanism known as diapedesis.
- Assuming amoeba-like behavior, an adhered leukocyte pushes a long, narrow projection through a capillary pore.
- Outside the vessel, the leukocyte moves in an amoeboid fashion toward the site of tissue damage and bacterial invasion.
- Phagocytic cells are guided in their direction of migration by attraction to certain chemical mediators, or chemotaxins.
- This process is referred to as chemotaxis.
- Within a few hours after the onset of the inflammatory response, the number of neutrophils in the blood may increase up to four or five times that of normal
- Phagocytosis involves the engulfment and intracellular degradation of foreign particles and tissue debris.
- Dead tissue and many foreign materials have surface characteristics that differ from normal body cells.
- Foreign particles are deliberately marked for phagocytic ingestion by being coated with chemical mediators (opsonins) generated by the immune system.
- Microbe-stimulated phagocytes release many chemicals, which function as mediators of the inflammatory response.
- Macrophages secrete nitric oxide which is toxic to nearby microbes.
- Neutrophils secrete lactoferrin, a protein that lightly binds with iron, making it unavailable for use by invading bacteria.
- Phagocytic secretions stimulate the release or histamine from mast cells.
- Phagocytic secretions trigger both clotting and anti-clotting systems.

- Phagocytic secretions split kininogens into active kinins.
- Kinins stimulate several key steps in the complement system.
- Kinins reinforce the vascular changes induced by histamine.
- Kinins activate nearby pain receptors.
- Kinins act as powerful chemotaxins to induce phagocyte migration to the affected area.
- Phagocytic secretions induce the development of fever by the release of endogenous pyrogen.
- Macrophages secrete a chemical mediator known as leukocyte endogenous mediator.
- Leukocyte endogenous mediator (LEM) causes a decrease in plasma concentration of iron by altering iron metabolism.
- The LEM stimulates granulopoiesis, the synthesis and release of neutrophils and other granulocytes by the bone marrow.
- The LEM also stimulates the release of acute-phase proteins from the liver.
- Macrophages secrete interleukin 1, which enhances the proliferation and differentiation of both B and T lymphocytes.
- The ultimate purpose of the inflammatory process is to isolate and destroy injurious agents and to clear the area for tissue repair.

*Interferon transiently inhibits multiplication of viruses in most cells.*

- A nonspecific defense mechanism is the release of interferon from virus-infected cells.
- Interferon binds with receptors on the plasma membranes of neighboring cells, signaling these cells to prepare for the possibility of impending viral attack.
- Interferon triggers the production of viral-blocking enzymes by potential host cells.
- Binding with interferon induces these other cells to synthesize enzymes that can break down viral messenger RNA.
- Interferon exerts its effect by activating a heretofore unknown signal transduction pathway, involving a newly identified class of intracellular chemicals known as Janus kinases within the cells to which it binds.
- Interferon markedly enhances the actions of cell-killing cells.
- Interferon also slows cell division and suppresses tumor growth.
- Through recombinant DNA technology, bacteria can be turned into interferon factories for large-scale commercial production of this valuable immune agent.
- Interferon has been approved for some forms of cancer, including one previously fairly rare form of leukemia and the AIDS-associated Kaposi's sarcoma, as well as for genital warts caused by papillomavirus.
- Interferon is also the first drug approved for treating multiple sclerosis.

*Natural killer cells destroy virus-infected cells and cancer cells upon first exposure to them.*

- Natural killer cells are naturally occurring, lymphocyte-like cells that nonspecifically destroy virus-infected cells and cancer cells by directly lysing their membranes.

*The complement system punches holes in microorganisms.*

- The system derives its name from the fact that it "complements" the action of antibodies.
- The complement system consists of plasma proteins that are produced by the liver and circulate in the blood in inactive form.
- Once the first component is activated, it activates the next component, and so on.
- The final five components assemble into a large protein complex, the membrane-attack complex (MAC), which attacks the surface membrane of nearby microorganisms by embedding itself, so that a large channel is created through the microbial surface membrane.
- The osmotic flux of water into the victim cell causes it to swell and burst.
- Complement-induced lysis is the major means of directly killing microbes without phagocytizing them.
- Complement components augment the inflammatory response by: (1) serving as chemotaxins, (2) acting as opsonins, (3) promoting vasodilation and increased vascular permeability, (4) stimulating the release of histamine, and (5) activating kinins.

## Adaptive Immunity: General Concepts

*Adaptive immune responses include antibody-mediated immunity and cell-mediated immunity.*

- There are two classes of specific immune responses: antibody-mediated, or humoral, immunity involving the production of antibodies by B lymphocyte derivatives known as plasma cells, and cell-mediated immunity, involving the production of activated T lymphocytes, which directly attack unwanted cells.
- During fetal life and early childhood, some of the immature lymphocytes migrate through blood to the thymus, where they undergo further processing to become T lymphocytes.
- Lymphocytes that mature without benefit of thymic education become B lymphocytes.

- The thymus gradually atrophies and becomes less important as the individual matures.
- The thymus does, however, continue to produce thymosin, a hormone important in maintaining the T cell lineage.

*An antigen induces an immune response against itself.*
- Both B and T cells must be able to specifically recognize unwanted cells and other material to be destroyed or neutralized as being distinct from the body's own normal cells.
- The presence of antigens enables them to make this distinction.
- Foreign proteins are the most common antigens.
- Haptens are low-molecular weight organic substances that are not antigens by themselves but can become antigens if they attach to body proteins.

## B Lymphocytes: Antibody-Mediated Immunity

*Antibodies stimulate B cells to convert into plasma cells that produce antibodies.*
- In the case of B cells, binding with an antigen induces the cell to differentiate into a plasma cell, which produces antibodies that are able to combine with the specific antigen that stimulated the antibodies' production.
- Antibodies are secreted into the blood or lymph where they are known as gamma globulins or immunoglobulins.

*Antibodies are Y-shaped and classified according to properties of their tail portion.*
- Antibodies are grouped into five subclasses.
- Antibody proteins in all five subclasses are composed of four interlinked polypeptide chains—arranged in the shape of a Y.
- Characteristics of the arm regions of the Y determine the specificity of the antibody.
- An antibody has two identical antigen-binding sites, one at the tip of each arm.
- These antigen-binding fragments (Fab) are unique for each different antibody.
- The tail portion of every antibody within each immunoglobulin subclass is identical.

*Antibodies largely amplify innate immune responses to promote antigen destruction.*
- Immunoglobulins cannot directly destroy foreign organisms or other unwanted materials upon binding with antigens on their surfaces.
- Antibodies exert their protective influence in one of two general ways: physical hindrance of antigens and amplification of nonspecific immune responses.
- Antibodies mark or identify foreign material as targets for actual destruction by the complement system, phagocytes, or killer cells.
- Occasionally, an overzealous antigen-antibody response can inadvertently cause damage to normal cells as well as to invading foreign cells, causing an immune-complex disease.

*Clonal selection accounts for the specificity of antibody production.*
- Each B lymphocyte is preprogrammed to respond to only one of the millions of different antigens.
- The clonal selection theory proposes that diverse B lymphocytes are produced during fetal development, each capable of synthesizing an antibody against a particular antigen before ever being exposed to it.

*Selected clones differentiate into active plasma cells and dormant memory cells.*
- B cells remain dormant, not actually secreting their particular antibody product until they come in contact with the appropriate antigen.
- Antigen binding causes the activated B cell clone to multiply and differentiate into two cell types: plasma cells and memory cells.
- Plasma cells switch to the production of IgG antibodies, which are secreted rather than remaining membranebound.
- In the blood, the secreted antibodies combine with invading free antigen, marking it for destruction by the complement system, phagocytic ingestion, or other means.
- A small proportion of the new B lymphocytes become memory cells, which do not participate in the current immune attack against the antigen but instead remain dormant and expand the specific clone.
- During initial contact with a microbial antigen, the antibody response is delayed for several days until plasma cells are formed and does not reach its peak for a couple of weeks.
- This response is known as the primary response.
- After reaching the peak, the antibody levels gradually decline over a period of time.
- Long-term protection against the same antigen is primarily attributable to the memory cells.
- If the same antigen ever reappears, the long-lived memory cells launch a more rapid, more potent and longer-lasting secondary response.
- This is the basis of long-term immunity against a specific disease.
- During vaccination, the individual is deliberately exposed to a pathogen that has been stripped of its

disease-inducing capability, but can still induce antibody formation against it.

*The huge repertoire of B cells is built by reshuffling a small set of gene fragments.*

- Only a relatively small number of gene fragments code for antibody synthesis.
- These fragments are cut, reshuffled, and spliced in a vast number of different combinations during B cell development.

*Active immunity is self-generated; passive immunity is borrowed.*

- The production of antibodies as a result of exposure to an antigen is referred to as active immunity against that antigen.
- A second way in which an individual can acquire antibodies is buy direct transfer of antibodies actively formed by another person (or animal).
- The transfer of antibodies is known as passive immunity.

*Blood types are a form of natural immunity.*

- Antibodies associated with blood types are the classic example of "natural antibodies. "
- Accordingly, the plasma of type A blood contains anti-B antibodies, type B blood contains anti-A antibodies, and both anti-A and anti-B antibodies are present in type O blood.
- High levels of these antibodies are found in the plasma of individuals who have never been exposed to a different type of blood.
- It is now known whether individuals are unknowingly exposed at an early age to small amounts of A-and B-like antigens associated with common intestinal bacteria.
- Because type O individuals do not have any A or B antigens, they are considered to be universal donors.
- Type O individuals can receive only type O blood.
- Type AB individuals are called universal recipients.
- Lacking both anti-A and anti-B antibodies, they can accept donor blood of any type.
- Individuals who possess the Rh factor (an erythrocyte antigen) are said to have Rh positive blood.
- Those lacking the Rh factor are considered to be Rh-negative.
- No naturally occurring antibodies develop against the Rh factor.

*Lymphocytes respond only to antigens presented to them by antigen-presenting cells.*

- Invading organisms or other antigens are first engulfed by macrophages.
- During phagocytosis, the macrophage processes the raw antigen intracellularly and then exposes the processed antigen on the outer surface of the macrophage's plasma membrane in such a way that the adjacent B cells can recognize and be activated by it.
- When a macrophage engulfs a foreign microbe, it digests the microbe into antigenic peptides.
- Each antigenic peptide is then bound to an MHC molecule.
- An MHC molecule has a deep groove into which a variety of antigenic peptides can bind.
- Loading of the antigenic peptide onto an MHC molecule takes place in a newly discovered specialized organelle within antigen-presenting cells, the compartment for peptide loading or CPL.
- The MHC molecule then transports the bound antigen to the cell surface where it is presented to passing lymphocytes.
- Macrophages secrete interleukin 1 that enhances the differentiation and proliferation of the now-activated B cell clone.
- Interleukin 1 is also largely responsible for the fever and malaise accompanying many infections.
- Many antigens are similarly presented to T cells.
- Helper T cells help B cells upon being activated by macrophage-presented antigen.
- The helper T cells secrete a chemical mediator, B cell growth factor, which further contributes to B cell function in concert with the interleukin 1 secreted by macrophages.
- Dendritic cells are specialized antigen-presenting cells found in sentinels in almost every tissue.

## T LYMPHOCYTES: CELL-MEDIATED IMMUNITY

*T cells bind directly with their targets.*

- Unlike B cells, T cells do not secrete antibodies.
- T cells must be in direct contact with their targets, a process known as cell-mediated immunity.
- Like B cells, T cells are clonal and exquisitely antigen-specific.
- T cells are activated by foreign antigen only when it is present on the surface of a cell that also carries a marker of the individual's own identity.

*The two types of T cells are cytotoxic T cells and helper T cells.*

- There are two subpopulations of T cells, depending on their roles when activated by antigens: (1) cytotoxic T cells, and (2) helper T cells.
- Like B cells, not all activated T cell progeny become effector T cells.
- A small proportion of them remain dormant, serving as a pool of memory T cells that are

primed and ready to respond should the same foreign antigen ever reappear within a body cell.
- The targets of cytotoxic T cells most frequently are host cells infected with viruses.

*Cytotoxic T cells secrete chemicals that destroy target cells.*
- One means by which cytotoxic T cells and natural killer cells destroy a targeted cell is by releasing perforin molecules, which penetrate into the target cell's surface membrane and join together to form porelike channels.
- Cytotoxic T cells also induce these virus-infected cells to self-destruct, a process know as apoptosis.
- The virus released upon destruction of the host cell is then directly destroyed in the extracellular fluid by phagocytic cells, neutralizing antibodies, and the complement system.
- Virus-infected neurons are spared from extermination by the immune system.
- Antibodies not only target viruses for destruction in the extracellular fluid but can also eliminate viruses inside neurons.

*Helper T cells secrete chemicals that amplify the activity of other immune cells.*
- In contrast to cytotoxic T cells, helper T cells are not killer cells.
- Helper T cells secrete B-cell growth factor, which enhances the antibody-secreting ability of the activated B-cell clone.
- Helper T cells secrete T cell growth factor, which augments the activity of cytotoxic T cells, suppressor T cells, and even other helper T cells responsive to the invading antigen.
- Some chemicals secreted by T cells act as chemotaxins to lure more neutrophils and macrophages-to-be to the invaded area.
- Macrophage-migration inhibition factor, another important cytokine released from helper T cells, keeps these large phagocytic cells in the region by inhibiting their outward migration.
- The AIDS virus selectively invades helper T cells, destroying or incapacitating the cells that normally orchestrate much of the immune response.
- The AIDS virus also invades macrophages.
- Recent studies have demonstrated the existence of two subsets of helper T cells, T helper 1 (TH1) cells and T helper 2 (TH2) cells.
- TH1 cells rally a cell-mediated response, whereas TH2 cells promote humoral immunity by B cells and rev up eosinophil activity.
- Helper T cells produced in the thymus are in a naive state until they encounter the antigen they are primed to recognize.
- Whether a naive helper T cell becomes a TH1 or TH2 cell depends on which cytokines are secreted by the macrophage as it presents the antigen to the naive T cell.
- Interleukin 12 drives a naive T cell specific for the antigen to become a TH1 cell, whereas interleukin 4 favors the development of a naive cell into a TH2 cell.
- Suppressor T cells limit the response of all other immune cells.

*The immune system is normally tolerant of self-antigens.*
- Suppressor T cells probably also play an important role in preventing the immune system from attacking the person's own tissues, a phenomenon known as tolerance.
- At least five different mechanisms appear to be involved in tolerance: (1) clonal deletion, (2) clonal anergy, (3) inhibition by suppressor T cells, (4) antigen sequestering, and (5) granting of immune privilege.

*Autoimmune diseases arise from loss of tolerance to self-antigens.*
- A condition in which the immune system fails to recognize and tolerate self-antigens associated with particular tissues is known as an autoimmune disease, of which myasthenia gravis is an example.
- Autoimmune disease may arise from a number of different causes: (1) a reduction in suppressor T cell activity or an imbalance in the ratio of suppressor to helper T cells specific for self-antigens; (2) normal self-antigens may be modified so that they are no longer recognized and tolerated by the immune system; (3) exposure to normally inaccessible self-antigens sometimes induces an immune attack against these antigens; and (4) exposure of the immune system to a foreign antigen structurally almost identical to a self-antigen.

*The major histocompatibility complex is the code for self-antigens.*
- The self-antigens, which the immune system learns to recognize as markers of a person's own cells, are plasma membrane-bound glycoproteins known as MHC molecules, because their synthesis is directed by a group of genes called the major histocompatibility complex or MHC.
- These are the same MHC molecules that escort engulfed foreign antigen to the cell surface for presentation by antigen-presenting cells.
- More than 100 different MHC molecules have been identified in human tissue, but each individual has a code for only three to six of these possible antigens.

- The exact pattern of MHC molecules varies from one individual to another much like a biochemical fingerprint.
- T cells bind with MHC self-antigens only when they are in association with a foreign antigen.
- T cells do bind with MHC antigens present on the surface of transplanted cells in the absence of foreign viral antigen.
- The ensuing destruction of the transplanted cells is responsible for rejection of transplanted or grafted tissues.
- To minimize the rejection phenomenon, the tissues of donor and recipient are matched according to MHC antigens.
- New therapeutic agents have become extremely useful in selectively depressing T cell rejection.
- The natural function of MHC antigens lies in their ability to direct the responses of T cells, not in their artificial role in the rejection of transplanted tissue.
- Each individual has two main classes of MHC-encoded glycoproteins that are differentially recognized by cytotoxic T and helper T cells.
- Cytotoxic T cells are able to respond to foreign antigen only in association with class I MHC glycoproteins, which are found on the surface of virtually all nucleated cells.
- Class II MHC glycoproteins, which are recognized by helper T cells, are restricted to the surface of a few special types of immune cells, such as B cells, cytotoxic T cells, and macrophages.
- The class I and II markers serve as signposts to guide cytotoxic and helper T cells to precise cellular locations where there immune capabilities can be most effective.
- Essentially all cells display class I MHC glycoproteins, enabling cytotoxic T cells to attack any invaded host cell.
- Helper T cells can bind with foreign antigen only when it is found on the surfaces of immune cells bearing class II glycoproteins.
- These include macrophages, which present antigen to helper T cells, and B and T cells, whose activity is enhanced by helper T cells.
- Specific binding requirements for the various T cells help ensure the appropriate T cell responses.

*Immune surveillance against cancer cells involves an interplay among immune cells and interferon.*

- Another important function generally attributed to the T cell system is its role in recognizing and destroying newly arisen, potentially cancerous tumor cells before they have a chance to multiply and spread – a process known as immune surveillance.
- Any normal cell may be transformed into a cancer cell if mutations occur within its genes responsible for controlling cell division and growth.
- Such mutations frequently occur by exposure to carcinogenic factors.
- A cell that has been transformed into a tumor cell defies the normal controls on its proliferation and position.
- Unrestricted multiplication of a single tumor cell results in a tumor.
- If the mass is slow growing and does not infiltrate into the surrounding tissue, it is considered a benign tumor.
- Invasive tumors are known as malignant tumors, commonly referred to as cancer.
- Often some of the cancer cells break away from the parent tumor and are transported through the blood to new territories.
- Metastasis is the term applied to this spread of cancer to other parts of the body.
- Cancer cells typically remain immature and do not become specialized, often resembling embryonic cells instead.
- Such dedifferentiated malignant cells lack the ability to perform the specialized functions of the normal cell type from which they mutated.
- Most mutations do not result in malignancy.
- Only a fraction of the mutations involve loss of control over the cell's growth and multiplication.
- A single mutation generally is not sufficient for a cell to become cancerous.
- Potentially cancerous cells that do arise are usually destroyed by the immune system early in their development.
- Immune surveillance against cancer depends on an interplay among three types of immune cells: cytotoxic T cells, natural killer cells, and macrophages.
- All three of these immune cell types secrete interferon.
- Interferon inhibits multiplication of cancer cells and increases the killing ability of the immune cells.
- Natural killer cells do not require prior exposure and sensitization to a cancer before being able to launch a lethal attack.
- Cytotoxic T cells are believed to be especially important in defending against the few kinds of virus-induced cancers.
- Natural killer cells and cytotoxic T cells release perforin and other toxic chemicals that destroy the targeted mutant cell.
- Macrophages are able to engulf and destroy cancer cells intracellularly.
- B cells, upon viewing a mutant cancer cell as an alien to normal self, may produce antibodies against it.

- The blocking antibodies are able to bind with the antigenic sites on the cancer cell, hiding these sites from recognition by cytotoxic T cells.

*A regulatory loop links the immune system with the nervous and endocrine systems.*

- There are important links between the immune system and the body's two major control systems, the nervous and endocrine systems.
- The immune system both influences and is influenced by the nervous and endocrine systems.
- Interleukin 1 can turn on the stress response by activating a sequence of nervous and endocrine events that result in the secretion of cortisol.
- Cortisol mobilizes the body's nutrient stores so that metabolic fuel is readily available to keep pace with the body's energy demand at a time when the person is sick.
- Cortisol mobilizes amino acids to repair any tissue damage.
- Cytokines released by immune cells enhanced the neurally and hormonally controlled stress response, while cortisol suppresses the immune system.

## Immune Diseases

*Immune diseases result from insufficient immune responses.*

- Abnormal functioning of the immune system can lead to immune diseases in two general ways: deficiency diseases and inappropriate immune attacks.
- Deficiency diseases occur when the immune system fails to respond adequately to foreign invasion.
- The most recent and tragically the most common acquired immune deficiency disease is AIDS, which is caused by HIV, a virus that invades and incapacitates the critical helper T cells.

*Allergies are inappropriate immune attacks against harmless environmental substances.*

- The other category of immune diseases involves inappropriate specific immune attacks that cause reactions harmful to the body.
- These include: (1) autoimmune responses, in which the immune system turns against one of the body's own tissues; (2) immune-complex diseases, which involve overexuberant antibody responses that spill over and damage normal tissues; and (3) allergies.
- An allergy is the acquisition of an inappropriate specific immune reactivity, or hypersensitivity, to a normally harmless environmental substance.
- In immediate hypersensitivity, the allergic response appears within about 20 minutes after the sensitized individual is exposed to an allergen, whereas in delayed hypersensitivity, the reaction is not generally manifested until a day or so following exposure.
- The most common allergens that provoke immediate hypersensitivities are pollen grains, bee stings, penicillin, certain foods, molds, dust, feathers and animal fur.
- These allergens bind to and elicit the synthesis of IgE antibodies rather than IgG antibodies.
- When an individual with an allergic tendency is first exposed to a particular allergen, compatible helper T cells secrete interleukin 4, which prods compatible B cells to synthesize IgE antibodies specific for the allergen.
- IgE antibodies do not freely circulate.
- Instead, their tail portions attach to mast cells and basophils.
- Binding of an appropriate allergen with the outreached arm regions of the IgE antibodies that are lodged tail-first in a mast cell or basophil triggers the rupture of the cell's preformed granules.
- As a result, histamine and other chemical mediators spew forth into the surrounding tissue.
- The following are among the most important chemicals released during immediate allergic reactions: (1) histamine, (2) slow-reactive substance of anaphylactic, and (3) eosinophil chemotactic factor.
- If the reaction is limited to the upper respiratory passages, the released chemicals bring about the symptoms of hay fever.
- If the reaction is concentrated primarily within the bronchioles, asthma results.
- Localized swelling in the skin because of allergy-induced histamine release causes hives.
- When large amounts of these chemical mediators gain access to the blood, the extremely serious systemic reaction known as anaphylactic shock occurs, which is frequently fatal.
- Shared characteristics of the immune reactions to allergens and parasitic worms include the production of IgE antibodies and increased basophil and eosinophil activity.
- The inflammatory response in the skin could wall off parasitic worms attempting to burrow in.
- Coughing and sneezing could expel worms that migrated to the lungs.
- Diarrhea could help flush out worms before they could penetrate or attach to the digestive tract lining.
- Some allergens invoke delayed hypersensitivity, a T cell-mediated immune response.

- Among these allergens are poison ivy toxin and certain chemicals to which the skin is frequently exposed, such as cosmetics and household cleaning agents.

EXTERNAL DEFENSES

*The skin consists of an outer protective epidermis and an inner connective tissue dermis.*

- The epidermis consists of numerous layers of epithelial cells.
- Epidermal cells are tightly bound together by spot desmosomes, which interconnect with intracellular keratin filaments to form a strong, cohesive covering.
- The dermis is a connective tissue layer that contains many elastin fibers and collagen fibers, as well as an abundance of blood vessels and specialized nerve endings.
- Special infoldings of the epidermis into the underlying dermis form the skin's exocrine glands—the sweat glands and the sebaceous glands—as well as the hair follicles.
- Melanocytes produce the pigment melanin.
- The amount and type of melanin are responsible for the different shades of skin color of the various races.
- Melanin is produced through complex biochemical pathways in which the melanocyte enzyme tyrosinase plays a key role.
- Two genetic factors prevent tyrosinase from functioning at full capacity: (1) much of the tyrosinase produced is in an inactive form, and (2) various inhibitors block tyrosinase action.
- Keratinocytes are specialists in keratin production.
- As they die, keratinocytes form the outer protective keratinized layer and are also responsible for generating hair and nails.
- Langerhans cells present antigen to helper T cells.
- Granstein cells interact with suppressor T cells, probably serving as a brake or skin-activated immune responses.
- The epidermis synthesizes vitamin D in the presence of sunlight.

*Protective measures within body cavities discourage pathogen invasion into the body.*

- Saliva secreted into the mouth at the entrance of the digestive system contains an enzyme that lyses certain bacteria.
- Many of the surviving bacteria that are swallowed are killed by the strongly acidic gastric juice that they encounter in the stomach.
- Some bacteria do manage to survive and reach the large intestine.
- These harmless resident flora competitively suppress the growth or potential pathogens that have escaped the antimicrobial measures or earlier parts of the digestive tract.
- Within the genitourinary system, would-be invaders encounter hostile conditions in the acidic urine and acidic vaginal secretions.
- The genitourinary organs also produce a sticky mucus, which, like flypaper, entraps small invading particles.
- Large airborne particles are filtered out of the inspired air by hairs at the entrance of the nasal passages.
- The respiratory airways are coated with a layer of thick, sticky mucus secreted by epithelial cells within the airway lining.
- This mucus sheet, laden with any inspired particulate debris is constantly moved upward to the throat by ciliary action.
- Also contributing to defense against respiratory infections are antibodies secreted in the mucus.
- In addition, an abundance of phagocytic specialists called the alveolar macrophages scavenge within the air sacs of the lungs.
- Cigarette smoking suppresses these normal respiratory defenses.

## *Key Terms*

ABO system
aids
active immunity
acquired immune deficiency syndrome (AIDS)
acute-phase proteins
adaptive (acquired) immune system
agglutination
allergen
allergy
alternate pathway
alveolar macrophage
angry macrophages
anaphylactic shock
antibody-mediated (humoral immunity)
antigen
antigen-binding fragment
antiserum (antitoxin)
asthma
autoimmune disease
bacteria
benign tumor
B cell growth factor
B lymphocyte
cancer
carcinogenic
cell-mediated immunity
chemotaxis

classical pathway
complement system
clonal anergy
clonal selection theory
clone
constant (Fc) region
cytokines
cytotoxic T cells (CD 8 cells)
dendritic cells
diapedesis
endogenous pyrogen
erythroblastosis fetalis (hemolytic disease of the newborn)
gut associated lymphoid tissue
gamma globulins (immunoglobulins)
Granstein cells
granuloma
granulopoiesis
gut-associated lymphoid tissue (GALT)
haptens
hay fever
helper T cells (CD4 cells)
histamine
host cell
humoral immunity
hypersensitivity
immune privilege
immune system
immunity
immune complex disease
immune surveillance
inflammation
innate immune system
interferon
interleukin 1
interleukin 2
kallikrein
keratinized layer
killer cells
kininogens
lactoferrin
langerhans cells
leukocyte endogenous mediator (LEM)
lymphoid cell
lymphoid tissue
margination
major histocompatibility complex (MHC)
mast cells
melanin
melanocytes
membrane attack complex (MAC)
metastasis
memory cells
MHC molecule
mucous escalator
naïve lymphocytes
natural killer cells
neutralization
opsonins
passive immunity
pathogens
Peyer's patches
perforin
phagocytosis
plasma cell
precipitation
primary immune response
psychoneuroimmunology
Rh factor
recombinant DNA technology
secondary immune response
severe combined immunodeficiency
serum sickness
scar tissue
skin-associated lymphoid tissue (SALT)
suppressor t cells
sweat glands
T lymphocyte
tolerance
toll-like receptors (TLRs)
transfusion reaction
thymus
thymosin
universal donor
universal recipient
virulence
viruses

## Review Exercises

*Answers are in the appendix.*

**True/False**

_____ 1. Lymphocytes that mature without benefit of "thymic education" become T lymphocytes.

_____ 2. During fetal development some immature lymphocytes migrate to the thymus, where they undergo further processing to become T lymphocytes.

_____ 3. There are four classes of specific immune responses.

_____ 4. Foreign proteins are the most common antigens.

_____ 5. The thymus becomes less important as the individual matures.

_____ 6. Antibodies are secreted into the blood or lymph.

_____ 7. Antibodies are grouped into six subclasses.

_____ 8. IgD is present on the surface of many T cells, but its function is uncertain.

_____ 9. Immunoglobulins cannot directly destroy foreign organisms.

_____ 10. Each B lymphocyte is preprogrammed to respond to only one of the millions of different antigens.

_____ 11. An individual can acquire antibodies by the direct transfer of antibodies actively formed by another person or animal.

_____ 12. There are two naturally occurring antibodies developed against the Rh factor.

_____ 13. The most common allergens that provoke immediate hypersensitivities are pollen grains, bee stings, penicillin, certain foods, molds, dust, feathers, and animal fur.

_____ 14. Severe hypotension can lead to circulatory failure.

_____ 15. Immune complex disease is when the immune system turns against one of the body's own tissues.

_____ 16. In immediate hypersensitivity, the allergic response appears within about 10 minutes after being exposed to an allergen.

_____ 17. A single mast cell may be coated with a number of different IgC antibodies.

_____ 18. T cells do not secrete antibodies.

_____ 19. Antibodies not only target viruses for destruction in the extracellular fluid but can also eliminate viruses inside neurons.

_____ 20. Helper T cells secrete B-cell growth factor.

_____ 21. The AIDS virus selectively invades cytotoxic T cells.

_____ 22. Self-antigens are plasma-membrane-bound glycoproteins.

_____ 23. B cells typically bind with HLA self-antigens only when they are in association with a foreign antigen.

_____ 24. Most body cells undergo mutations that result in malignancy.

_____ 25. Epidermal cells are tightly bound together by spot desmosomes.

_____ 26. The epidermis has very little direct blood supply.

_____ 27. The keratinized layer is airtight, fairly waterproof, and impervious to most substances.

_____ 28. Epidermal enzymes are not able to convert many potential carcinogens into harmless compounds.

_____ 29. The dermal blood vessels have no control in temperature regulation.

_____ 30. Langerhans cells present antigen to helper T cells.

_____ 31. The epidermis synthesizes vitamin A in the presence of sunlight.

_____ 32. T lymphocytes secrete antibodies that indirectly lead to the destruction of foreign material.

_____ 33. Natural killer cells are a family of proteins that nonspecifically defend against viral infections.

_____ 34. Released histamine increases the capillaries permeability.

_____ 35. Neutrophils are the first to arrive at the site of an injury.

_____ 36. Phagocytic secretions stimulate the release of kininogens from the liver.

_____ 37. Salicylates reduce fever by inhibiting the production of prostaglandins.

**Fill in the Blank**

38. Antibodies are secreted into the blood or lymph, where they are known as ____________________ or ____________________.

39. ____________________ is the term applied to the process in which foreign cells, such as bacteria or mismatched transfused red blood cells, bind together a clump.

40. Antibodies mark or identify foreign material as targets for actual destruction by the ____________________, ____________________, or ____________________.

41. ____________________ remain dormant, not actually secreting their particular antibody product until they come into contact with the appropriate ______________________________.

42. The production of antibodies as a result of exposure to an antigen is referred to as ____________________ against that antigen.

43. Antigen-presenting macrophages secrete ____________________, a multipurpose chemical mediator that enhances the differentiation and proliferation of the now activated B-cell clone.

44. T-cells must be in direct contact with their targets, a process known as ____________________.

45. ____________________ destroy host cells bearing foreign antigen, such as body cells invaded by viruses, cancer cells, and transplanted cells.

46. Natural killer cells release ____________________ molecules, which penetrate into the target cell's surface membrane and join together to form porelike channels.

47. Helper T cells secrete T cell growth factor, also known as ____________________.

48. ____________________ are by far the most numerous of the T cells.

49. ____________________ is an example of an autoimmune disease.

50. The most recent and tragically the most commonly acquired immune deficiency disease is ____________________.

51. A(n) ____________________ is the acquisition of an inappropriate specific immune reactivity, or ____________________ to a normally harmless environmental substance.

52. When large amounts of chemical mediators or allergens gain access to the blood, the extremely serious systemic reaction known as ____________________ occurs.

53. ____________________ refers to the body's ability to resist or eliminate potentially harmful foreign materials or abnormal cells.

54. ____________________ refer collectively to the tissues that store, produce, or process lymphocytes.

55. Localized vasodilation is primarily induced by ____________________ that has been released in the area of tissue damage from mast cells.

56. Upon exposure to tissue thromboplastin in the injured tissue and to specific chemicals secreted by phagocytes on the scene, ____________________, the final factor in the clotting system, is converted into ____________________.

57. Neutrophils secrete ____________________, a protein that tightly binds with iron, making it unavailable for use by invading bacteria.

58. A more-or-less permanently walled-off structure in which nondestructible offending material is imprisoned is known as a(n) ____________________.

59. Numerous drugs can suppress the inflammatory process; the most effective are the ____________________ and related compounds and ____________________.

60. There are two classes of specific immune responses: ____________________ and ____________________.

61. The ____________________ gradually atrophies and becomes less important as the individual matures. It does, however, continue to prove important in maintaining the T cell lineage.

62. ____________________ are low-molecular-weight organic substances that are not antigenic by themselves but can become antigenic if they attach to body proteins.

63. In ____________________, the allergic response appears within about 20 minutes after a sensitized individual is exposed to an allergen.

64. Epidermal cells are tightly bound together by ____________________.

65. The cells of the ____________________ produce an oily secretion known as ____________________.

66. ____________________ release a dilute salt solution through small openings, ____________________ onto the surface of the body.

67. ____________________ produce the pigment ____________________.

68. The most abundant epidermal cells are the ____________________.

69. ____________________ present an antigen to helper T cells, thereby facilitating their responsiveness to skin-associated antigens.

70. ____________________ interact with suppressor T cells, probably serving as a brake on skin-activated immune responses.

71. The ____________________ synthesizes vitamin D in the presence of sunlight.

72. Phagocytic specialists called ____________________ scavenge within the air sacs of the lungs.

**Matching**
*Match the lymphoid tissue to the function.*

a. thymus
b. spleen
c. bone marrow
d. lymph nodes, tonsils, adenoids, appendix, gutassociated lymphoid tissue

_____ 73. Exchanges lymphocytes with blood.

_____ 74. Origin of all blood cells.

_____ 75. Stores a small percentage of red blood cells.

_____ 76. Secretes the hormone thymosin.

_____ 77. Exchanges lymphocytes with lymph.

_____ 78. Site of maturational processing for T lymphocytes.

_____ 79. Resident lymphocytes produce antibodies and sensitized T cells, which are released into the blood.

_____ 80. Site of maturational processing for B lymphocytes.

_____ 81. Resident macrophages remove microbes and other particulate debris from lymph.

_____ 82. Resident lymphocytes produce antibodies and sensitized cells, which are released into the lymph.

_____ 83. Resident macrophages remove microbes and other particulate debris, most notably worn-out red blood cells, from the blood.

**Multiple Choice**

84. Which of the following substances causes an inappropriate immune response?
   a. perforins
   b. opsonins
   c. endogenous pyrogens
   d. haptens
   e. allergens

85. Which of the following substances induces the development of fever?
   a. perforins
   b. opsonins
   c. endogenous pyrogens
   d. haptens
   e. allergens

86. Which of the following substances is a secretory product of helper T cells?
   a. kallikrein
   b. interleukin 1
   c. haptens
   d. interleukin 2
   e. allergens

87. Which of the following substances make bacteria more susceptible to phagocytosis?
   a. allergens
   b. opsonins
   c. interleukin 1
   d. interleukin 2
   e. haptens

88. Which of the following substances are not antigenic by themselves but can become antigenic if they attach to body proteins?
   a. opsonins
   b. allergens
   c. perforins
   d. histamines
   e. haptens

89. Which of the following substances is secreted by neutrophils and binds tightly with iron?
   a. ferrokinogen
   b. lactoferrin
   c. endogenous pyrogen
   d. interleukin 1
   e. interferon

90. Which of the following substances is a secretory product of macrophages?
   a. interleukin 1
   b. interferon
   c. kallikrein
   d. interleukin 2
   e. perforin

91. Which of the following substances brings about vasodilation and increased capillary permeability?
   a. interferon
   b. histamine
   c. perforin
   d. interleukin 2
   e. kallikrein

92. Which of the following substances triggers the production of viral-blocking enzymes by potential host cells?
   a. perforin
   b. interleukin 2
   c. interleukin 1
   d. ferrokinogen
   e. interferon

93. Which of the following substances is released by cytotoxic T cells and binds to a target cell's surface membrane?
   a. perforin
   b. interferon
   c. interleukin 1
   d. interleukin 2
   e. ferrokinogen

**Modified Multiple Choice**

*Indicate whether the characteristic applies to active immunity or passive immunity by writing the appropriate letter in the blank using the following answer code.*

A = applies to active immunity
B = applies to passive immunity
C = applies to both active and passive immunity

_____ 94. Exposure to antigen required for immunity to develop
_____ 95. Antibodies produced by a source other than ones own body
_____ 96. Often confers long-lasting immunity
_____ 97. Resistance to antigen exposure immediate upon injection of antibodies
_____ 98. Natural immunity is actually a special case of this type of immunity

*Indicate which type of T cell is being described by writing the appropriate letter in the blank using the answer code below.*

A = applies to cytotoxic T cells
B = applies to helper T cells
C = applies to suppressor T cells
D = applies to both helper T cells and suppressor T cells
E = applies to all three types of T cells

_____ 99. destroys host cells bearing foreign antigen
_____ 100. suppresses both T-cell and B-cell activity
_____ 101. secretes B cell growth factor
_____ 102. enhances the development of antigen-stimulated B cells into antibody-secreting cells
_____ 103. called regulatory T cells
_____ 104. secretes interleukin 2
_____ 105. releases perforin
_____ 106. secretes macrophage-migration inhibition factor
_____ 107. attacked by the AIDS virus
_____ 108. most numerous of the T cells
_____ 109. serves to limit immune reactions
_____ 110. believed to play a role in tolerance
_____ 111. recognizes class I MHC glycoproteins
_____ 112. recognizes class II MHC glycoproteins

## Points to Ponder

1. With the current social views in this country, do you feel that the eradication of AIDS is possible?

2. In that part of the body defenses involving B cells and T cells, cellular memory plays a major role. When and how is this memory lost?

3. Do animal rights activists hinder our fight against viral diseases?

4. How does exercise strengthen the immune system?

5. If you have an allergy, will your children inherit it? Explain why or why not.

6. A drug has been discovered that destroys all mast cells. How might this drug help prevent allergies? Would there be any negative side affects by taking this drug?

7. How does an elevated body temperature help an infection caused by a virus?

8. What is the difference between a nonspecific host defense mechanism and a specific defense mechanism? Give examples.

9. How does the clonal selection theory explain the ability of the immune response to distinguish between self and nonself?

10. What is the basis of immunological memory?

## Clinical Perspectives

1. What is the significance of a booster shot?

2. This summer you will direct a biological field trip to the remote areas of the Chihuahuan Desert. You and the students will spend a month living in tents as you study the desert. Based on this chapter, what precautions should you take before you begin this trip?

3. Why don't we become immune to the cold virus?

4. Why do most children outgrow common food allergies?

5. Why are cortisone and its analogues used clinically in patients who have a transplanted organ?

6. While painting the house, you get stung by a bee. This is the first time you have ever been stung by a bee. You develop an itchy rash, which is not relieved by antihistamines prescribed by your physician. The physician then tries cortisone, which alleviates the rash. Several weeks later, you get stung again by a bee; however, unlike the first time, you respond well to the antihistamines. Why did this occur?

7. The tuberculin skin test and poison ivy are both type IV hypersensitivities. What does this mean? Explain the common mechanism involved.

8. How does hemolytic disease of the newborn develop?

9. What is the immunological basis behind blood transfusions?

10. In desensitization procedures, the allergist injects more of the same allergen to which the person is allergic. How can this be beneficial?

## PhysioEdge Activities

**Related to Text:**
Media Exercise 12.1: The Body's Defenses.
Media Exercise 12.2: Inflammation.
Media Exercise 12.3: Basics of Specific Immunity.

**Related to Figures:**
*Figure 12.1*. For an interaction related to this figure, see Media Exercise 12.1: The Body's Defenses.
*Figure 12.3*. For an interaction related to this figure, see Media Exercise 12.2: Inflammation.
*Figure 12.6*. For an interaction related to this figure, see Media Exercise 12.1: The Body's Defenses.
*Figure 12.7*. For an interaction related to this figure, see Media Exercise 12.3: Basics of Specific Immunity.
*Figure 12.9*. For an interaction related to this figure, see Media Exercise 12.3: Basics of Specific Immunity.

## Media Resources

**PhysioEdge CD-ROM**
For a visual review of concepts in this chapter, check out:

Media Exercise 12.1: The Body's Defenses
Media Exercise 12.2: Inflammation
Media Exercise 12.3: Basics of Specific Immunity

**Book Companion Website**
The website for this book contains a wealth of helpful study aids, as well as many ideas for further reading and research. Log on to:

**http://www.brookscole.com/sherwoodhp6**

Select Chapter 12 from the drop-down menu, or click on one of the many resources areas, including **Case Histories**, which introduce clinical aspects of human physiology. For this chapter check out Case History 18: "It's Nothing – Just a Bee Sting"; Case History 19: A Close Call; Case History 20: David – Life in a Germ-Free World; Case History 21: AIDS; Case History 22: Acne Can Be Controlled.

For Suggested Readings, consult **InfoTrac® College Edition**, your online research library, at:

**http://infotrac.thomsonlearning.com**

# 13

# The Respiratory System

## Chapter Overview

Respiration involves the sum of the processes that accomplish ongoing passive movement of oxygen from the atmosphere to the tissues to support cellular metabolism, as well as the continual passive movement of metabolically produced carbon dioxide from the tissues to the atmosphere. The homeostatic role of the respiratory system overlaps that of the circulatory system: supplying oxygen to the cells of the body and removing carbon dioxide, a byproduct of cellular metabolism. While these are not the only functions of the respiratory system they are surely the most important. The entire ventilation-transport process, from air outside the body to the gaseous exchange at the systemic cellular level, involves diffusion down gradients. For the most part, they are pressure gradients.

The mechanics of ventilation rely heavily upon the physical properties of the alveolar wall and surface tension. Ventilation is controlled by the nervous system at both conscious and subconscious levels.

The bonding of oxygen and carbon dioxide to hemoglobin is important for transport by the blood. Oxygen bonds to the heme portion and carbon dioxide attaches to the globin fraction of the hemoglobin molecule. This is, however, only one step in the process of respiration.

The path from alveolar air to the mitochondria of the various cells involves many barriers through which oxygen and carbon dioxide must traverse. Crossing the numerous barriers, oxygen and carbon dioxide diffuse down pressure gradients involving differences in their respective partial pressures. Using quite sophisticated control mechanisms, the respiratory system plays a major role in maintaining homeostasis.

## Chapter Outline

INTRODUCTION

*The respiratory system does not participate in all steps of respiration.*

- The primary function of the respiratory system is to obtain oxygen for use by the body's cells and to eliminate the carbon dioxide the cells produce.
- Internal or cellular respiration refers to the intracellular metabolic processes carried out within the mitochondria, which use oxygen and produce carbon dioxide during the derivation of energy from nutrient molecules.
- External respiration refers to the entire sequence of events in the exchange of oxygen and carbon dioxide between the external environment and the cells of the body.
- External respiration, the topic of this chapter, encompasses four steps: (1) air is alternately moved in and out of the lungs so that exchange of air can occur between the atmosphere and the alveoli of the lungs; (2) oxygen and carbon dioxide are exchanged between air in the alveoli and blood within the pulmonary capillaries by the process of diffusion; (3) oxygen and carbon dioxide are transported by the blood between the lungs and the tissues; and (4) the exchange of oxygen and carbon dioxide takes place between the tissues and the blood by the process of diffusion across the systemic capillaries.
- The respiratory system additionally performs the following nonrespiratory functions: (1) it provides a route for water loss and heat elimination; (2) it enhances venous return; (3) it enables speech, singing, and other vocalization; (4) it defends against inhaled foreign matter; (5) it removes, modifies, activates, or inactivates various materials passing through the pulmonary circulation; (6) the nose serves as the organ of smell; and (7) it contributes to the maintenance of normal acid-base balance.

*The respiratory airways conduct air between the atmosphere and alveoli.*

- The respiratory system includes the respiratory airways leading into the lungs, the lungs themselves, and the structures of the thorax involved in producing movement of air through the airways into and out of the lungs.

- The airways include: (1) nasal passages, (2) pharynx, (3) larynx, (4) trachea, (5) bronchi, and (6) bronchioles.
- The term "vocal chords" has been replaced by the more descriptive term vocal folds.

*The gas-exchanging alveoli are thin-walled, inflatable air sacs encircled by pulmonary capillaries.*

- The alveoli are clusters of thin walled, inflatable, grapelike sacs at the terminal branches of the conducting airways.
- The alveolar walls consist of a single layer of flattened Type I alveolar cells.
- The interstitial space between an alveolus and the surrounding capillary network forms an extremely thin barrier.
- In addition to the Type I cells, the alveolar epithelium also contains Type II alveolar cells, which secrete pulmonary surfactant.
- The minute pores of Kohn in the alveolar walls permit airflow between adjacent alveoli, a process known as collateral ventilation.

*The lungs occupy much of the thoracic cavity.*

- There is no muscle within the alveolar walls to cause them to inflate or deflate during the breathing process.
- The lungs, heart and associated vessels, esophagus, thymus, and some nerves occupy the thoracic cavity.
- The outer chest wall is formed by 12 pairs of curved ribs, which join the sternum anteriorly and the thoracic vertebrae posteriorly.
- The diaphragm forms the floor of the thoracic cavity.

*A pleural sac separates each lung from the thoracic wall.*

- Separating each lung from the thoracic wall and other surrounding structures is a doubled-walled, closed sac called the pleural sac.
- The surfaces of the pleura secrete a thin intrapleural fluid, which lubricates the pleural surfaces.

## RESPIRATORY MECHANICS

*Interrelationships among pressures inside and outside the lungs are important in ventilation.*

- Air tends to move from a region of higher pressure to a region of lower pressure down a pressure gradient.
- Three different pressure considerations are important in ventilation: (l) atmospheric pressure, (2) intra-alveolar pressure, and (3) intrapleural pressure.

*The lungs are normally stretched to fill the larger thorax.*

- The intrapleural fluid's cohesiveness and the transmural pressure gradient hold the lungs and thoracic wall in tight apposition, even though the lungs are smaller than the thorax.
- The polar water molecules in the intrapleural fluid resist being pulled apart because of their attraction to each other.
- The resultant cohesiveness of the intrapleural fluid tends to hold pleural surfaces together.
- An even more important reason that the lungs follow the movements of the chest wall is the transmural pressure gradient that exists across the lung wall.
- The intra-alveolar pressure is greater than the intrapleural pressure: so a greater pressure is pushing outward than is pushing inward across the lung wall.
- A similar transmural pressure gradient exists across the thoracic wall.
- The atmosphere pressure pushing inward on the thoracic wall is greater than the intrapleural pressure pushing outward on the same wall.
- The stretched lungs have a tendency to pull inward away from the thoracic wall, whereas the compressed thoracic wall tends to move away from the lungs.
- If the intrapleural pressure were ever to equilibrate with the atmospheric pressure, the lungs and thorax would separate and assume their own inherent dimensions.

*Flow of air into and out of the lungs occurs because of cyclical changes in intra-alveolar pressure.*

- Because air flows down a pressure gradient, the intra-alveolar pressure must be less than atmospheric pressure for air to flow into the lungs during inspiration.
- At the onset of inspiration, the inspiratory muscles—the diaphragm and external intercostal muscles—are stimulated to contract, resulting in enlargement of the thoracic cavity.
- The intra-alveolar pressure is now less than atmospheric pressure, air flows into the lungs down the pressure gradient from higher to lower pressure.
- During inspiration, the intrapleural pressure falls.
- The resultant increase in the transmural pressure gradient during inspiration ensures that the lungs are stretched to fill the expanded thoracic cavity.
- At the end of inspiration, the inspiratory muscles relax.
- The chest wall and the stretched lungs recoil to their preinspiratory size because of their elastic properties.

- In a resting expiration, the intra-alveolar pressure increases.
- Air now leaves the lungs down a pressure gradient from high intra-alveolar pressure to lower atmospheric pressure.
- Expiration is normally a passive process.
- Inspiration is always an active process.
- To produce active expiration, the expiratory muscles must contract to further reduce the volume of the thoracic cavity and lungs.

*Airway resistance influences airflow rates.*

- The primary determinant to resistance to airflow is the radius of the conducting airway.
- Normally, modest adjustments in airway size can be accomplished by autonomic nervous system regulation to suit the body's needs.
- Parasympathetic stimulation promotes bronchiolar smooth muscle contraction, which increases airway resistance by producing bronchoconstriction.
- Sympathetic stimulation and its associated hormone, epinephrine, bring about bronchodilation and decreased airway resistance by promoting bronchiolar smooth muscle relaxation.
- Resistance becomes an extremely important impediment to airflow when airway lumens become narrowed as a result of disease.

*Airway resistance is abnormally increased with chronic obstructive pulmonary disease.*

- Chronic obstructive pulmonary disease is a group of lung diseases characterized by increased airway resistance resulting from the narrowing of the lumen of the lower airways.
- Chronic obstructive pulmonary disease encompasses three chronic diseases: asthma, chronic bronchitis, and emphysema.
- In asthma, airway obstruction is due to: (1) profound constriction of the smaller airways caused by allergy-induced spasm of the smooth muscle in the walls of these airways; (2) plugging of airways by excess secretion of very thick mucus; and (3) thickening of the walls of the airways due to inflammation and histamine-induced edema.
- Chronic bronchitis is a long-term inflammation of the lower respiratory airways, generally triggered by irritating cigarette smoke, polluted air, or allergens.
- Emphysema is characterized by collapse of the smaller airways and a breakdown of alveolar walls.
- When airway resistance is increased as a result of chronic obstructive lung disease of any type, expiration is more difficult to accomplish than inspiration.

*Elastic behavior of the lungs is due to elastic connective tissue and alveolar surface tension.*

- Two interrelated concepts are involved in pulmonary elasticity: (1) elastic recoil, and (2) compliance.
- Elastic recoil refers to how readily the lungs rebound after having been stretched.
- Compliance refers to how much effort is required to stretch or distend the lungs.
- A highly compliant lung stretches further for a given increase in the pressure difference than does a less compliant lung.
- Pulmonary elastic behavior depends mainly on two factors: highly elastic connective tissue in the lungs and alveolar surface tension.
- Pulmonary connective tissue contains large quantities of elastin fibers arranged into a meshwork that amplifies their elastic behavior.
- Surface tension is responsible for a twofold effect.
- First, the liquid layer resists any force that increases its surface area.
- Second, the liquid surface area tends to become as small as possible because the surface water molecules try to get as close together as possible.

*Pulmonary surfactant decreases surface tension and contributes to lung stability.*

- The tremendous surface tension of pure water is normally counteracted by secretion of pulmonary surfactant by the Type II alveolar cells.
- By lowering the alveolar surface tension, pulmonary surfactant provides two important benefits: (1) it increases pulmonary compliance, thus reducing the work of inflating the lungs; and (2) it reduces the lungs' tendency to recoil so that they do not collapse readily.
- Pulmonary surfactant's role in reducing the alveoli's tendency to recoil is important in helping maintain lung stability.
- A second factor that contributes to alveolar stability is the interdependence of neighboring alveoli.
- If an alveoli starts to collapse, the surrounding alveoli are stretched.
- By recoiling in resistance to being stretched, these neighboring alveoli exert expanding forces on the collapsing alveolus and thereby help keep it open.
- This phenomenon is termed interdependence.
- A deficiency of pulmonary surfactant is responsible for newborn respiratory distress syndrome.
- In an infant born prematurely, pulmonary surfactant may be insufficient to reduce the alveolar surface tension to manageable levels.

- The resultant collection of symptoms that develop is referred to as newborn respiratory distress syndrome.
- It is more difficult to expand a collapsed alveolus by a given volume than to increase an already partially expanded alveolus by the same volume.
- With newborn respiratory distress syndrome, lung expansion may require considerably larger transmural pressure gradients.
- The problem is compounded by the fact that the newborn's muscles are still weak.

*The work of breathing normally requires only about 3 percent of total energy expenditure.*
- Normally, the lungs are highly compliant and the airway resistance is low, so only about 3 percent of the total energy expended by the body is used to accomplish quiet breathing.
- The work of breathing may be increased in four different situations: (1) when pulmonary compliance is decreased, (2) when airway resistance is increased, (3) when elastic recoil is decreased, and (4) when there is a need for increased ventilation.
- Because total energy expenditure by the body is increased up to fifteen- to twenty-fold during heavy exercise, the energy used to accomplish the increased ventilation still represents only about 5 percent of total energy expended.

*The lungs normally operate at about "half full."*
- On average, the maximum amount of air that the lungs can hold is about 5.7 liters in males and 4.2 liters in females.
- Anatomical build, age, the distensibility of the lungs, and the presence or absence of respiratory disease affect this total lung capacity.
- At the end of a normal quiet expiration, the lungs still contain about 2,200 ml of air.
- About 500 ml of air are inspired, and the same quantity is expired during quiet breathing.
- Using a spirometer, the following lung volumes and lung capacities can be determined: (1) tidal volume, (2) inspiratory reserve volume, (3) inspiratory capacity, (4) expiratory reserve volume, (5) residual volume, (6) functional residual capacity, (7) vital capacity, (8) total lung capacity, and (9) forced expiratory volume in one second.
- Two general categories of respiratory dysfunction yield abnormal results during spirometry: obstructive and restrictive lung disease.
- To determine what abnormalities are present, the diagnostician relies on a variety or respiratory function tests in addition to spirometry, including X-ray examination, blood gas determinations, and tests to measure the diffusion capacity of the alveolar capillary membrane.

*Alveolar ventilation is less than pulmonary ventilation because of dead space.*
- Not all of the inspired air gets down to the site of gas exchange in the alveoli.
- Part of the air remains in the conducting airways, where it is not available for gas exchange.
- This volume is considered to be anatomic dead space.
- Alveolar ventilation—the volume of air exchanged between the atmosphere and the alveoli per minute—is more important than pulmonary ventilation.

*Local controls act on smooth muscle of the airways and arterioles to maximally match blood flow to airflow.*
- If an alveolus is receiving too little airflow in comparison to its blood flow, carbon dioxide levels will increase in the alveolus.
- This local increase in carbon dioxide acts directly on the bronchiolar smooth muscle involved to induce the airway supplying the under-aerated alveolus to relax.
- The result is an increased airflow to the involved alveolus, so its airflow now matches its blood supply.
- A localized decrease in carbon dioxide associated with an alveolus that is receiving too much air for its blood supply directly increases contractile activities.
- The result is a reduction in airflow to the over-aerated alveolus.
- If the blood flow is greater than the airflow in a given alveolus, the oxygen level in the alveolus and surrounding tissue will fall below normal.
- The local decrease in oxygen concentration causes vasoconstriction of the pulmonary arteriole.
- The result is a reduction in blood flow to match the smaller airflow.
- An increase in alveolar oxygen concentration caused by a mismatched large airflow and small blood flow brings about pulmonary vasodilation, which increases blood flow to match the larger airflow.

## Gas Exchange

*Gases move down partial pressure gradients.*
- The ultimate purpose of breathing is to provide a continual supply of fresh oxygen for pickup by the blood and to constantly remove carbon dioxide unloaded from the blood.
- Gas exchange at both the pulmonary-capillary and tissue-capillary levels involves simple passive

diffusion of oxygen and carbon dioxide down partial pressure gradients.
- If, as in the case with oxygen, the alveolar partial pressure of a gas is higher than the partial pressure of that gas in the blood entering the pulmonary capillaries, the higher alveolar partial pressure drives more oxygen into the blood.
- Conversely, if the alveolar partial pressure of a gas is lower than its partial pressure in the entering blood—the situation that exists for carbon dioxide—the lower alveolar partial pressure permits some of the carbon dioxide to escape from solution in the blood.
- A gas always diffuses down its partial pressure gradient from the area of higher partial pressure to the area of lower partial pressure.

*Oxygen enters and carbon dioxide leaves the blood in the lungs passively down partial pressure gradients.*
- The alveolar partial pressure of oxygen remains relatively constant throughout the respiratory cycle.
- The partial pressure of oxygen in the blood likewise remains fairly constant at the same value.
- A similar situation in reverse exists for carbon dioxide.
- The appropriate partial pressure gradients between the alveoli and blood are maintained to ensure that oxygen enters the blood and carbon dioxide leaves the blood.
- The amount of oxygen picked up in the lungs matches the amount extracted and used by the tissues.
- The amount of carbon dioxide given up to the alveoli from the blood matches the amount of carbon dioxide picked up at the tissues.

*Factors other than the partial pressure gradient influence the rate of gas transfer.*
- According to Fick's law of diffusion, the rate of diffusion of a gas depends on the surface area and thickness of the membrane and on the diffusion coefficient of the gas.
- During exercise, the surface area available for exchange can be physiologically increased to enhance the rate of gas transfer.
- Many of the previously closed pulmonary capillaries are forced open.
- The alveolar membranes are stretched further than normal during exercise because of deeper breathing.
- Such stretching increases the alveolar surface area and decreases the thickness of the alveolar membrane.
- Collectively these changes expedite gas exchange during exercise.
- Surface area is reduced in emphysema.
- Inadequate gas exchange can also occur when the thickness of the barrier separating the air and blood is pathologically increased.
- The thickness increases in: (1) pulmonary edema, (2) pulmonary fibrosis, and (3) pneumonia.
- The rate of gas transfer is directly proportional to the diffusion coefficient.
- The diffusion coefficient for carbon dioxide is 20 times that of oxygen because carbon dioxide is much more soluble in body tissues than oxygen is.
- In a diseased lung in which diffusion is impeded because the surface area is decreased or the blood-air barrier is thickened, oxygen transfer is usually more impaired than carbon dioxide because of the larger carbon dioxide diffusion coefficient.

*Gas exchange across the systemic capillaries also occurs down a partial pressure gradient.*
- Just as they do at the pulmonary capillaries, oxygen and carbon dioxide move between the systemic capillary blood and the tissue cells by simple passive diffusion down partial pressure gradients.
- The amount of oxygen transferred to the cells and the amount of carbon dioxide carried away from the cells depend on the rate of cellular metabolism.

## GAS TRANSPORT

*Most oxygen in the blood is transported bound to hemoglobin.*
- Oxygen is present in the blood in two forms: (1) physically dissolved, and (2) chemically bound to hemoglobin.
- Very little oxygen is physically dissolved in the plasma water because oxygen is poorly soluble in body fluids.
- Of the oxygen transported by the blood, 98.5 percent is transported in combination with hemoglobin.
- Hemoglobin, an iron-bearing protein molecule contained within the red blood cells, has the ability to form a loose, easily reversible combination with oxygen.

*The partial pressure of oxygen is the primary factor determining the percent of hemoglobin saturation.*
- Each hemoglobin molecule can carry up to four molecules of oxygen.
- The most important factor determining the percent of hemoglobin saturation is the partial pressure of oxygen of the blood, which in turn is related to the concentration of oxygen physically dissolved in the blood.
- When the partial pressure or oxygen in the blood is increased, as it is in the pulmonary capillaries,

the reaction is driven toward an increase in the formation of oxyhemoglobin.
- When the partial pressure of oxygen is decreased, as it is in the systemic capillaries, the reaction is driven toward an increase in the formation of reduced hemoglobin.

*Hemoglobin promotes the net transfer of oxygen at both the alveolar and tissue levels.*
- Net diffusion of oxygen from alveoli to blood occurs continuously until the hemoglobin becomes saturated with oxygen as completely as it can be at that particular partial pressure.
- At a normal partial pressure for oxygen, hemoglobin is 97.5 percent saturated.
- Not until hemoglobin can store no more oxygen does all of the oxygen transferred into the blood remain dissolved and directly contribute to the partial pressure.
- Once the partial pressure of oxygen in the blood equilibrates with the alveolar partial pressure, no further oxygen transfer can take place.
- The reverse situation occurs at the tissue level.
- When the partial pressure of oxygen in the blood falls, hemoglobin is forced to unload some of its stored oxygen because the percent of hemoglobin saturation is reduced.
- Only when hemoglobin is no longer able to release any more oxygen into solution can the partial pressure of oxygen in the blood become as low as in the surrounding tissue.
- At this time further transfer of oxygen ceases.
- Hemoglobin plays an important role in the total quantity of oxygen that the blood can pick up in the lungs and drop off in the tissues.

*Factors at the tissue level promote the unloading of oxygen from hemoglobin.*
- An increase in the partial pressure of carbon dioxide reduces the amount or oxygen and hemoglobin that can be combined.
- An increase in acidity also reduces the amount of oxygen and hemoglobin that can be combined.
- The influence of carbon dioxide and acid on the release of oxygen is known as the Bohr effect.
- Local elevation in temperature enhances oxygen release from hemoglobin for use by the more active tissues.
- Thus, increases in carbon dioxide, acidity, and temperature at the tissue level enhances the effect of a drop in the partial pressure of oxygen in facilitating the release of oxygen and hemoglobin.
- These effects are largely reversed at the pulmonary level, where the extra acid-forming carbon dioxide is blown off and the local environment is cooler.
- Inside the red blood cell, 2,3-diphosphoglycerate is produced during cellular metabolism.
- This erythrocyte constituent can bind reversibly with hemoglobin and reduce its affinity for oxygen.

*Hemoglobin has a much higher affinity for carbon monoxide than for oxygen.*
- Carbon monoxide and oxygen compete for the same binding sites on hemoglobin, but the affinity of hemoglobin for carbon monoxide is 240 times that of the bond strength between hemoglobin and oxygen.
- Fortunately, carbon monoxide is not a normal constituent of inspired air.
- If carbon monoxide is being produced in a closed environment, it can reach lethal levels without the victim ever being aware of the danger.

*The majority of carbon dioxide is transported in the blood as bicarbonate.*
- Carbon dioxide is transported in the blood in three ways: (l) physically dissolved, (2) bound to hemoglobin, and (3) as bicarbonate.
- The amount of carbon dioxide physically dissolved in the blood is only 10 percent of the blood's total carbon dioxide content at normal systemic venous partial pressure levels.
- Another 30 percent of the carbon dioxide combines with hemoglobin to form carbamino-hemoglobin.
- Carbon dioxide binds with the globin portion of hemoglobin.
- By far the most important means of carbon dioxide transport is as bicarbonate.
- Carbon dioxide combines with water to form carbonic acid.
- This reaction proceeds swiftly within red blood cells because of the presence of the erythrocyte enzyme carbonic anhydrase which catalyzes the reaction.
- The one carbon and two oxygen atoms of the original carbon dioxide molecule are thus present in the blood as an integral part of the bicarbonate ion.
- Bicarbonate and hydrogen start to accumulate within the red blood cells in the systemic capillaries.
- Bicarbonate diffuses down its concentration gradient out of erythrocytes into the plasma.
- Chloride ions, the dominant plasma anions, diffuse into the red blood cells down the electrical gradient to restore neutrality.
- This inward shift of chloride ions in exchange for the outflux of carbon dioxide-generated bicarbonate ions is known as the chloride shift.

- Reduced hemoglobin has a greater affinity for hydrogen ions than does oxyhemoglobin.
- The fact that removal of oxygen from hemoglobin increases the ability of hemoglobin to pick up carbon dioxide and carbon dioxide-generated hydrogen is known as the Haldane effect.
- The Haldane effect and Bohr effect work in synchrony to facilitate oxygen liberation and the uptake of carbon dioxide and carbon dioxide-generated hydrogen ions.

*Various respiratory states are characterized by abnormal blood gas levels.*

- Hypoxia refers to insufficient oxygen at the cellular level.
- There are four general categories of hypoxia: (1) hypoxic hypoxia, (2) anemic hypoxia, (3) circulatory hypoxia, and (4) histotoxic hypoxia.
- Hypercapnia refers to excess carbon dioxide in the arterial blood.
- Hypercapnia is caused by hypoventilation.

## CONTROL OF RESPIRATION

*Respiratory centers in the brain stem establish a rhythmic breathing pattern.*

- Respiratory control centers housed in the brain stem are responsible for generating the rhythmic pattern of breathing.
- The generation of respiratory rhythm is now widely believed to lie in the preBotzinger complex.
- The primary respiratory control center is the medullary respiratory center.
- There are two other respiratory centers higher in the brain stem in the pons—the apneustic center and the pneumotaxic center.
- The medullary respiratory center consists of two neuronal clusters known as the dorsal respiratory group and the ventral respiratory group.
- The dorsal respiratory group consists mostly of inspiratory neurons whose descending fibers terminate on the motor neurons that supply the inspiratory muscles.
- This ventral respiratory group is composed of inspiratory neurons and expiratory neurons.
- The ventral respiratory group is called into play by the dorsal respiratory group as an "overdrive" mechanism during periods when demands for ventilation are increased.
- The generation of respiratory rhythm is now widely believed to lie in the rostral ventromedial medulla, a region located near the upper end of the ventral respiratory group.
- The pontine centers exert "fine-tuning" influences over the medullary center to help produce normal, smooth inspirations and expirations.
- The pneumotaxic center sends impulses to the dorsal respiratory group that help "switch off" the inspiratory neurons, thereby limiting the duration of inspiration.
- The apneustic center prevents the inspiratory neurons from being switched off.
- The pneumotaxic center is dominant over the apneustic center.
- When the tidal volume is large, as during exercise, the Hering-Breuer reflex is triggered to prevent overinflation of the lungs.
- Action potentials from the pulmonary stretch receptors travel through afferent nerve fibers to the medullary center and inhibit the inspiratory neurons.

*Decreased arterial partial pressure of oxygen increases ventilation only as an emergency mechanism.*

- The medullary respiratory center receives inputs that provide information about the body's needs for gas exchange.
- The arterial partial pressure of oxygen is monitored by peripheral chemoreceptors known as the carotid bodies and aortic bodies, which are located at the bifurcation of the common carotid arteries and in the arch of the aorta, respectively.
- The peripheral chemoreceptors are not sensitive to modest reductions in the arterial partial pressure of oxygen.
- Reflex stimulation of respiration by the peripheral chemoreceptors serves as an important emergency mechanism in dangerously low arterial partial pressures of oxygen conditions.

*Carbon dioxide-generated hydrogen ions in the brain are normally the primary regulator of ventilation.*

- The arterial partial pressure of carbon dioxide is the most important input regulating the magnitude of ventilation under resting conditions.
- An increase in the arterial partial pressure of carbon dioxide reflexly stimulates the respiratory center, with the resultant increase in ventilation promoting elimination of excess carbon dioxide.
- Conversely, a fall in arterial partial pressure of carbon dioxide reflexly reduces respiratory drive.
- There are no important receptors that monitor the arterial partial pressure of carbon dioxide, per se.
- Located in the medulla in the vicinity of the respiratory center, central chemoreceptors are sensitive to changes in carbon dioxide-induced hydrogen ion concentration in the brain extracellular fluid.
- Increased partial pressure of carbon dioxide within the brain extracellular fluid causes a corresponding increase in the concentration of hydrogen ions.

*Adjustments in ventilation in response to changes in arterial hydrogen ions are important in acid-base balance.*

- An elevation in the hydrogen ion concentration in the brain extracellular fluid directly stimulates the central chemoreceptors, which in turn increase ventilation by stimulating the respiratory center.
- A decline in arterial partial pressure of carbon dioxide below normal is paralleled by a fall in partial pressure of carbon dioxide and hydrogen ions in the brain extracellular fluid, the result of which is a central chemoreceptor-mediated decrease in ventilation.
- The powerful influence of the central chemoreceptors on the respiratory center is responsible for your inability to deliberately hold your breath for more than about a minute.
- Carbon dioxide induced changes in the hydrogen ion concentration in the arterial blood are detected by the peripheral chemoreceptors.
- The result is reflex stimulation of ventilation in response to an increase in arterial hydrogen ions and depression of ventilation is association with a decrease in arterial hydrogen ion concentration.

*Exercise profoundly increases ventilation, but the mechanisms involved are unclear.*

- The cause of increased ventilation during exercise is still largely speculative.

*Factors that may increase ventilation during exercise.*

- Researchers have suggested that a number of other factors, including the following, play a role in the ventilatory response to exercise: (1) reflexes originating from body movements, (2) increase in body temperature, (3) epinephrine release, and (4) impulses from the cerebral cortex.

*Ventilation can be influenced by factors unrelated to the need for gas exchange.*

- Protective reflexes such as sneezing and coughing temporarily govern respiratory activity in an effort to expel irritant materials from the respiratory passages.
- Inhalation of particularly noxious agents frequently triggers immediate cessation of ventilation.
- Voluntary control of breathing is accomplished by the cerebral cortex, which does not act on the respiratory center in the brain stem but instead sends impulses directly to the motor neurons in the spinal cord that supply the respiratory muscles.

*During apnea, a person "forgets to breathe," during dyspnea, a person feels "short of breath."*

- Apnea is the transient cessation of ventilation with the expectation that breathing will resume spontaneously.
- The condition is called respiratory arrest.
- In exaggerated cases of sleep apnea, the victim may be unable to recover from an apneic period, and death results.
- This is the case of sudden infant death syndrome.
- Most evidence suggests that the baby "forgets to breathe" as a result of the immaturity of the respiratory control mechanisms.
- Certain risk factors make babies more vulnerable to sudden infant death syndrome (SIDS).
- Among them are sleeping position (a higher incidence of SIDS is associated with sleeping on the abdomen rather than on the back or side) and exposure to nicotine during fetal life or after birth.
- People who have dyspnea have the subjective sensation that they are not getting enough air.

## Key Terms

alveolar surface tension
alveoli
alveolar surface tension
alveolar ventilation
anatomical dead space
aortic bodies
apneusis
apneustic center
apnea
asthma
atelectasis
atmospheric pressure
2.3-bisphosphoglycerate
Bohr effect
Boyle's law
breathing (ventilation)
bronchi
bronchioles
bronchoconstriction
bronchodilation
carbaminohemoglobin
carbonic anhydrase
carbon monoxide
carboxyhemoglobin
carotid bodies
cellular respiration (internal)
central chemoreceptors
chloride shift
chronic bronchitis
chronic obstructive pulmonary disease (COPD)
collateral ventilation
compliance
dorsal respiratory group
dyspnea
elastic recoil
emphysema
expiratory reserve volume
forced expiratory volume in one second ($FEV_1$)
functional residual capacity
Haldane effect

hypercapnia
hyperoxia
hyperpnea
hyperventilation
hypoxia
intrapleural fluid
intrapulmonary pressure
inspiration
inspiratory capacity
inspiratory reserve volume
internal (cellular) respiration
LaPlace's law
larynx
law of mass action
lungs
newborn respiratory distress syndrome
oxygen-hemoglobin dissociation curve
oxyhemoglobin
partial pressure
partial pressure gradient
percent (%Hb) hemoglobin saturation
pleural sac
pleurisy
pneumotaxic center
pneumothorax
preBotzinger complex
pressure gradient
pores of Kohn
pulmonary (minute) respiration
pulmonary stretch receptors
pulmonary surfactant
pulmonary ventilation
residual hemoglobin
residual volume
respiratory airways
respiratory arrest
respiration
respiratory quotient (R.Q.)
sleep apnea
spirometer
sudden infant death syndrome
total lung capacity
transmural pressure gradient
tidal volume
thorax
Type I alveolar cells
Type II alveolar cells
ventilation (breathing)
ventral respiratory group
vital capacity
vocal folds

## Review Exercises

*Answers are in the appendix*

**True/False**

_____ 1. Pulmonary connective tissue contains large quantities of elastin fibers.

_____ 2. The work of breathing may be increased when pulmonary compliance is increased.

_____ 3. The energy used to accomplish the increased ventilation still represents only about 3 percent of total energy expended.

_____ 4. The maximum amount of air that the lungs can hold is about 4.2 liters in males.

_____ 5. The lungs still contain about 2,200 ml of air at the end of a normal quiet expiration.

_____ 6. Alveolar ventilation is more important than pulmonary ventilation.

_____ 7. A gas always diffuses down its partial pressure gradient from the area of higher to the area of lower partial pressure.

_____ 8. The alveolar partial pressure of oxygen remains relatively constant at about 150 mm Hg throughout the respiratory cycle.

_____ 9. Surface area is reduced in emphysema.

_____ 10. The rate of gas transfer is directly proportional to the diffusion coefficient.

_____ 11. The amount of oxygen transferred to the cells and the amount of carbon dioxide carried away from the cells depend on the rate of cellular metabolism.

_____ 12. The net diffusion of oxygen occurs first between the blood and the tissues.

_____ 13. Respiratory control centers housed in the brain stem are responsible for generating the rhythmic pattern of breathing.

_____ 14. VRG is generally regarded as being responsible for the basic rhythm of ventilation.

_____ 15. The pneumotaxic center is dominant over the apneustic center.

_____ 16. Arterial partial pressure of carbon dioxide is the most important input regulating the magnitude of ventilation under resting conditions.

_____ 17. A fall in arterial partial pressure of carbon dioxide reflexly reduces the respiratory drive.

_____ 18. A decrease in hydrogen ion concentrations in the brain ECF directly stimulates the central chemoreceptors.

_____ 19. The cause of increased ventilation during exercise is still largely speculative.

_____ 20. In sudden infant death syndrome (SIDS) babies two to five months are victims of poorly developed carotid bodies.

_____ 21. Only 2.5 percent of the carbon dioxide in the blood is dissolved.

_____ 22. Each hemoglobin molecule can carry up to four molecules of oxygen.

_____ 23. In the systemic capillaries the blood partial pressure of carbon dioxide is increased.

_____ 24. At a normal partial pressure of oxygen of 100 mm Hg, hemoglobin is 98.5 percent saturated.

_____ 25. Local elevation in temperature enhances oxygen release from hemoglobin for use by the more active tissues.

_____ 26. Carbon dioxide binds with the globin portion of hemoglobin.

_____ 27. Histotoxic hypoxia arises when too little oxygenated blood is delivered to the tissues.

_____ 28. There are no muscles within the alveolar walls.

_____ 29. Air tends to move from a region of lower pressure to a region of higher pressure.

_____ 30. The intra-pleural pressure is greater than the intra-alveolar pressure.

_____ 31. The stretched lungs have a tendency to pull inward.

_____ 32. During inspiration, the intrapleural pressure increases.

_____ 33. At resting expiration, the intra-alveolar pressure decreases.

_____ 34. Expiration is normally a passive process.

_____ 35. If the blood flow is greater than the airflow to a given alveolus, the oxygen level in the alveolus and surrounding tissues will fall below normal.

**Fill in the Blank**

36. ____________________ oxygen delivery to the tissues is normal, but the cells are unable to use the oxygen available to them

37. ____________________ refers to excess carbon dioxide in the arterial blood that is caused by hypoventilation.

38. The relationship between partial pressure of oxygen and percent hemoglobin saturation is depicted by an S-shaped curve known as the ____________________.

39. The influence of carbon dioxide and acid on the release of oxygen is known as the _______________.

40. The combination of carbon monoxide and hemoglobin is known as ____________________.

41. The fact that removal of oxygen from hemoglobin increases the ability of hemoglobin to pick up $C0_2$ and $C0^{2-}$ generated H is known as the ____________________.

42. ____________________ is characterized by a low arterial blood partial pressure of oxygen accompanied by inadequate hemoglobin saturation.

43. ____________________ is characterized by inflammatory fluid accumulation within or around the alveoli.

44. The individual pressure exerted independently by a particular gas within a mixture of gases is known as its ____________________.

45. A difference in partial pressure between pulmonary blood and alveolar air is known as ____________________.

46. Across systemic capillaries: oxygen partial pressure gradient from blood to tissue cell is ______________ and carbon dioxide partial pressure gradient from tissue cell to blood is ____________________.

47. ____________________ is the volume of air in the lungs at the end of a normal passive expiration.

48. Two general categories of respiratory dysfunction yield abnormal results during spirometry: ____________________ and ____________________ lung disease.

49. Not all of the inspired air gets down to the site of gas exchange in the alveoli. Part of it remains in the conducting airways, where it is not available for gas exchange. The volume is considered to be ____________________.

50. ____________________ is the pressure exerted by the weight of the air in the atmosphere on objects on the Earth's surface.

51. ____________________ refers to decreased airway resistance by promoting bronchiolar smooth muscle relaxation.

52. ____________________ is characterized by collapse of the smaller airways and a breakdown of alveolar walls.

53. ____________________ refers to how much effort is required to stretch or distend the lungs.

54. A second factor that contributes to alveolar stability is the ____________________ of neighboring alveoli.

55. ____________________ refers to the magnitude of the inward-directed collapsing pressure directly proportional to the surface tension and inversely proportional to the radius of the bubble.

56. ____________________ is the volume of air entering or leaving the lung during a single breath.

57. The primary respiratory control center is the ____________________.

58. There are two other respiratory centers higher in the brain stem in the pons: (1)___________________ and (2) ___________________.

59. The ___________________ consists mostly of inspiratory neurons.

60. ___________________ is when the breathing pattern consists of prolonged inspiratory gasps abruptly interrupted by very brief expirations.

61. Arterial partial pressure of oxygen is monitored by peripheral chemoreceptors known as the ___________________ and ___________________.

62. ___________________ is the transient cessation of ventilation with the expectation that breathing will resume spontaneously.

63. People who have ___________________ have the subjective sensation that they are not getting enough air.

64. ___________________ involves the sum of the processes that accomplish ongoing passive movement of oxygen from the atmosphere to the tissues to support cellular metabolism.

65. Two tubes lead from the pharynx: the ___________________ through which air is conducted to the lungs, and the ___________________, the tube through which food passes to the stomach.

66. ___________________ are two bands of elastic tissue that lie across the opening of the larynx, which can be stretched and positioned in different shapes by laryngeal muscles.

67. The alveolar walls consist of a single layer of flattened ___________________.

68. ___________________ secrete pulmonary surfactant.

69. ___________________ are present in the alveolar walls to permit airflow between adjacent alveoli.

**Matching**

*Match the description of the lung volume or lung capacity to the name of lung volume or capacity.*

a. total lung capacity
b. tidal volume
c. vital capacity
d. functional residual capacity
e. inspiratory reserve volume
f. residual volume
g. inspiratory capacity
h. expiratory reserve volume

_____ 70. the vital capacity plus the residual volume

_____ 71. the minimum volume of air remaining in the lungs even after maximal expiration

_____ 72. the extra volume of air that can be maximally inspired over and above the typically resting tidal volume

_____ 73. the maximum volume of air that the lungs can hold

_____ 74. the tidal volume plus the inspiratory reserve volume

_____ 75. the volume of air in the lungs at the end of a normal passive expiration

_____ 76. the volume of air entering or leaving the lungs during a single breath

_____ 77. the extra volume of air that can be actively expired by maximal contraction of the expiratory muscles beyond that normally passively expired at the end of a typical resting tidal volume

_____ 78. the tidal volume plus the inspiratory reserve volume plus the expiratory reserve volume

_____ 79. the expiratory reserve volume plus the residual volume

_____ 80. the maximum volume of air that can be inspired at the end of a normal quiet expiration

_____ 81. the maximum volume of air that can be moved out during a single breath following a maximal inspiration

**Multiple Choice**

82. Which condition is characterized by the person "forgetting" to breathe?
 a. newborn respiratory distress syndrome
 b. atelectasis
 c. sudden infant death syndrome
 d. dyspnea
 e. hypercapnia

83. Which of the following conditions is characterized by air entering the pleural cavity?
 a. apnea
 b. pleurisy
 c. dyspnea
 d. asthma
 e. pneumothorax

84. Which of the following conditions refers to insufficient oxygen at the cellular level?
 a. hypoxia
 b. hypercapnia
 c. hyperpnea
 d. emphysema
 e. atelectasis

85. Which of the following conditions indicates that the partial pressure of carbon dioxide is above normal in arterial blood?
 a. hyperoxia
 b. hypercapnia
 c. hyperpnea
 d. atelectasis
 e. hypercarbonium

86. Which of the following is classified as a chronic obstructive pulmonary disease?
 a. pneumothorax
 b. pleurisy
 c. atelectasis
 d. hypoxia
 e. asthma

87. Which of the following conditions is characterized by the lungs collapsing to its unstretched size?
 a. pleurisy
 b. atelectasis
 c. emphysema
 d. dyspnea
 e. pneumothorax

88. Which of the following refers to an inflammation of the pleural sac?
   a. pleuritis
   b. pneumothorax
   c. pleurisy
   d. emphysema
   e. hyperpnea

89. Which of the following is classified as a chronic obstructive pulmonary disease?
   a. pleuritis
   b. pleurisy
   c. sleep apnea
   d. pneumothorax
   e. emphysema

90. Which of the following conditions refers to a situation in alveolar surface tension due to insufficient pulmonary surfactant?
   a. sudden infant death syndrome
   b. newborn respiratory distress syndrome
   c. pleurisy
   d. pleuritis
   e. pulmonary insurfactantosis

91. Which of the following conditions is characterized by transient cessation of ventilation?
   a. apnea
   b. hypocapnia
   c. hyperoxia
   d. hyperdypseia
   e. atelectasis

92. Which of the following conditions refers to increased ventilation to meet increased oxygen needs?
   a. hypopnea
   b. apnea
   c. dyspnea
   d. hyperpnea
   e. hypoxia

93. Which of the following conditions is characterized by inflammation of the lower respiratory airways?
   a. dyspnea
   b. chronic bronchitis
   c. asthma
   d. pleuritis
   e. pleurisy

94. Which of the following conditions is a subjective sensation of not getting enough air?
   a. dyspnea
   b. emphysema
   c. apnea
   d. hyperpnea
   e. pleurisy

95. At high altitudes:
   a. The alveolar $P_{O_2}$ is higher than normal.
   b. The alveolar $P_{O_2}$ is lower than normal.
   c. The alveolar $P_{CO_2}$ is higher than normal.
   d. Both a and c above are correct.
   e. Both b and c above are correct.

96 The receptors that are stimulated by a large drop in the blood partial pressure of oxygen are located where?
   a. in the respiratory center of the brain.
   b. in the carotid and aortic bodies.
   c. in the tissue capillaries.
   d. two of the above are correct.
   e. all of the above (a-c) are correct.

97. The apneustic center:
   a. Is located in the medulla.
   b. Stimulates the inspiratory neurons.
   c. Inhibits inspiratory activity.
   d. Both a and b above are correct.
   e. Both a and c above are correct.

98. Hypercapnia:
   a. Refers to excess carbon dioxide in the arterial blood.
   b. Occurs when carbon dioxide is blown off to the atmosphere at a rate faster than it is being produced by the tissues.
   c. Always accompanis hypoxia.
   d. Two of the above are correct.
   e. All of the above (a-c) are correct.

99. Which of the following conditions exists at high altitudes?
   a. histotoxic hypoxia
   b. hypoxic hypoxia
   c. anemic hypoxia
   d. hypocapnia
   e. none of the above are correct.

100. The normal percent saturation of hemoglobulin in venous blood is:
   a. 97 percent.
   b. 75 percent.
   c. 50 percent.
   d. 40 percent.
   e. 10 percent.

**Modified Multiple Choice**
*Indicate which lung volume or capacity is being described in the column by filling in the appropriate letter in the blank. There is only one correct answer for each question and each answer may be used more than once.*

a. vital capacity
b. respiratory rate
c. $FEV_1$
d. tidal volume
e. residual volume
f. total lung capacity
g. functional residual capacity
h. alveolar ventilation
i. pulmonary ventilation
j. inspiratory reserve volume
k. expiratory reserve volume
l. inspiratory capacity
m. anatomic dead space volume

_____ 101. respiratory rate X (tidal volume – dead space volume)
_____ 102. maximum volume of air that the lungs can hold
_____ 103. the volume of air entering or leaving the lungs in a single breath during quiet breathing
_____ 104. the minimum volume of air remaining in the lungs after maximal expiration
_____ 105. the extra volume of air that can be maximally inspired over and above the tidal volume
_____ 106. amount of air breathed in and out in one minute
_____ 107. maximum volume of air that can be moved in and out during a single breath
_____ 108. volume of air that can be expired during the first second of expiration in a vital-capacitydetermination

_____ 109. the maximum volume of air that can be inspired at the end of a normal expiration
_____ 110. inspiratory reserve volume + tidal volume + expiratory reserve volume
_____ 111. vital capacity + residual volume
_____ 112. volume of air in the respiratory airways
_____ 113. the extra volume of air that can be actively expired by contraction of the expiratory muscles
_____ 114. respiratory rate X tidal volume
_____ 115. volume of air in the lungs at the end of a normal passive expiration
_____ 116. amount of air that is available for exchange of gases with the blood per minute
_____ 117. breaths/minute

*Indicate which type of hypoxia would be present in each of the circumstances listed below by writing the appropriate letter in the blank using the following answer code.*

A = anemic hypoxia
B = circulatory hypoxia
C = histotoxic hypoxia
D = hypoxic hypoxia

_____ 118. cyanide poisoning
_____ 119. high altitude
_____ 120. carbon monoxide poisoning
_____ 121. emphysema
_____ 122. hemoglobin deficiency
_____ 123. congestive heart failure

## Points to Ponder

1. The lung of a frog is quite different when compared to that of a human. The frog lung is a thin cylindrical tube much like a finger on a rubber glove. How does this very simple structure adequately serve the frog whereas humans require millions of alveoli to survive?

2. How would you explain the position of the trachea being ventral to the esophagus–the food must pass over the trachea to be swallowed?

3. What should you do when a small child declares that he will hold his breath until he dies if his demands are not met?

4. Why do some athletes breathe pure oxygen on the sidelines?

5. Why do we yawn?

6. Why is it important for the respiratory system to have a dual blood supply?

7. Why is it important that the capillaries surrounding the alveoli have a small diameter?

8. Under normal conditions, why is the first breath of life difficult?

9. After driving from sea level to a trail high in the Sierras, you get out of your SUV and feel dizzy. What do you suppose is causing your dizziness?

10. Nicotine from cigarette smoke causes the buildup of mucus and paralyzes the ciliated epithelial cells that line the bronchioles. How might these conditions affect pulmonary function tests?

## Clinical Perspectives

1. How do you avoid carbon monoxide in a closed environment?
2. How would you relate drowning to this chapter?
3. How does taking a dive in a hyperbaric chamber enhance treatment against certain anaerobic bacteria?
4. You have just come upon an accident victim. After summoning help you see that the victim has a pneumothorax. How could you possibly help this person until medical help arrives?
5. Your college roommate has just returned from a football game and is very hoarse from all of the cheering he did. He knows you are taking a physiology course and wants to know why he is hoarse. How would you explain to him why yelling makes us hoarse?
6. Are common colds spread mostly through sneezes? Explain why or why not.
7. Why is the disease asthma often treated with glucocorticoid drugs?
8. When or under what conditions would the use of hyperbaric oxygen therapy be useful?
9. People who hyperventilate during psychological stress are sometimes told to breathe into a paper bag? What is the physiological basis for this act?
10. Your friend asks you to explain to her what "mountain sickness" is. How would you explain this to her?

## Experiments of the Day

1. Exhale normally, then exhale a little more and a little more until you can no longer exhale. From which of the lung volumes or capacities are you exhaling? Repeat using inhalation until you can no longer inhale. Into which lung volumes or capacities are you inhaling?
2. Do some exercise until you are ventilating rapidly. Place a bag over your mouth and nose. What are the results and why?

## PhysioEdge Activities

**Related to Text:**
Media Exercise 13.1:Anatomy of the Respiratory System.
Media Exercise 13.2: Mechanics of Ventilation.
Media Exercise 13.3: Gas Transport and Exchange.
Media Exercise 13.4: Control of Ventilation, Lung Volumes, and Terms.

**Related to Figures:**
*Figure 13.2*. For an interaction related to this figure, see Media Exercise 13.1: Anatomy of the Respiratory System.
*Figure 13.5*. For an interaction related to this figure, see Media Exercise 13.2: Mechanics of Ventilation.
*Figure 13.6*. For an animation of this figure, click the Role of Pressure tab (p.2) in the Respiratory tutorial.
*Figure 13.11*. For an animation of this figure, click the Ventilation tab (p.1) in the Respiratory System tutorial. For an interaction related to this figure, see Media Exercise 13.2: Mechanics of Ventilation.
*Figure 13. 13*. For an animation of this figure, click the Role of Pressure tab in the Respiratory System tutorial.
*Figure 13.14*. For an interaction related to this figure, see Media Exercise 13.2: Mechanics of Ventilation.
*Figure 13.18*. For an interaction related to this figure, see Media Exercise 13.42: Control of Ventilation, Lung Volumes and Terms.
*Figure 13.22*. For an animation of this figure, click the Ventilation tab (p.2) in the Respiratory tutorial.

*Figure 13.25*. For an animation of this figure, click the Gas Exchange tab in the Respiratory tutorial.
*Figure 13.27*. For an animation of this figure, click the Gas Transport tab (p.4) in the Respiratory tutorial. For an interaction related to this figure, see Media Exercise 13.3: Gas Transport and Exchange.
*Figure 13.28*. For an animation of this figure, click the Gas Exchange tab in the Respiratory tutorial.
*Figure 13.29* For an interaction related to this figure, see Media Exercise 13.3: Gas Transport and Exchange.
*Figure 13.30*. For an animation of this figure, click the Gas Transport tab (p.4) in the Respiratory tutorial. For an interaction related to this figure, see Media Exercise 13.3: Gas Transport and Exchange.
*Figure 13.32*. For an interaction related to this figure, see Media Exercise 13.3: Gas Transport and Exchange.

## Media Resources

**PhysioEdge CD-ROM**
For a visual review of concepts in this chapter, check out:

Tutorial: The Respiratory System
Media Exercise 13.1: Anatomy of the Respiratory System
Media Exercise 13.2: Mechanics of Ventilation
Media Exercise 13.3: Gas Transport and Exchange
Media Exercise 13.4: Control of Ventilation, Lung Volumes and Terms

**Book Companion Website**
The website for this book contains a wealth of helpful study aids, as well as many ideas for further reading and research. Log on to:

**http://www.brookscole.com/sherwoodhp6**

Select Chapter 13 from the drop-down menu, or click on one of the many resources areas, including **Case Histories**, which introduce clinical aspects of human physiology. For this chapter check out Case History 1: Fighting for Every Breath; Case History 2: Asthma and Influenza: A Dangerous Combination; Case History 3: When Help Becomes Harm.

For Suggested Readings, consult **InfoTrac® College Edition**, your online research library, at:

**http://infotrac.thomsonlearning.com**

# 14

# The Urinary System

## Chapter Overview

The survival and proper functioning of cells depend on the maintenance of stable concentrations of salt, acids, and other electrolytes in the internal environment. Cell survival also depends on the continual removal of toxic metabolic wastes produced by the cells as they perform life-sustaining chemical reactions. The urinary system, more specifically, the kidneys are major contributors to the phenomenon of multicellular organisms. The existence of multicellular organisms depends upon the presence of the extracellular fluid. Since virtually all cells interface with the extracellular fluid, the homeostasis of this fluid is of utmost importance. For the most part, it is the role of the kidney to provide this homeostasis.

To fulfill this role the kidneys regulate most extracellular fluid ions; maintain pH, plasma volume, water balance, and osmolarity; eliminate waste products of metabolism and many foreign compounds; and in a secondary capacity, produce erythropoietin and convert vitamin D into its active form.

While the functions of the kidneys can be listed quite simply in a short paragraph the physiology of the kidney is extremely complex. The kidneys are true guardians of homeostasis in the multicellular organism.

## Chapter Outline

INTRODUCTION

*The kidneys perform a variety of functions aimed at maintaining homeostasis.*

- The simplest forms of life live in an external environment of fixed composition, the sea.
- The individual cells of complex multicellular organisms are able to function and survive only in a fluid environment of essentially constant composition similar to the sea.
- This salty internal fluid environment is the extracellular fluid that bathes all the cells of the body and must be homeostatically maintained.
- Terrestrial animals are able to live on dry land independent of the sea because of their kidneys, the organs that, in concert with the hormonal and neural inputs that control their function, are primarily responsible for maintaining the stability of extracellular fluid volume and electrolyte composition.
- The kidneys are the primary route for elimination of potentially toxic metabolic wastes and foreign compounds from the body.
- The kidneys make adjustments in urinary output of water, salt, and other electrolytes to compensate for their abnormal losses through heavy sweating, vomiting, diarrhea, or hemorrhage.
- The following are specific functions performed by the kidneys: (1) maintain water balance in the body, (2) regulate the quality and concentration of most ECF ions, (3) maintain proper plasma volume, (4) assist in maintaining the proper acid-base balance, (5) maintain proper osmolarity, (6) excrete end products of bodily metabolism, (7) excrete many foreign compounds, (8) secrete erythropoietin, (9) secrete rennin, and (10) convert vitamin D into its active form.

*The kidneys form urine; the rest of the urinary system carries the urine to the outside.*

- The urinary system consists of the urine-forming organs—the kidneys—and the structures that carry the urine from the kidneys to the outside for elimination from the body.
- The kidney acts on the plasma flowing through it to produce urine, conserving materials to be retained in the body and eliminating unwanted materials into the urine.

*The nephron is the functional unit of the kidney.*

- Each kidney is composed of about one million microscopic functional units known as nephrons.
- A nephron is the smallest unit capable of urine formation.
- The arrangement of nephrons within the kidneys gives rise to two distinct regions: an outer granular-appearing renal cortex, and an inner region, the renal medulla.

- Each nephron consists of a vascular component and a tubular component.
- The dominant portion of the vascular component is the glomerulus.
- On entering the kidney, the renal artery systematically subdivides to ultimately form many small vessels known as afferent arterioles, one of which supplies each nephron.
- The afferent arteriole delivers blood to the glomerular capillaries, which rejoin to form another arteriole, the efferent arteriole.
- No oxygen or nutrients are extracted from the blood for use by the kidney tissues nor are waste products picked up from the surrounding tissue.
- The efferent arteriole quickly divides into a second set of capillaries, the peritubular capillaries, which supply the renal tissue with blood and are important in exchanges between the tubular system and blood during conversion of the filtered fluid into urine.
- The peritubular capillaries rejoin to form venules that ultimately drain into the renal vein, by which the blood leaves the kidney.
- The tubular component begins with Bowman's capsule.
- From Bowman's capsule, the filtered fluid passes into the proximal tubule, which lies entirely in the cortex.
- The next segment, the loop of Henle, dips into the renal medulla.
- The descending limb of Henle's loop plunges from the cortex into the medulla; the ascending limb traverses back up into the cortex.
- The ascending limb passes through the fork formed by the afferent and efferent arterioles.
- Both the tubular and vascular cells at this point are specialized to form the juxtaglomerular apparatus.
- Beyond the juxtaglomerular apparatus, the tubule once again becomes highly coiled to form the distal tubule.
- The distal tubule empties into a collecting tubule.
- Each collecting duct or tubule plunges down through the medulla to empty its fluid contents into the renal pelvis.
- There are two types of nephrons: (1) cortical nephrons, and (2) juxtamedullary nephrons.

*The three basic renal processes are glomerular filtration, tubular reabsorption, and tubular secretion.*

- Glomerular filtration is the first step in urine formation.
- The filtrate flows through the tubules.
- Selective movement of substances from inside the tubule into the blood is referred to as tubular reabsorption.
- Tubular secretion refers to the selective transfer of substances from the peritubular capillary blood into the tubular lumen.
- Urine excretion refers to the elimination of substances from the body in the urine and is the result of the first three processes.

## Glomerular Filtration

*The glomerular membrane is considerably more permeable than capillaries elsewhere.*

- Fluid filtered from the glomerulus into Bowman's capsule must pass through three layers that make up the glomerular membrane: (1) the wall of the glomerular capillaries, (2) an acellular gelatinous layer known as the basement membrane, and (3) the inner layer of Bowman's capsule.
- The glomerular capillary wall is perforated by many large pores or fenestrae.
- A basement membrane composed of collagen and glycoproteins is sandwiched between the glomerulus and Bowman's capsule.
- The final layer consists of podocytes, octopuslike cells that encircle the glomerular tuft.

*Glomerular capillary blood pressure is the major force that induces glomerular filtration.*

- No active transport mechanisms or local energy expenditures are involved in moving fluid from the plasma across the glomerular membrane.
- Passive physical forces are responsible for glomerular filtration.
- Three physical forces are involved in glomerular filtration: (1) glomerular capillary blood pressure, (2) plasma-colloid osmotic pressure, and (3) Bowman's capsule hydrostatic pressure.
- The glomerular capillary blood pressure is higher than capillary blood pressure elsewhere.
- Plasma-colloid osmotic pressure is caused by the unequal distribution of plasma proteins across the glomerular membrane.
- The resultant tendency for water to move by osmosis down its own concentration gradient from Bowman's capsule into the glomerulus opposes glomerular filtration.
- The fluid in Bowman's capsule exerts a hydrostatic pressure.
- This pressure, which tends to push fluid out of Bowman's capsule, opposes glomerular filtration.
- The total force favoring filtration is attributable to the glomerular capillary blood pressure.
- The total of the two forces opposing filtration is less.
- The net difference favoring filtration is referred to as the net filtration pressure.

*Changes in the GFR occur primarily as a result of changes in glomerular capillary blood pressure.*

- The glomerular filtration rate (GFR) is controlled by two mechanisms: (1) autoregulation, which is aimed at preventing spontaneous changes in the GFRI, and (2) extrinsic sympathetic control, which is aimed at long-termed regulation of arterial blood pressure.
- Presently, two intrarenal mechanisms are thought to contribute to autoregulation: (1) a myogenic mechanism, which responds to changes in pressure within the nephron's vascular component, and (2) a tubuloglomerular feedback mechanism, which senses changes in flow through the nephron's tubular component.
- Arteriolar vascular smooth muscle contracts inherently in response to the stretch accompanying increased pressure within the vessel.
- Conversely, inherent relaxation of an unstretched afferent arteriole when pressure within the vessel is reduced increases blood flow into the glomerulus despite the fall in arterial pressure.
- The tubuloglomerular mechanism involves the juxtaglomerular apparatus.
- The macula densa cells detect changes in the rate at which fluid is flowing past them through the tubule.
- In response, the macula densa cells trigger the release of vasoactive chemicals from the juxtaglomerular apparatus.
- The CFR can be changed on purpose by extrinsic-control mechanisms that override the autoregulatory responses.
- Extrinsic control of GFR, which is mediated by sympathetic nervous system input to the afferent arterioles, is aimed at the regulation or arterial blood pressure.
- The parasympathetic nervous system does not exert any influence on the kidneys.
- No new mechanism is required to decrease the GFR.
- It is reduced as a result of the baroreceptor reflex response to a fall in blood pressure.
- The afferent arterioles constrict as a result of increased sympathetic activity.
- The resultant decrease in GFR in turn leads to a reduction in urine volume.
- When the baroreceptors detect a rise in blood pressure, sympathetic vasoconstrictor activity to the arterioles, including the renal afferent arterioles, is reflexly reduced, allowing afferent arteriolar vasodilation to occur.
- As more blood enters the glomeruli through the dilated afferent arterioles, glomerular capillary blood pressure rises, increasing the GFR.

*The GFR can be influenced by changes in the filtration coefficient.*

- The rate of glomerular filtration, however, is dependent on the filtration coefficient as well as on the net filtration pressure.
- Both factors on which the filtration coefficient depends—the surface area and the permeability of the glomerular membrane—can be modified by contractile activity within the membrane.
- Contraction of the mesangial cells closes off a portion of the filtering capillaries, thereby reducing the surface area available for filtration.
- Contractile activity of the podocytes can decrease or increase the number of filtration slits available in the inner membrane of Bowman's capsule, which in turn affects permeability and the filtration coefficient.

*The kidneys normally receive 20 percent to 25 percent of the cardiac output.*

- Most of the blood goes to the kidneys, not to supply the renal tissue but to be adjusted and purified by the kidneys.
- Only by continuously processing such a large proportion of the blood are the kidneys able to precisely regulate the volume and electrolyte composition of the internal environment and adequately eliminate large quantities of metabolic waste products that are constantly produced.

## Tubular Reabsorption

*Tubular reabsorption is tremendous, highly selective, and variable.*

- Tubular reabsorption is a highly selective process.
- The tubules have a high reabsorptive capacity for substances needed by the body and a poor or no reabsorptive capacity for substances of no value.
- Only a small percentage of filtered plasma constituents that are useful to the body are present in the urine, most having been reabsorbed and returned to the blood.
- Only excess amounts of essential materials such as electrolytes are excreted in the urine.
- As water and other valuable constituents are reabsorbed, the waste products remaining in the tubular fluid become highly concentrated.

*Tubular reabsorption involves transepithelial transport.*

- To be reabsorbed, a substance must traverse five distinct barriers: (1) It must leave the tubular fluid by crossing the luminal membrane of the tubular cell. (2) It must pass through the cytosol from one side of the tubular cell to the other. (3) It must traverse the basolateral membrane of the tubular

cell to enter the interstitial fluid. (4) It must diffuse through the interstitial fluid. (5) It must penetrate the capillary wall to enter the blood plasma.
- This entire sequence is known as transepithelial transport.
- There are two types of tubular reabsorption: passive reabsorption and active reabsorption.

*An active sodium-potassium ATPase pump in the basolateral membrane is essential for sodium reabsorption.*
- Sodium reabsorption in the proximal tubule plays a pivotal role in the reabsorption of glucose, amino acids, water, chloride ions, and urea.
- Sodium reabsorption in the loop of Henle, along with chloride ion reabsorption, plays a critical role in the kidneys' ability to produce urine of varying concentrations and volumes, depending on the body's need to conserve or eliminate water.
- Sodium reabsorption in the distal portions of the nephron is variable and subject to hormonal control, being important in the regulation of ECF volume.
- Sodium reabsorption is also linked in part with potassium and hydrogen ion secretion.

*Aldosterone stimulates sodium reabsorption in the distal and collecting tubules.*
- The most important and best known hormonal system involved in the regulation of sodium is the renin-angiotensin-aldosterone system, which stimulates sodium reabsorption in the distal and collecting tubules.
- The granular cells of the juxtaglomerulus apparatus secrete a hormone, renin, into the blood.
- Renin acts as an enzyme to activate angiotensinogen into angiotensin I.
- On passing through the lungs, angiotensin I is converted to angiotensin II.
- Angiotensin II is the primary stimulus for the secretion of the hormone aldosterone.
- Aldosterone increases sodium reabsorption by the distal and collecting tubules.
- The renin-angiotensin-aldosterone system thus promotes salt retention, and a resultant water retention and elevation of arterial blood pressure.
- Renin-angiotensin-aldosterone activity is also responsible in part for the fluid retention and edema accompanying congestive heart failure.
- Diuretics are therapeutic agents that cause diuresis.
- ACE inhibitor drugs block the action of angiotensin-converting enzyme.

*Atrial natriuretic peptide inhibits sodium reabsorption.*
- Atrial natriuretic peptide is released from the cardiac atria when the ECF volume is expanded.
- The primary action of atrial natriuretic peptide is to inhibit sodium reabsorption in the distal parts of the nephron.

*Glucose and amino acids are reabsorbed by sodium-dependent secondary active transport.*
- Glucose and amino acids are transferred by means of secondary active transport; a specialized cotransport carrier simultaneously transfers both sodium and the specific organic molecule from the lumen into the cell.
- Glucose and amino acids get a "free ride" at the expense of energy already used in the reabsorption of sodium.

*In general, actively reabsorbed substances exhibit a tubular maximum.*
- The tubular maximum in the kidney is the maximum amount of a substance that tubular cells can actively transport within a given time period.
- The quantity of any substance filtered per minute is known as the filtered load.

*Glucose is an example of an actively reabsorbed substance that is not regulated by the kidneys.*
- At a constant GFR, the glucose filtered load is directly proportional to the plasma glucose concentration.
- The plasma glucose concentration can become extremely high in diabetes mellitus.
- Consequently, although glucose does not normally appear in the urine, it is found in the urine of persons with diabetes when the plasma glucose concentration exceeds the renal threshold.

*Phosphate is an example of an actively reabsorbed substance that is regulated by the kidneys.*
- The kidneys do directly contribute to the regulation of many electrolytes such as calcium and phosphate.
- Our diets are generally rich in phosphates, but because the tubules can reabsorb up to the normal plasma concentration's worth of phosphate and no more, the excess ingested phosphate is quickly spilled into the urine.
- Unlike the reabsorption or organic nutrients, the reabsorption of phosphate and calcium is also subject to hormonal control.
- Parathyroid hormone can alter the renal thresholds for phosphate and calcium, thus adjusting the quantity of these electrolytes conserved depending on the body's momentary needs.

*Active sodium reabsorption is responsible for the passive reabsorption of chloride ions, water, and urea.*

- The negatively charged chloride ions are passively reabsorbed down the electrical gradient created by the active reabsorption of positively charged sodium ions.
- Water is passively reabsorbed by osmosis throughout the length of the tubule.
- Water passes through recently discovered water channels, or aquaporins, formed by specific plasma membrane proteins in the tubular cells.
- The passive reabsorption of urea is also indirectly linked to active sodium reabsorption.

*In general, unwanted waste products are not reabsorbed.*

- Urea molecules, being the smallest of waste products, are the only wastes able to be passively reabsorbed as a result of the concentrating effect of the kidneys.
- The waste products, failing to be reabsorbed, generally remain in the tubules and are excreted in the urine in a highly concentrated form.

## TUBULAR SECRETION

*Hydrogen ion secretion is important in acid-base balance.*

- The most important substances secreted by the tubules are hydrogen ions, potassium ions, and organic anions and cations, many of which are compounds foreign to the body.
- Renal hydrogen ion secretion is extremely important in the regulation of acid-base balance in the body.
- Hydrogen ions can be added to the filtered fluid by being secreted by the proximal, distal, and collecting tubules.

*Potassium secretion is controlled by aldosterone.*

- Potassium is actively reabsorbed in the proximal tubule and actively secreted in the distal and collecting tubules.
- Potassium-ion secretion in the distal and collecting tubules is coupled to sodium ion reabsorption by means of the energy-dependent basolateral sodium-potassium pump.
- Increased aldosterone secretion always promotes simultaneous sodium reabsorption and potassium secretion.

*Organic anion and cation secretion helps efficiently eliminate foreign compounds from the body.*

- The proximal tubule contains two distinct types of secretory carriers, one for the secretion or organic anions and a separate system for the secretion of organic cations.
- The ability of the organic-ion secretory systems to eliminate many foreign compounds from the body is most important.
- Nonselectivity permits the organic-ion secretory systems to hasten the removal of many foreign organic chemicals, including food additives, environmental pollutants, drugs, and other non-nutritive organic substances that have gained entrance to the body.

## URINE EXCRETION AND PLASMA CLEARANCE

- On the average, one milliliter of urine is excreted per minute.
- Typically, of the 125 ml or plasma filtered per minute, 124 ml/min are reabsorbed, so the final quantity of urine formed averages 1 ml/min.
- Thus, 1.5 liters of urine per day are excreted, out of the 180 liters per day filtered.
- A relatively small change in the quantity of filtrate reabsorbed can bring about a large change in the volume of urine formed.

*Plasma clearance is the volume of plasma cleared of a particular substance per minute.*

- If a substance is filtered but not reabsorbed or secreted, its plasma clearance rate equals the GFR.
- If a substance is filtered and reabsorbed but not secreted, its plasma clearance rate is always less than the GFR.
- If a substance is filtered and secreted but not reabsorbed, its plasma clearance rate is always greater than the GFR.
- If a substance is filtered but not reabsorbed or secreted, its plasma clearance rate equals the GFR.
- If a substance is filtered and reabsorbed but not secreted, its plasma clearance rate is always less than the GFR.
- If a substance is filtered and secreted but not reabsorbed, its plasma clearance is always greater than the GFR.

*The kidneys can excrete urine of varying concentrations depending on the body's state of hydration.*

- The ECF osmolarity depends on the relative amount of water compared to solute.
- At normal fluid balance and solute concentration, the body fluids are said to be isotonic.
- If there is too much water relative to the solute load, the body fluids are hypotonic.
- If a water deficit exists relative to the solute load, the body fluids are too concentrated or are hypertonic.

*The medullary vertical osmotic gradient is established by means of countercurrent multiplication.*

- A large vertical osmotic gradient is uniquely maintained in the interstitial fluid of the medulla of each kidney.
- The concentration of the interstitial fluid progressively increases from the cortical boundary down through the depth of the renal medulla until it reaches a maximum of 1,200 mosm/liter in humans at the junction with the renal pelvis.
- The presence of this gradient enables the kidneys to produce urine that ranges in concentration from 100 to 1,200 mosm/liter.
- When the body is in ideal fluid balance, isotonic urine is formed.
- When the body is overhydrated, the kidneys are able to produce a large volume of dilute urine, thus eliminating the excess water into the urine.
- Conversely, the kidneys are able to put out a small volume of concentrated urine when the body is dehydrated.
- The juxtamedullary nephrons' long loops of Henle establish the vertical gradient. Their vasa recta prevent the dissolution of this gradient while providing blood to the renal medulla.

*Vasopressin-controlled, variable water reabsorption occurs in the final tubular segments.*

- The collecting tubules of all nephrons use the gradient, in conjunction with the hormone vasopressin, to produce urine of varying concentrations.
- Collectively, this entire functional organization is known as the medullary countercurrent system.
- The long loops of Henle of juxtamedullary nephrons establish the medullary vertical osmotic gradient by means of countercurrent multiplication.
- The descending limb is highly permeable to water but does not actively extrude sodium.
- The ascending limb actively transports NaCl out of the tubular lumen into the surrounding interstitial fluid and is always impermeable to water, so salt leaves the tubular fluid without water osmotically following along.
- The vertical osmotic gradient in the medullary interstitial fluid is used by the collecting ducts to concentrate the tubular fluid so that urine more concentrated than normal body fluids can be excreted.
- The fact that the fluid is hypotonic as it enters the distal portions of the tubule enables the kidneys to excrete urine more dilute than normal body fluids.
- The active salt pump in the ascending limb is able to transport sodium chloride out of the lumen until the surrounding interstitial fluid is 200 mosm/liter more concentrated than the tubular fluid in this limb.
- When the ascending limb pump starts actively extruding salt, the medullary interstitial fluid becomes hypertonic.
- Water cannot follow osmotically from the ascending limb because this limb is impermeable to water.
- Net diffusion of water does occur from the descending limb into the interstitial fluid.
- The tubular fluid entering the descending limb from the proximal tubule is isotonic.
- Because the descending limb is highly permeable to water, net diffusion of water occurs by osmosis out of the descending limb into the more concentrated interstitial fluid.
- The tubular fluid entering the loop of Henle immediately starts to become more concentrated as it loses water.
- The concentration of tubular fluid is progressively increasing in the descending limb and progressively descending in the ascending limb.
- The fluid in the descending limb becomes progressively more hypertonic until it reaches a maximum concentration of 1200 mosm/liter at the bottom of the loop.
- The tubular fluid becomes hypotonic as it leaves the ascending limb to enter the distal tubule at a concentration of 100 mosm/liter.

*Countercurrent exchange within the vasa recta conserves the medullary vertical osmotic gradient.*

- The medullary vertical osmotic gradient permits excretion of urine of differing concentrations by means of vasopressin-controlled, variable water reabsorption from the final tubular segments.
- In order for water reabsorption to occur across a segment of the tubule, two criteria must be met: (1) an osmotic gradient must exist across the tubule, and (2) the tubular segment must be permeable to water.
- The distal and collecting tubules are impermeable to water except in the presence of vasopressin, also known as antidiuretic hormone.
- Vasopressin is produced in the hypothalamus and is stored in the posterior pituitary gland.
- The hypothalamus controls the release of vasopressin from the posterior pituitary into the blood.
- In negative-feedback fashion, vasopressin secretion is stimulated by a water deficit when water must be conserved for the body and inhibited by a water excess when surplus water must be eliminated in the urine.
- Vasopressin reaches the basolateral membrane of the tubular cells lining the distal and collecting

tubules through the circulatory system, whereupon it binds with receptors specific for it.

- This binding activates the cyclic AMP second messenger system within the tubular cells, which increases the permeability of the opposite luminal membrane to water by increasing the number of water channels.
- The tubular response to vasopressin is graded; the more vasopressin present, the more water channels inserted, and the greater the permeability of the distal and collecting tubules to water.
- The channels are retrieved when vasopressin secretion decreases.
- When vasopressin secretion is increased in response to a water deficit and the permeability of the distal and collecting tubules to water is increased, the hypotonic tubular fluid entering the distal tubules is able to lose progressively more water by osmosis into the interstitial fluid as the tubular fluid first flows through the isotonic cortex and then is exposed to the ever-increasing osmolarity of the medullary interstitial fluid as it plunges toward the renal pelvis.
- Under the influence of maximum levels of vasopressin, it is possible to concentrate the tubular fluid up to 1,200 mosm/liter by the end of the collecting tubules.
- No further modification of the tubular fluid occurs beyond the collecting tubule, so what remains in the tubules at this point is urine.
- The minimum volume of urine that is required to excrete wastes is 500 ml/day.
- When a person consumes large quantities of water, the excess water must be removed from the body without simultaneously losing solutes that are critical to the maintenance of homeostasis.
- Under these circumstances, no vasopressin is secreted, so the distal and collecting tubules remain impermeable to water.
- None of the water remaining in the tubules can leave the lumen to be reabsorbed even though the tubular fluid is less concentrated than the surrounding interstitial fluid.
- Urine flow may be increased up to 25 ml/min in the absence of vasopressin, compared to the normal urine production of 1 ml/min.
- The loop of Henle plays a key role in allowing the kidneys to excrete urine that ranges in concentration from 100 to 1,200 mosm/liter.
- The extent of reabsorption varies directly with the amount of vasopressin secreted.
- Vasopressin influences water permeability only in the distal and collecting tubules.

*Water and solute reabsorption are only partially coupled.*

- In tubular segments that are permeable to water, solute reabsorption is always accompanied by comparable water reabsorption because of osmotic considerations.
- Solute excretion is always accompanied by comparable water excretion because of osmotic considerations.
- A loss or gain of pure water that is not accompanied by comparable solute deficit or excess in the body leads to changes in extracellular fluid osmolarity.
- Urea recycling in the renal medulla contributes to medullary hypertonicity and helps concentrate urea in the urine.
- The early portion of the collecting tubule is impermeable to urea.
- Vasopressin increases the permeability of the late collecting tubule to urea.
- Urea reabsorption is beneficial for two reasons: (1) reabsorbed urea contributes to medullary hypertonicity, and (2) urea recycling provides a mechanism for concentrating urea in the excreted fluid while economizing on the simultaneous loss of water.
- Countercurrent exchange within the vasa recta enables the medulla to be supplied with blood while conserving the medullary vertical osmotic gradient.
- It is important that circulation of blood through the medulla does not disturb the vertical gradient of hypertonicity established by the loops of Henle.
- The hairpin construction of the vasa recta, by looping back through the concentration gradient in reverse, allows the blood to leave the medulla and enter the renal vein essentially isotonic to incoming arterial blood.
- As blood passes down the descending limb of the vasa recta, it picks up salt and loses water until it is hypertonic by the bottom of the loop.
- Then, as blood flows up the ascending limb, salt diffuses back out into the interstitium, and water reenters that vasa recta as progressively decreasing concentrations are encountered.
- This passive exchange of solutes and water between the two limbs of the vasa recta and the interstitial fluid is known as countercurrent exchange.
- The countercurrent exchange does not establish the concentration gradient.
- The countercurrent exchange prevents the dissolution of the gradient.

*Renal failure has wide-ranging consequences.*

- When the functions of both kidneys are disrupted to the point that they are unable to perform their regulatory and excretory functions sufficiently to maintain homeostasis, renal failure is said to exist.
- Among the causes of renal failure are: (1) infectious organisms, (2) toxic agents, (3) inappropriate immune responses, (4) obstruction of urine flow, and (5) an insufficient renal blood supply.

*Urine is temporarily stored in the bladder, from which it is emptied by micturition.*

- Peristaltic contractions of the smooth muscles within the ureteral wall propel the urine forward from the kidneys to the bladder.
- As the bladder fills, the ureteral ends within its walls are compressed closed.
- Urine can still enter because ureteral contractions generate sufficient pressure to overcome the resistance.
- As is characteristic of smooth muscle, bladder muscle is able to stretch tremendously without a build up in bladder wall tension.

*The role of the bladder.*

- The bladder smooth muscle is richly supplied by parasympathetic fibers, stimulation of which causes bladder contraction.
- The exit from the bladder is guarded by two sphincters: the internal urethral sphincter and the external urethral sphincter.
- The internal urethral sphincter is under involuntary control.
- The external urethral sphincter and pelvic diaphragm are under voluntary control.
- Micturition, the process of bladder emptying, is governed by two mechanisms: the micturition reflex and voluntary control.

## Key Terms

ace inhibitor drugs
active reabsorption
afferent arterioles
aldosterone
angiotensinogen
angiotensinogen i
angiotensinogen ii
angiotensin-converting enzyme
aquaporins (water channels)
atrial natriuretic peptide (ANP)
autoregulation
blood urea nitrogen (BUN)
bowman's capsule
collecting tubule (duct)
cortical nephrons
countercurrent exchange
countercurrent multiplication
distal tubule
diuretics
efferent arterioles
filtration coefficient ($K_f$)
filtered load
filtration fraction
filtration slits
glomerular filtration
glomerular filtration rate (GFR)
glomerular membrane
glomerulus
granular cells
hypertonic
hypotonic
inulin
isotonic
juxtaglomerular apparatus
juxtamedullary nephrons
kidneys
loop of Henle
macula densa
medullary countercurrent system
mesangial cells
micturition (urination)
micturition reflex
myogenic mechanism
net filtration pressure
nephrons
net filtration pressure
osmotic diuresis
para-aminohippuric acid (PAH)
passive reabsorption
peritubular capillaries
plasma clearance
podocytes
proximal tubule
renal cortex
renal failure
renal medulla
renal pyramids
renal threshold
renin
renin-angiotensin-aldosterone system
sodium load
tubuloglomerular feedback mechanism
tubular maximum ($T_m$)
tubular reabsorption
tubular secretion
urea
ureter
urethra
urine excretion
urinary bladder
urinary system
urinary incontinence
vasopressin (antidiuretic hormone)
vasa recta
water channels (aquaporins)
water diuresis

## *Review Exercises*

*Answers are in the appendix.*

**True/False**

_____ 1. The afferent arteriole delivers blood to the glomerular capillaries.

_____ 2. The efferent arteriole carries venous blood to the renal vein to exit the kidney.

_____ 3. The peritubular capillaries supply the renal tissue with oxygen.

_____ 4. The glomerulus in a juxtamedullary nephron is located in the medulla of the kidney.

_____ 5. Podocytes are found in the membranes of the distal tubule.

_____ 6. The GFR is dependent upon the net filtration pressure only.

_____ 7. Macula densa cells are found in the glomerulus.

_____ 8. Aldosterone stimulates sodium reabsorption in the collecting tubules.

_____ 9. Atrial natriuretic peptide is released from the juxtaglomerular apparatus.

_____ 10. The filtered load is known as the level at which the tubular maximum has been reached and the particular substance begins to appear in the urine.

_____ 11. The negatively charged chloride ions are passively reabsorbed in the proximal tubule.

_____ 12. Urea is passively reabsorbed in the distal and collecting tubule.

_____ 13. Organic cations are secreted by the proximal tubules.

_____ 14. Sodium is passively reabsorbed in the proximal tubule.

_____ 15. The most important substances secreted by the distal and collecting tubules are sodium and potassium.

_____ 16. Potassium is actively reabsorbed in the proximal tubule

_____ 17. Potassium is actively secreted in the distal and collecting tubules.

_____ 18. Urine excretion is about 10.5 liters per day.

_____ 19. If a substance is filtered and reabsorbed but not secreted, its plasma clearance is always greater than the GFR.

_____ 20. If a substance is filtered and secreted but not reabsorbed, its plasma clearance is always less than the GFR.

_____ 21. If a substance is filtered but not secreted, its plasma clearance rate equals the GFR.

_____ 22. Vasopressin is also known as antidiuretic hormone.

**Fill in the Blank**

23. ____________________ is a waste product resulting from the breakdown of proteins.

24. The passive reabsorption of urea is also directly linked to active ____________________ reabsorption.

25. The parathyroid hormone can alter renal thresholds for ____________________.

26. Atrial natriuretic peptide is released from the cardiac atria when the ECF volume _______________.

27. Fluid retention and edema that accompany congestive heart failure are brought about by ____________________ activity.

28. Renin acts as an enzyme to activate angiotensinogen into ____________________.

29. ____________________ is the primary stimulus for the secretion of aldosterone.

30. ____________________ increases sodium reabsorption by the distal and collecting tubules.

31. Tubular reabsorption involves ____________________ transport.

32. Each tuft of glomerular capillaries is held together by ____________________.

33. The rate of glomerular filtration is dependent on the ____________________ as well as on the ____________________.

34. From Bowman's capsule, the filtered fluid passes into the tubule.

35. Solute excretion is always accompanied by comparable water excretion because of ____________________ considerations.

36. The early portion of the collecting tubule is ____________________ to urea.

37. Causes of renal failure are: a. ____________________, b. ____________________, c. ____________________, d. ____________________, e. ____________________.

38. ____________________ contractions of the smooth muscle within the ureteral wall propel urine forward from the kidneys to the bladder.

39. The exit from the bladder is guarded by two sphincters: ____________________ and ____________________.

40. The passive exchange of solutes and water between the two limbs of the vasa recta and the interstitial fluid is known as ____________________.

41. Body fluids are said to be isotonic at an osmolarity of __________ mosm/l.

42. Potassium-ion secretion in the distal and collecting tubules is coupled to sodium reabsorption by means of the energy-dependent basolateral ____________________.

43. Increased aldosterone secretion always promotes simultaneous sodium reabsorption and ____________________.

**Matching**

*Using the code below match the reabsorption or secretion function to all the tubules that perform the function.*

a. proximal tubule
b. distal tubule
c. collecting tubule

_____ 44. organic-ion secretion; not subject to control

_____ 45. variable water reabsorption, controlled by vasopressin

_____ 46. variable hydrogen ion secretion, depending on acid-base status of the body

_____ 47. all filtered glucose and amino acids reabsorbed by secondary active transport, not subject to control

_____ 48. variable sodium reabsorption, controlled by aldosterone; chloride follows passively

_____ 49. all filtered potassium reabsorbed; not subject to control

_____ 50. fifty percent of filtered urea passively reabsorbed; not subject to control

_____ 51. variable amounts of phosphate ions and other electrolytes reabsorbed; subject to control

_____ 52. variable potassium secretion, depending on acid-base status of the body

_____ 53. sixty-seven percent of filtered sodium actively reabsorbed; not subject to control; chloride follows passively

_____ 54. sixty-five percent of filtered water osmotically reabsorbed; not subject to control

**Multiple Choice**

55. Which of the following substances is secreted by the kidneys and stimulates red blood cell production?
a. aldosterone
b. vasopressin
c. para-aminohippuric acid
d. renin
e. erythropoietin

56. Which substance is secreted by the hypothalamus, stored in the posterior pituitary, and stimulates the reabsorption of water in the collecting tubules?
a. aldosterone
b. angiotensin I
c. vasopressin
d. para-aminohippuric acid
e. inulin

57. Which of the following substances is used to clinically ascertain the GFR?
a. aldosterone
b. inulin
c. renin
d. atrial natriuretic peptide
e. vasopressin

58. Which of the following substances is the primary stimulus for the secretion of the hormone aldosterone?
    a. angiotensin I
    b. renin
    c. inulin
    d. angiotensin II
    e. atrial natriuretic peptide

59. Which substance inhibits sodium reabsorption in the distal parts of the nephron?
    a. Atrial natriuretic peptide
    b. Para-aminohippuric acid
    c. Vasopressin
    d. Aldosterone
    e. Renin

60. Which of the following substances is a hormone of the adrenal gland?
    a. renin
    b. aldosterone
    c. inulin
    d. vasopressin
    e. erythropoietin

61. Which substance functions as an enzyme to activate angiosinogen into angiotensin I?
    a. aldosterone
    b. vasopressin
    c. inulin
    d. renin
    e. para-hippuric acid

62. Which of the following substances can be used to measure renal plasma flow?
    a. angiotensin I
    b. para-aminohippuric acid
    c. angiotensin 11
    d. aldosterone
    e. vasopressin

63. Which of these substances must pass through the lungs before they can become a stimulus for hormone release?
    a. angiotensin I
    b. angiotensin II
    c. angiotensinogen
    d. inulin
    e. aldosterone

64. Which of the following substances increases sodium reabsorption by the distal tubule?
    a. renin
    b. angiotensin I
    c. para-aminohippuric acid
    d. atrial natriuretic peptide
    e. aldosterone

65. What is the concentration of the tubular fluid as it leaves the ascending limb and enters the distal tubule?
    a. 200 mosm/liter
    b. 600 mosm/liter
    c. 100 mosm/liter
    d. 1,200 mosm/liter
    e. 800 mosm/liter

66. With maximum levels of vasopressin, what is the concentration of the tubular fluid at the end of the collecting tubules?
    a. 200 mosm/liter
    b. 600 mosm/liter
    c. 100 mosm/liter
    d. 1,200 mosm/liter
    e. 800 mosm/liter

67. What is the minimum volume of urine required to excrete wastes in each day?
    a. 1 ml
    b. 50 ml
    c. 500 ml
    d. 600 ml
    e. 1,200 ml

68. In which tubules or renal structures does vasopressin influence water permeability?
    a. proximal tubules and loop of Henle
    b. descending and ascending limbs of loop of Henle
    c. descending limb of loop of Henle and distal tubule
    d. loop of Henle only
    e. distal and collecting tubules

69. Which of the following is not a function of the kidneys?
    a. excrete the end products of bodily metabolism.
    b. maintain proper plasma volume.
    c. secrete aldosterone to regulate sodium balance in the body.
    d. maintain proper osmolarity of body fluids.
    e. assist in maintaining the proper acid-base balance of the body.

70. The functional unit of the kidney is:
    a. the renal medulla
    b. the nephron
    c. the countercurrent system
    d. the loop of Henle
    e. the glomerulus

71. Which of the following is not part of the nephron?
    a. the glomerulus
    b. the promimal tubule
    c. the renal pelvis
    d. the collecting duct
    e. bowman's capsule

72. The peritubular capillaries:
    a. Supply nutrients and oxygen to the tubular cells.
    b. Take up the substances that are reabsorbed by the tubules.
    c. Supply substances that are secreted by the tubules.
    d. All of the above are correct.
    e. None of the above are correct.

73. The blood that flows through the kidneys:
    a. Is normally about 20-25 percent of the total cardiac output.
    b. Is all filtered through the glomeruli.
    c. Is all used to supply the renal tissue with oxygen and nutrients.
    d. Both a and b are correct.
    e. All of the above (a-c) are correct.

74. The glomerular filtration rate:
    a. Averages 125 ml/min.
    b. Averages 180 L/day.
    c. Represents about 20-25 percent of the total cardiac output.
    d. Both a and b above are correct.
    e. All of the above (a-c) are correct.

75. The glomerular filtrate as it enters Bowman's capsule:
    a. is a protein-free plasma
    b. is identical in composition to urine
    c. contains only substances that are not needed by the body
    d. is formed as a result of active forces
    e. is formed at a constant rate under all circumstances

76. Which of the following does not normally appear in the glomerular filtrate?
    a. plasma proteins
    b. glucose
    c. sodium
    d. urea
    e. calcium

77. The filtration coefficient:
    a. Is a measure of the surface area and permeability of the glomerular membrane.
    b. Is a constant value.
    c. Can be varied by contraction of the podocytes and mesangial cells.
    d. Both a and b above are correct.
    e. Both a and c above are correct.

78. Filtered substances do not pass through which of the following as they move across the glomerular membrane?
    a. Podocytes.
    b. Filtration slits.
    c. None of the above is correct.

79. Which of the following forces oppose glomerular filtration?
    a. Blood-colloid osmotic pressure.
    b. Bowman's capsule hydrostatic pressure.
    c. Glomerular capillary blood pressure.
    d. Both a and b above are correct.
    e. Both a and c above are correct.

80. Bowman's capsule:
    a. Filters water and solute from the blood.
    b. Exerts a hydrostatic pressure that opposes filtration.
    c. Exerts a hydrostatic pressure that favors filtration.
    d. Both a and b above are correct.
    e. Both a and c above are correct.

81. Which of the following statements concerning the process of glomerular filtration is correct?
    a. Bowman's capsule hydrostatic pressure opposes filtration.
    b. The glomerular filtration rate is limited by a tubular max.
    c. All the plasma that enters the glomerulus is filtered.
    d. Two of the above are correct.
    e. All of the above (a-c) are correct.

82. The macula densa:
    a. Consists of specialized tubular cells in the juxtaglomerular apparatus.
    b. Consists of specialized arteriolar smooth-muscle cells in the juxtaglomerular apparatus.
    c. Secretes renin
    d. Both a and c above are correct.
    e. Both b and c above are correct.

83. Tubular reabsorption:
    a. Refers to the movement of a substance from the peritubular capillary blood into the tubular fluid.
    b. Occurs by either active or passive transport.
    c. Involves the process of transepithelial transport.
    d. Both a and c above are correct.
    e. All of the above (a-c) are correct.

84. The vessels that substances enter during tubular reabsorption are the:
    a. afferent arterioles
    b. efferent arterioles
    c. peritubular capillaries
    d. glomerular capillaries
    e. collecting tubules

85. The proximal tubule:
    a. Reabsorbs about 65 percent of the filtered water.
    b. Is the site of action of aldosterone.
    c. Is the location where glucose is reabsorbed.
    d. Both a and c above are correct.
    e. All of the above (a-c) are correct.

86. Which of the following plasma constituents is not regulated by the kidneys?
    a. glucose
    b. sodium
    c. hydrogen
    d. phosphate
    e. water

87. Which of the following substances exhibits a tubular max?
    a. sodium
    b. chloride
    c. amino acids
    d. urea
    e. water

88. The juxtaglomerular apparatus:
    a. Is a combination of specialized tubular and vascular cells.
    b. Secretes aldosterone.
    c. Secretes renin.
    d. Both a and b above are correct.
    e. Both a and c above are correct.

89. Aldosterone:
    a. Stimulates sodium reabsorption in the distal and collecting tubules.
    b. Is secreted by the juxtaglomerular apparatus.
    c. Stimulates potassium secretion in the distal tubules.
    d. Both a and b above are correct.
    e. Both a and c above are correct.

90. Aldosterone secretion:
    a. Occurs in the kidneys.
    b. Is stimulated by angiotensin II.
    c. Is controlled by the plasma concentration of chloride ion.
    d. All of the above (a-c) are correct.
    e. None of the above are correct.

91. Sodium reabsorption in the distal portions of the nephron is stimulated by:
    a. atrial natriuretic peptide
    b. vasopressin
    c. aldosterone
    d. renin
    e. angiotensin II

92. Which of the following does not play a role in sodium reabsorption?
    a. renin
    b. vasopressin
    c. angiotensin
    d. aldosterone
    e. atrial natritic peptide

93. The distal and collecting tubules:
    a. Are the site of the co-transport carriers for glucose and amino acid reabsorption.
    b. Are the site of the organic ion secretory systems.
    c. Are the site of aldosterone and vasopressin action.
    d. Both a and c above are correct.
    e. Both b and c above are correct.

94. Water reabsorption is under the control of vasopressin:
    a. along the entire length of the nephron.
    b. only in the loop of Henle.
    c. only in the distal and collecting tubules.
    d. only in the proximal tubule.
    e. only in the glomerulus.

95. Water reabsorption:
    a. Occurs passively by osmosis in the proximal tubule.
    b. Is under control of vasopressin in the distal and collecting tubules.
    c. Occurs by active transport in the distal and collecting tubules.
    d. Both a and b above are correct.
    e. All of the above (a-c) are correct.

96. Water reabsorption:
    a. Cannot occur from any portion of the nephron in the absence of vasopressin.
    b. Occurs to the greatest extent in the proximal convoluted tubule.
    c. Is under vasopressin control in the proximal tubule.
    d. Is under vasopressin control in the distal and collecting tubules.
    e. Both b and c above are correct.

97. Urea:
    a. Is the waste product with the smallest molecular size in the glomerular filtrate.
    b. Is in greater concentration at the end of the proximal tubule than in other body fluids.
    c. Clearance rate is less than the GFR.
    d. Both a and b above are correct.
    e. All of the above (a-c) are correct.

98. Urea:
    a. Reabsorption occurs by active transport.
    b. Is only 50% reabsorbed in the proximal tubule because the carrier has a low tubular max.
    c. Is a waste product of protein metabolism
    d. Both b and c above are correct.
    e. All of the above (a-c) are correct.

99. Secretion of foreign substances such as drugs generally occurs in the:
    a. distal tubule
    b. loop of Henle
    c. proximal tubule
    d. collecting duct
    e. glomerulus

100. Potassium:
    a. Is actively reabsorbed in the proximal tuble.
    b. Is actively secreted in the distal and collecting tubules.
    c. Secretion is controlled by aldosterone.
    d. Both b and c above are correct.
    e. All of the above are correct.

**Modified Multiple Choice**

*Indicate whether the first item in the statement increases, decreases, or has no effect on the second item by filling in the appropriate letter using the following code.*

A = increases
B = decreases
C = has no effect on

101. Increased osmolarity of body fluids _____ vasopressin secretion.
102. Decreased vasopressin secretion _____ water reabsorption.
103. Decreased sodium in body fluids (sodium depletion) _____ renin secretion.
104. Increased renin secretion _____ angiotensin I activation.
105. Increased vasopressin secretion _____ urinary output.
106. Increased angiotensin II activation _____ aldosterone secretion.
107. Increased aldosterone secretion _____ sodium reabsorption.
108. Increased vasopressin secretion _____ sodium reabsorption.

*Indicate which substance in the right column undergoes the process in the left column by writing the appropriate letter in the blank.*

_____ 109. filtered and actively reabsorbed
_____ 110. filtered and passively reabsorbed
_____ 111. filtered and secreted, bit not reabsorbed
_____ 112. filtered and both actively reabsorbed and actively secreted
_____ 113. filtered but not reabsorbed or secreted
_____ 114. not filtered

A. potassium
B. glucose
C. inulin
D. plasma protein
E. urea
F. hydrogen ion

## Points to Ponder

1. What is the advantage of sweat containing salt and other ions? Would water work just as well?

2. What makes the urine yellow?

3. Mammals excrete urea, uric acid, and creatinine as liquids while other vertebrates excrete these waste products as solids. Is there an advantage to this?

4. What would happen if we cut one ureter and inserted the proximal end into a vein so that the urine flows directly back into the blood? The other kidney would function normally in this scenario.

5. How often does the average adult urinate each day? A baby? How can you explain this difference?

6. Why does drinking beer stimulate the need to urinate?

7. How and when does the autonomic nervous system regulate kidney function?

8. Why does the glomerular ultrafiltrate have a low protein concentration?

9. What is the functional significance of the vasa recta and the countercurrent exchange mechanism?

10. What happens to urinary bicarbonate excretion when a person hyperventilates?

## Clinical Perspective

1. A favorite fraternity pledge prank is to put methyl blue in the coffee of your big brother Explain why the urine turns blue.

2. Why does urinary incontinence occur more often in the elderly?

3. How would you explain to your fraternity brother how kidney stones form?

4. How would you explain to your classmate what diabetes insipidus is and why these patients drink a lot of fluids?

5. Why is creatinine used clinically as an index of kidney function?

6. Based on what the kidney does to penicillin, why must large amounts of penicillin be given to an individual in order to be effective?

7. Why does removal of both adrenal glands cause hyperkalemia?

8. One of the complications of diuretics is hypokalemia. Why does this occur?

9. The drug acetazolamide is used to treat acute mountain sickness. Its pharmacology involves inhibiting renal carbonic anhydrase. How does this drug work in the kidney?

10. Suppose a student has a history of polycystic kidney disease. This student will also have protenuria and an elevated blood creatinine and reduced inulin clearance. What do these physiological parameters indicate?

11. Mammals excrete urea, uric acid, and creatinine as liquids while other vertebrates excrete these waste products as solids. Is there an advantage to this?

## Experiment of the Day

1. Eat some asparagus. Note the time of ingestion. Drink plenty of fluids. During the next 24 hours check your urine for any noticeable changes. Note the time that the changes are first observed. Explain the changes should you become aware of any. If you observe no changes, ask a classmate for their changes.

## PhysioEdge Activities

**Related to Text:**
No tutorials or media exercises directly related to text.

**Related to Figures:**
*Figure 14.1*. For an animation of this figure, click the Anatomy of the Kidney tab in the Renal Physiology tutorial. For an interaction related to this figure, see Media Exercise 14.2: The Kidney.
*Figure 14.3*. For an interaction related to this figure, see Media Exercise 14.3: The Nephron.
*Figure 14.6*. For an animation of this figure, click the Urine Formation tab in the Renal Physiology tutorial.
*Figure 14.7*. For an interaction related to this figure, see Media Exercise 14.1: The Urinary System.
*Figure 14.16*. For an interaction related to this figure, see Media Exercise 14.1: The Urinary System.
*Figure 14.23*. For an animation of this figure, click the Urine Formation tab in the Renal Physiology tutorial.
*Figure 14.25*. For an animation of this figure, click the Vertical Osmotic Gradient and Countercurrent Multiplication tab in the Renal Physiology tutorial.
*Figure 14.27*. For an animation of this figure, click the Reabsorption and Urine Concentration tab in the Renal Physiology tutorial.

## Media Resources

**PhysioEdge CD-ROM**
For a visual review of concepts in this chapter, check out:

Tutorial: Renal Physiology
Media Exercise 14.1: The Urinary System
Media Exercise 14.2: The Kidney
Media Exercise: 14.3: The Nephron

**Book Companion Website**
The website for this book contains a wealth of helpful study aids, as well as many ideas for further reading and research. Log on to:

**http://www.brookscole.com/sherwoodhp6**

Select Chapter 14 from the drop-down menu, or click on one of the many resources areas, including **Case Histories**, which introduce clinical aspects of human physiology. For this chapter check out Case History 11: Congestive Heart Failure; Case History 12: Urinary Tract Infection.

For Suggested Readings, consult **InfoTrac College Edition**, your online research library, at:

**http://infotrac.thomsonlearning.com**

15

# Fluid and Acid-Base Balance

## Chapter Overview

Homeostasis depends on maintaining a balance between the input and output of all constituents in the internal fluid environment. As changes occur, homeostasis is maintained by body systems responding to those changes. This chapter deals with two very important areas of change: (1) water (control of solute concentration and salt balance), and (2) pH.

Water is located primarily in two major compartments: (1) extracellular fluid (plasma and interstitial fluid), and (2) intracellular fluid. All of these have different amounts of water and ions. To regulate and maintain these fluid balances, adjustments are made in the extracellular volume and the extracellular osmolarity.

Acid-base balance is very critical in the survival of cells. The somewhat narrow pH range is maintained by four buffer systems: carbonic acid/carbonate buffer system, protein buffer system, hemoglobin buffer system, and the phosphate buffer system.

The buffers are the first line of defense against pH changes. The second line is the lungs. By removing carbon dioxide less carbonic acid is added to body fluids. The last line of defense in the prevention of change in pH is the kidney. The big role of the kidney is to excrete the necessary amounts of hydrogen ions, bicarbonate ions, and ammonium ions. Maintaining a relatively constant pH is truly remarkable considering the great variability of substances brought into and produced by the body.

## Chapter Outline

### BALANCE CONCEPT

*The internal pool of a substance is the amount of that substance in the ECF.*

- The quantity of any particular substance in the ECF is considered to be a readily available internal pool.
- The ECF pool can further be altered by transferring a particular constituent into storage within the cells or bones.

*To maintain stable balance of an ECF constituent, its input must equal its output.*

- If the quantity of a substance is to remain stable within the body, its input by means of ingestion or metabolic production must be balanced by an equal output by means of excretion or metabolic consumption.

### FLUID BALANCE

- Water is the most abundant component of the human body.
- Known as the balance concept, this relationship is extremely important in the maintenance of homeostasis.
- If the body as a whole has a surplus or a deficit of a particular stored substance, the storage site can be expanded or partially depleted to maintain the extracellular fluid concentration of the substance within homeostatically prescribed limits.
- When total body input of a particular substance equals its total body output, a stable balance exists.
- When the gains via input exceed its losses via output, a positive balance exists.
- When the losses of a substance exceed its gains, a negative balance exists.

*Body water is distributed between the ICF and ECF.*

- The intracellular fluid compartment comprises about two-thirds of the total body water.
- The remaining one-third of the body water found in the extracellular fluid is further subdivided into plasma and interstitial fluid.
- Two other major categories are included in the extracellular fluid compartment: lymph and transcellular fluid.
- Transcellular fluid consists of a number of small specialized fluid volumes: cerebrospinal fluid, intraocular fluid, synovial fluid, pericardial fluid, pleural fluid, peritoneal fluid, and the digestive juices.
- Although these fluids are extremely important functionally, they represent an insignificant fraction of the total body water.

*The plasma and interstitial fluid are similar in composition but the ECF and ICF differ markedly.*

- Plasma and interstitial fluid are nearly identical in composition, except that interstitial fluid lacks plasma proteins.
- Any change in one of these ECF compartments is quickly reflected in the other compartment because they are constantly mixing.
- The composition of the ECF differs considerably from that of the ICF.
- Among the major differences are: (1) the presence of cellular proteins in the ICF that are unable to permeate the enveloping membranes to leave the cells, and (2) the unequal distribution of sodium and potassium and their attendant anions as a result of the action of the membrane-bound sodium-potassium ATPase pump that is present in all cells.
- Sodium is the primary ECF cation and potassium is primarily found in the ICF.
- In the extracellular fluid, sodium is accompanied primarily by the anion chloride and to a lesser extent by bicarbonate.
- The major intracellular anions are phosphate and the negatively charged proteins trapped within the cell.
- The movement of water between the plasma and the interstitial fluid across capillary walls is governed by relative imbalances between capillary blood pressure and colloid osmotic pressure.
- The net transfer of water between the interstitial fluid and the intracellular fluid across the cellular plasma membranes occurs as a result of osmotic effects alone.

*Fluid balance is maintained by regulating ECF volume and osmolarity.*

- Any control mechanism that operates on the plasma in effect regulates the entire extracellular fluid.
- The ICF, in turn, is influenced by changes in the ECF to the extent permitted by the permeability of the membrane barriers surrounding the cells.
- The factors regulated to maintain fluid balance in the body are ECF volume and ECF osmolarity.
- Extracellular fluid volume must be closely regulated to help maintain blood pressure.
- Extracellular fluid osmolarity must be closely regulated to prevent swelling or shrinking of the cells.

*Control of ECF volume is important in the long-term regulation of blood pressure.*

- Two compensatory measures come into play to transiently adjust the blood pressure until the ECF volume can be restored to normal: (1) baroreceptor reflex mechanisms alter both cardiac output and total peripheral resistance through autonomic nervous system effects on the heart and blood vessels, and (2) fluid shifts occur temporarily and automatically between the plasma and interstitial fluid.
- A reduction in plasma volume is partially compensated for by a shift of fluid out of the interstitial compartment into the blood vessels, thereby expanding the circulating plasma volume at the expense of the interstitial compartment.
- Conversely, when the plasma volume is too large, much of the excess fluid is shifted into the interstitial compartment.
- Responsibility for long-term regulation of blood pressure rests with the kidneys and the thirst mechanism, which control urinary output and fluid intake.

*Control of salt balance is primarily important in regulating ECF volume.*

- The total mass of sodium salts in the ECF determines the ECF's volume, and, appropriately, regulation of the ECF volume depends primarily on controlling salt balance.
- To maintain salt balance at a set level, salt input must equal salt output.
- Since we typically consume salt in excess of our needs, it is obvious that salt intake in humans is not well controlled.
- The three avenues for salt output are obligatory loss of salt in sweat and feces and controlled excretion of salt in the urine.

*The amount of sodium filtered is controlled through regulation of the GFR.*

- By regulating the rate of urinary salt excretion, the kidneys keep the total sodium mass in the ECF constant despite any notable changes in dietary intake of salt or unusual losses through sweating, diarrhea, or other means.
- The amount of sodium excreted in the urine represents the amount of sodium that is filtered but is not subsequently reabsorbed.
- The kidneys accordingly adjust the amount of salt excreted by controlling two processes: (1) the glomerular filtration rate, and (2) the tubular reabsorption of sodium.
- The afferent arterioles that supply the renal glomeruli are constricted as part of the generalized vasoconstriction aimed at elevating a reduced blood pressure.
- As a result of reduced blood flow into the glomeruli, the glomerular filtration rate decreases and, accordingly, the amount of sodium and accompanying fluid that are filtered decreases.

- Excretion of salt and fluid are diminished.
- An elevation in ECF volume and arterial blood pressure is reflexly countered by a baroreceptor reflex response that leads to an increase in glomerular filtration rate, which in turn results in enhanced salt and fluid excretion.
- Baroreceptors that monitor fluctuations in blood pressure are responsible for bringing about adjustments in the amounts of sodium filtered and eventually excreted.

*The amount of sodium reabsorbed is controlled through the renin-angiotensin-aldosterone system.*

- The main factor controlling the extent of sodium reabsorption in the distal and collecting tubule is the powerful renin-angiotensin-aldosterone system.
- A fall in arterial blood pressure brings about a twofold effect in the renal handling of sodium: (1) a reflex reduction in the glomerular filtration rate to decrease the amount of sodium filtered, and (2) a hormonally adjusted increase in the amount of sodium reabsorbed.
- A rise in arterial blood pressure brings about: (1) increases in the amount of sodium filtered, and (2) a reduction in renin-angiotensin-aldosterone activity, which decreases salt and water reabsorption.

*Control of ECF osmolarity prevents changes in ICF volume.*

- The osmotic activity across the capillary wall is due to the unequal distribution of plasma proteins.
- Plasma proteins are present only in the plasma.
- Osmotic activity across the cellular plasma membranes is directly related to any differences in ionic concentration between the ECF and ICF.
- Plasma proteins play no role in osmosis of water across the cell membranes because they are absent in both the interstitial fluid and the ICF that are separated by cell membranes.
- Sodium and its attendant anions, being by far the most abundant solutes in the ECF in terms of numbers of particles, account for the vast majority of the ECF's osmotic activity.
- Potassium and its accompanying intracellular anions are responsible for the ICF's osmotic activity.
- Normally, the osmolarities of the ECF and ICF are the same.

*During ECF hypertonicity, the cells shrink as water leaves them.*

- Hypertonicity of the ECF, or the excessive concentration of ECF solutes, is usually associated with dehydration, or a negative free water balance.
- Dehydration with accompanying hypertonicity can be brought about in three major ways: (1) insufficient water intake, (2) excessive water loss, and (3) diabetes insipidus.
- Whenever the ECF compartment becomes hypertonic, water moves out of the cells by osmosis into the more concentrated ECF until the ICF osmolarity equilibrates with the ECF.
- The cells shrink as water leaves them.
- Shrinking of brain neurons causes disturbances in brain function.
- Circulatory problems may range from slight reduction in blood pressure to circulatory shock and death.

*During ECH hypotonicity, the cells swell as water enters them.*

- Hypotonicity of the ECF is usually associated with over hydration, that is, excess of free water is present.
- Hypotonicity can arise in three ways: (1) Patients with renal failure who are unable to excrete a dilute urine become hypotonic when they consume relatively more water than solutes. (2) Hypotonicity can occur transiently in healthy people if water is rapidly ingested to such an excess that the kidneys are unable to respond quickly enough to eliminate the extra water. (3) Hypotonicity can occur when excess water without solute is retained in the body as a result of inappropriate secretion of vasopressin.
- The resultant difference in osmotic activity between ECF and ICF induces water to move by osmosis from the more dilute ECF into the cells, with the cells swelling.
- Pronounced swelling in brain cells also leads to brain dysfunction and circulatory disturbances, including hypertension and edema.
- When an isotonic fluid is injected into the ECF compartment, the ECF volume increases, but the concentration of ECF solutes remains unchanged the ECF and ICF are still in osmotic equilibrium.
- In the case of an isotonic fluid loss such as occurs in hemorrhage, the loss is confined to the ECF with no corresponding loss of fluid from the ICF.

*Control of water balance by means of vasopressin is important in regulating ECF osmolarity.*

- Of the many sources of water input and output only two can be regulated to maintain water balance: (1) control of water input by thirst, and (2) control of water output in the urine by vasopressin.
- A thirst center is located in the hypothalamus in close proximity to the vasopressin-secreting cells.

- Thirst increases water input, whereas vasopressin, by reducing urine production, decreases water output.
- Vasopressin and thirst are both stimulated by a free water deficit and suppressed by a free water excess.
- The predominant excitatory input for both vasopressin secretion and thirst come from the hypothalamus osmoreceptors located near the vasopressin-secreting cells and thirst center.
- Left atrial volume receptors monitor the blood pressure, which is a reflection of the ECF volume.

*Vasopressin secretion and thirst are largely triggered simultaneously.*

- In response to a major reduction in ECF volume and arterial pressure, as during hemorrhage, the left atrial volume receptors reflexly stimulate both vasopressin secretion and thirst.
- The outpouring of vasopressin and increased thirst lead to decreased urine output and increased fluid intake.
- Vasopressin exerts a potent vasoconstrictor effect on arterioles in addition to having an effect on the kidney tubules.
- Vasopressin and thirst are both inhibited when the ECF/plasma volume and arterial blood pressure are elevated.
- Aldosterone controlled sodium reabsorption is the most important factor in regulating ECF volume, with the vasopressin and thirst mechanisms playing only a supportive role.
- Another stimulus for increasing both thirst and vasopressin is angiotensin II.
- Angiotensin II, in addition to stimulating aldosterone secretion, acts directly on the brain to give rise to the urge to drink and concurrently stimulates vasopressin to enhance renal water reabsorption.
- Several factors affect vasopressin secretion, but not thirst.
- Vasopressin is stimulated by stress-related inputs, such as pain, fear, and trauma, which have nothing directly to do with maintaining water balance.
- Alcohol inhibits vasopressin secretion and can lead to ECF hypertonicity by promoting excessive free water excretion.

## ACID-BASE BALANCE

*Acids liberate free hydrogen ions whereas bases accept them.*

- Acids are a special group of hydrogen-containing substances that dissociate when in solution to liberate free hydrogen ions.
- A base is a substance that can combine with a free hydrogen ion thus removing it from solution.

*The pH designation is used to express [H].*

- The pH equals the logarithm to the base 10 of the reciprocal of the hydrogen ion concentration.
- The greater the hydrogen ion concentration the lower the pH.
- Every unit change in pH actually represents a tenfold change in hydrogen ion concentration
- Solutions having a pH less than 7.0 are considered to be acidic.
- Solutions having a pH greater than 7.0 are considered to be basic or alkaline.

*Fluctuations in hydrogen-ion concentration alter nerve, enzyme, and potassium activity.*

- Changes in excitability of nerve and muscle cells are among the major clinical manifestations of pH abnormalities.
- Hydrogen-ion concentration exerts a marked influence on enzyme activity.
- Changes in hydrogen-ion concentration influence potassium levels in the body.

*Hydrogen ions are continually being added to the body fluids as a result of metabolic activities.*

- Normally hydrogen is continually being added to the body fluids from the three following sources: (1) carbonic acid formation, (2) inorganic acids produced during the breakdown of nutrients, and (3) organic acids resulting from intermediary metabolism.
- Three lines of defense against changes in hydrogen ion concentration operate to maintain the hydrogen ion concentration of body fluids at a nearly constant level despite unregulated input: (1) the chemical buffer systems, (2) the respiratory mechanism of pH control, and (3) the renal mechanism of pH control.

*Chemical buffer systems minimize changes in pH by binding with or yielding free hydrogen ions.*

- A chemical buffer system is a mixture in a solution of two chemical compounds that minimize pH changes when either an acid or a base is added or removed from the solution.
- There are four buffer systems in the body: (1) the carbonic acid/bicarbonate buffer system, (2) the protein buffer system, (3) the hemoglobin buffer system, and (4) the phosphate buffer system.

*The carbonic/bicarbonate buffer pair is the primary ECF buffer for non-carbonic acids.*

- The carbonic acid/bicarbonate buffer pair is the most important buffer system in the ECF for buffering pH changes brought about by causes other than fluctuations in carbon dioxide-generated carbonic acid.

*The protein buffer system is primarily important intracellularly.*

- The protein buffer system is primarily important intracellularly.

*The hemoglobin buffer system buffers hydrogen generated from carbonic acid.*

- The hemoglobin buffer system buffers hydrogen ion generated from carbonic acid.

*The phosphate buffer system is an important urinary buffer.*

- The phosphate buffer system is an important urinary buffer.
- The phosphate buffer system is composed of an acid phosphate salt and a basic phosphate salt.
- This system contributes significantly to intracellular buffering.
- The phosphate system serves as an excellent urinary buffer.

*Chemical buffer systems act as the first line of defense against changes in hydrogen ion concentration.*

- All chemical buffer systems act immediately, within fractions of a second, to minimize changes in pH.

*The respiratory system regulates hydrogen-ion concentration by controlling the rate of carbon dioxide removal.*

- When arterial hydrogen ion concentration increases, the respiratory center in the brain stem is reflexly stimulated to increase pulmonary ventilation.
- When arterial hydrogen ion concentration falls, pulmonary ventilation is reduced.
- Carbon dioxide diffuses from the cells into the blood faster than it is removed from the blood by the lungs, so higher-than-usual amounts of acid-forming carbon dioxide accumulate in the blood, thus restoring hydrogen ion concentration toward normal.

*The respiratory system serves as the second line of defense against changes in [H].*

- If a deviation in hydrogen ion concentration is not swiftly and completely corrected by the buffer systems, the respiratory system comes into action a few minutes later, thus serving as the second line of defense against changes in hydrogen ion concentration.

*The kidneys are a powerful third line of defense against changes in [H].*

- By simultaneously removing acid from and adding base to the body fluids, the kidneys are able to restore the pH toward normal more efficiently than the lungs.
- Also contributing to the kidneys' acid-base regulatory potency is their ability to return the pH almost exactly to normal.

*The kidneys adjust their rate of $H^+$ secretion.*

- The kidneys control the pH of the body fluids by adjusting three interrelated factors: (1) hydrogen excretion, (2) bicarbonate excretion, and (3) ammonia secretion.
- The magnitude of hydrogen secretion depends on a direct effect of the plasma acid-base status on the kidneys' tubular cells.

*The kidneys conserve or excrete bicarbonate ions depending on the plasma hydrogen ion concentration.*

- When the hydrogen ion concentration of the plasma passing through the peritubular capillaries is elevated above normal, the tubular cells respond by secreting greater-than-usual amounts of hydrogen from the plasma into the tubular fluid to be excreted in the urine.
- When the plasma hydrogen ion concentration is lower than normal, the kidneys conserve hydrogen by reducing their secretion and subsequent excretion in the urine.
- The kidneys regulate plasma bicarbonate ion concentration by two interrelated mechanisms: (1) variable reabsorption of the filtered bicarbonate back into the plasma, and (2) variable addition of new bicarbonate to the plasma.
- When the plasma hydrogen ion concentration is increased above normal during acidosis, renal compensation includes the following: (1) increased secretion and subsequent increased excretion of hydrogen in the urine, thereby eliminating the excess hydrogen and decreasing plasma hydrogen ion concentration, and (2) reabsorption of all filtered bicarbonate, plus addition of new bicarbonate to the plasma, resulting in increased plasma bicarbonate ion concentration.
- When plasma hydrogen ion concentration is reduced below normal during alkalosis, renal responses include the following: (1) decreased secretion and subsequent reduced excretion of hydrogen in the urine, resulting in conservation of hydrogen and increased plasma hydrogen ion concentration, and (2) incomplete reabsorption of filtered bicarbonate and subsequent increased

excretion of bicarbonate, resulting in reduction of plasma bicarbonate ion concentration.
- There are two important urinary buffers: (1) filtered phosphate buffers, and (2) secreted ammonia.
- Normally, secreted hydrogen is first buffered by the phosphate buffer system.

*The kidneys secrete ammonia during acidosis to buffer secreted $H^+$.*
- When acidosis exists, the tubular cells secrete ammonia into the tubular fluid once the normal urinary phosphate buffers are saturated.
- Ammonia enables the kidneys to continue secreting additional hydrogen ions because ammonia combines with free hydrogen in the tubular fluid to form ammonium ions.
- The tubular membranes are not very permeable to ammonium, so the ammonium ions remain in the tubular fluid and are lost in the urine, each one taking a hydrogen ion with it.

*Acid-base imbalances can arise from either respiratory dysfunction or metabolic disturbances.*
- Respiratory acidosis is the result of abnormal carbon dioxide retention arising from hypoventilation.
- The primary defect in respiratory alkalosis is excessive loss of carbon dioxide from the body as a result of hyperventilation.
- The following are the most common causes of metabolic acidosis: (1) severe diarrhea, (2) diabetes mellitus, (3) strenuous exercise, and (4) uremic acidosis.
- Metabolic alkalosis arises most commonly from the following: (1) vomiting and (2) ingestion of alkaline drugs

## Key Terms

acidic
acidosis
acids
alkalosis
ammonia
ammonium ions
balance concept
bases
basic (alkaline)
chemical buffer system
dehydration
diabetes insipidus
dissociation constant
extracellular fluid
fluid balance
Henderson-Hasselbach equation
hypothalamic osmoreceptors
insensible loss
intracellular fluid
interstitial fluid
left atrial volume receptors
metabolic acidosis
metabolic alkalosis
negative balance
osmolarity
overhydration
metabolic water
pH
plasma
positive balance
respiratory acidosis
respiratory alkalosis
salt balance
stable balance
thirst center
transcellular fluid
vasopressin
water balance
water intoxication

## Review Exercises

*Answers are in the appendix.*

**True/False**

_____ 1. When the gains via input exceed its losses via output, a negative balance exists.

_____ 2. The ICF compartment comprises about one-half of the total body water.

_____ 3. Plasma is a transcellular fluid.

_____ 4. The major intracellular anions are phosphate ions.

_____ 5. Diabetes insipidus is a disease characterized by a deficiency in vasopressin.

_____ 6. Vasopressin decreases water output by reducing urine production.

_____ 7. Vasopressin gets its name because it exerts a potent vasoconstrictor effect on arterioles.

_____ 8. Bases are a special group of hydrogen containing substances that dissociate when in solution to liberate free hydrogen

_____ 9. A base is a substance that can combine with free hydrogen and thus remove it from solution.

_____ 10. The greater the hydrogen ion concentration, the lower the pH.

_____ 11. The first line of defense against change in pH is the respiratory control of pH.

_____ 12. Tubular cells secrete ammonium ions that can combine with free hydrogen to form ammonia.

_____ 13. Metabolic acidosis can be caused by diabetes insipidus.

_____ 14. Metabolic alkalosis can be caused by vomiting.

**Fill in the Blank**

15. When total body input of a particular substance equals its total body output, a(n) ________________________ exists.

16. __________ and ____________________ are nearly identical in composition, except that ____________________ lacks plasma proteins.

17. The movement of water between the plasma and interstitial fluid across capillary walls is governed by relative imbalances between ________________________________ and ________________________________.

18. The three avenues for salt output are __________________________________________________, ____________, and controlled excretion of salt in the ___________.

19. The main factor controlling the extent of sodium reabsorption in the distal tubule and collecting tubule is the powerful _____________________________________ system.

20. Dehydration with accompanying hypertonicity can be brought about in three major ways: (1) _________________, (2) ____________________, and (3) _________________.

21. The predominant excitatory input for both vasopressin secretion and thirst comes from ______________________________.

22. The _______ equals the logarithm to the base 10 of the reciprocal of the hydrogen ion concentration.

23. The _________________________ serves as an excellent urinary buffer.

24. Respiratory acidosis is the result of abnormal carbon dioxide retention arising from _____________.

25. Uremic acidosis causes _________________________.

26. Vomiting causes _________________________.

27. Bicarbonate cannot buffer urinary hydrogen as it does in the ECF because bicarbonate is not excreted in the urine simultaneously with _______________.

28. The kidneys control the pH of the body fluids by adjusting three interrelated factors: (1) _________________, (2) ___________________, and (3) ___________________.

**Matching**

*Match the cause on the left to the acid-base condition on the right.*

_____ 29. diabetes mellitus
_____ 30. hyperventilation
_____ 31. vomiting
_____ 32. ingesting of alkaline drugs
_____ 33. strenuous exercise
_____ 34. uremic acidosis
_____ 35. hypoventilation
_____ 36. severe diarrhea

a. metabolic alkalosis
b. respiratory acidosis
c. metabolic acidosis
d. respiratory alkalosis

**Multiple Choice**

37. Which of the following disorders is characterized by a deficiency of vasopressin?
 a. diabetes mellitus
 b. hemorrhage
 c. hypotonicity
 d. colloidal isotonicity
 e. diabetes insipidus

38. Which of the following disorders or symptoms can be caused by hypotonicity?
 a. diabetes insipidus
 b. hypertension and edema
 c. hemorrhage
 d. diarrhea
 e. diabetes mellitus

39. Which of the following disorders or symptoms causes changes in the ECF but not in the ICF?
 a. diabetes insipidus
 b. hypertension and edema
 c. hemorrhage
 d. dehydration
 e. overhydration

40. Which of the avenues for salt output eliminates the most salt per day?
 a. tears
 b. feces
 c. hemorrhage
 d. sweat
 e. controlled excretion in urine

41. Which is the primary cation found in the ECF?
 a. bicarbonate
 b. phosphate
 c. sodium
 d. chloride
 e. potassium

42. Which is the primary cation found in the ICF?
 a. bicarbonate
 b. sodium
 c. chloride
 d. phosphate
 e. potassium

43. Which of the following organs is responsible for long-term regulation of blood pressure?
    a. lungs
    b. heart
    c. spleen
    d. kidneys
    e. intestines

44. Which fluid compartment has the greatest percentage of body fluid?
    a. lymph
    b. plasma
    c. ICF
    d. ECF
    e. transcellular fluid

45. Which fluid compartment represents the greatest percentage of body fluid within the ECF?
    a. lymph
    b. interstitial fluid
    c. synovial fluid
    d. plasma
    e. cerebrospinal fluid

46. When the plasma volume is too large, to which fluid compartment is the excess fluid shifted?
    a. lymph
    b. interstitial fluid
    c. transcellular fluid
    d. cerebrospinal fluid
    e. none of the above

**Modified Multiple Choice**

*Indicate which fluid imbalance is being described by writing the appropriate letter in the blank using the answer code below.*

A = overhydration
B = dehydration
C = both overhydration and dehydration
D = neither overhydration or dehydration

_____ 47. Symptoms include dry skin, parched tongue, and sunken eyeballs.
_____ 48. Water enters the cells by osmosis.
_____ 49. The body fluids have a lower concentration of solutes than normal.
_____ 50. ECF and ICF become hypertonic.
_____ 51. Cells become swollen.
_____ 52. No fluid shift occurs between the ECF and ICF.
_____ 53. Occurs as a consequence of water deprivation.
_____ 54. Cells shrink.
_____ 55. ECF and ICF become hypertonic.
_____ 56. Convulsions and coma may occur.
_____ 57. Occurs as a consequence of diabetes insipidus.
_____ 58. Occurs as a consequence of heavy vomiting.
_____ 59. Osmolarity of the body fluids is decreased.
_____ 60. Occurs as a consequence of excessive fluid intake.
_____ 61. Occurs as a consequence of excessive vasopressin secretion.
_____ 62. Vasopressin secretion is stimulated as a compensatory mechanism.
_____ 63. Increased urinary output as a compensatory mechanism.

*Indicate which type of acid-base imbalance might occur in each of the following situations by writing the appropriate letter in the blank.*

A = respiratory acidosis
B = respiratory alkalosis
C = metabolic acidosis
D = metabolic alkalosis

_____ 64. fever
_____ 65. excessive ingestion of alkaline drugs
_____ 66. aspirin poisoning
_____ 67. anxiety
_____ 68. severe exercise
_____ 69. uremia
_____ 70. damage to the respiratory center
_____ 71. severe diarrhea
_____ 72. pneumonia
_____ 73. excessive vomiting of gastric contents
_____ 74. diabetes mellitus

*When a person has diarrhea, s/he loses excessive salt and water from the body. This fluid loss results in sodium depletion, dehydration, a decreased extracellular fluid volume, a reduction in plasma volume, and a decreased systemic arterial blood pressure. The following refers to a sequence of events that occur to compensate for this fluid loss. Indicate whether each factor listed*

A = exhibits no change.
B = is increased to compensate for fluid loss.
C = is decreased to compensate for fluid loss.

_____ 75. sympathetic activity to the afferent arterioles of the nephrons
_____ 76. caliber of the afferent arterioles
_____ 77. glomerular capillary blood pressure
_____ 78. net filtration pressure
_____ 79. GFR
_____ 80. amount of sodium and water filtered
_____ 81. renin secretion
_____ 82. angiotensin I and II production
_____ 83. aldosterone secretion
_____ 84. amount of sodium reabsorbed
_____ 85. amount of sodium secreted
_____ 86. vasopressin secretion
_____ 87. permeability of distal and collecting tubules to water
_____ 88. amount of water reabsorbed
_____ 89. amount of water excreted
_____ 90. urinary volume
_____ 91. thirst
_____ 92. amount of potassium lost

*Indicate which acid-base abnormality is represented by the $[HCO_3^-] / [CO_2]$ ratio by writing the appropriate letter in the blank.*

A = uncompensated respiratory acidosis
B = uncompensated metabolic acidosis
C = uncompensated respiratory alkalosis
D = uncompensated metabolic alkalosis

_____ 93. 20/2
_____ 94. 40/1
_____ 95. 10/1
_____ 96. 20/0.5

## *Points to Ponder*

1. As you shop in your favorite grocery store you may select a meat product that has been "sugar cured." How does this process protect the meat and make it safe for you to eat?

2. At present one of the makers of a common antacid boasts that their product uses calcium instead of sodium and that this is better. Do you find any truth in this advertisement? Explain.

3. Most Americans are aware of the caffeine in popular drinks: coffee, tea, and most soft drinks. How are the other ingredients related to this chapter? Are these drinks diuretics? Are these beverages acidic or basic? How does your body handle these fluids?

4. What forces control the movement of water between the plasma and the interstitial fluid?

5. What are the two main forces that move water between compartments?

6. How is water intake regulated?

7. Why can't humans drink sea water to quench our thirst?

8. What is the role of thirst in regulating fluid intake?

9. What is meant by fluid balance?

10. Define the term body fluid.

## *Clinical Perspective*

1. The patient has lost a lot of blood due to the traumatic injuries received in an auto accident. Intravenous fluids are administered. Start with the hemorrhage and explain the sequence of events with reference to this chapter.

2. What causes a blister?

3. Why do high sodium diets tend to cause a person to gain weight?

4. For the sake of convenience, long-distance truck drivers often limit their intake of fluids, such as coffee. During this time, the urine becomes very dark in color. What makes it dark?

5. With respect to the above question (#4), if the truck driver drinks several glasses of water, the urine will be much lighter in color. Why does increased fluid intake lighten the urine?

6. How could a person go into hypochloremia? Hyponatremia?

7. How are alkalosis and acidosis compensated and treated?

8. What are the main physiological effects of acidosis and alkalosis?

9. What are some common causes of respiratory acidosis?

10. What are some characteristics of hypocalcemia?

## Experiment of the Day

Equipment:
3 glasses
3 dried prunes
sugar
salt

1. Place water in all three glasses. Place several spoons of salt into one glass and several spoons of sugar into still another. The third glass has plain water. Drop a prune into each glass. Note time. Explain the results.

## PhysioEdge Activities

**Related to Text:**
No tutorials or media exercises directly related to text.

**Related to Figures:**
*Figure 15.2.* For an interaction related to this figure, see Media Exercise 15.1: Fluid and Electrolyte Balance.
*Figure 15.3.* For an interaction related to this figure, see Media Exercise 15.1: Fluid and Electrolyte Balance.
*Figure 15.6.* For an interaction related to this figure, see Media Exercise 15.2: Basics of Acid-Base Balance.

## Media Resources

**PhysioEdge CD-ROM**
For a visual review of concepts in this chapter, check out:

Media Exercise 15.1: Fluid and Electrolyte Balance
Media Exercie 15.2: Basics of Acid\-Base Balance

**Book Companion Website**
The website for this book contains a wealth of helpful study aids, as well as many ideas for further reading and research. Log on to:

**http://www.brookscole.com/sherwoodhp6**

Select Chapter 15 from the drop-down menu, or click on one of the many resources areas, including **Case Histories**, which introduce clinical aspects of human physiology. For this chapter check out Case History 5: Eight Years Later.

For Suggested Readings, consult **InfoTrac® College Edition**, your online research library, at:

**http://infotrac.thomsonlearning.com**

# 16

# The Digestive System

## Chapter Overview

To maintain homeostasis, nutrient molecules used for energy production must continually be replaced by the acquisition of new, energy-rich nutrients. Cells need these nutrients, vitamins, and minerals to survive. The digestive system meets these cellular needs by removing the substances from the external environment and presenting them to the cells in a form that the cells can utilize. The food systematically passes through the mouth, pharynx, esophagus, stomach, small intestine, large intestine, and anus. The digestive tract receives enzymes and other secretions from the salivary glands, the exocrine pancreas, the liver, and exocrine cells in the digestive tract's epithelial lining. During this trip, carbohydrates, proteins and fats are broken down by various enzymes and other substances to yield monosaccharides, amino acids, and monoglycerides respectively. These smaller nutrient molecules are then absorbed into the small intestine and pass to the liver or lymphatic system. The liver performs the final modification on the nutrient molecules. Undigested or unutilized material, along with bacteria is passed to the outside through the anus. Nutrition prepared by the digestive system is another important step in homeostasis. The digestive system is under the control of the central nervous system and several hormones.

## Chapter Outline

INTRODUCTION

- The primary function of the digestive system is to transfer nutrients, water, and electrolytes from the food we eat into the body's internal environment.

*The digestive system performs four basic digestive processes.*

- There are four basic digestive processes: (1) motility, (2) secretion, (3) digestion, and (4) absorption.
- Propulsive movements push the contents forward through the digestive tract at varying speeds.
- Mixing movements serve a twofold function.
- First, by mixing food with the digestive juices, these movements promote digestion of the food.
- Second, they facilitate absorption by exposing all portions of the intestinal contents to the absorbing surfaces of the digestive tract.
- Each digestive secretion consists of water, electrolytes, and specific organic constituents that are important in the digestive process, such as enzymes, bile salts, or mucus.
- Digestion refers to the breaking-down process whereby the structurally complex foodstuffs of the diet are converted to smaller absorbable units by the enzymes produced within the digestive system.
- Digestion is accomplished by enzymatic hydrolysis.
- Through the process of absorption, the small absorbable units that result from digestion, along with water, vitamins, and electrolytes, are transferred from the digestive tract lumen into the blood or lymph.

*The digestive tract and accessory digestive organs make up the digestive system.*

- The accessory digestive organs include the salivary glands, the exocrine pancreas, and the biliary system (liver and gallbladder).
- The digestive tract includes the following organs: mouth, pharynx, esophagus, stomach, small intestine, large intestine, and anus.

*The digestive tract wall has four layers.*

- A cross section of the digestive tube reveals four tissue layers: (1) mucosa, (2) submucosa, (3) muscularis externa, and (4) serosa.
- The mucosa lines the luminal surface of the digestive tract and is divided into three layers: (1) mucous membrane, (2) lamina propria, and the (3) muscularis mucosa.

*Regulation of digestive function is complex and synergistic.*

- Four factors are involved in the regulation of digestive system function: (1) autonomous smooth muscle function, (2) intrinsic nerve plexuses, (3) extrinsic nerves, and (4) gastrointestinal hormones.

- The prominent type of self-induced electrical activity in digestive smooth muscle is slow-wave potentials, alternately referred to as the digestive tract's basic electrical rhythm or pacesetter potential.
- The interstitial cells of Cajal are the pacesetter cells responsible for instigating cyclic slow-wave activity.
- The second factor involved in the regulation of digestive tract function is the intrinsic nerve plexuses.
- Two major networks of nerves form the plexus of the digestive tract: (1) the myenteric plexus and (2) the submucous plexus.
- The two plexuses are often termed the enteric nervous system.
- Sensory neurons are present that possess receptors that respond to specific local stimuli in the digestive tract.
- Other neurons innervate the smooth muscle cells and exocrine and endocrine cells of the digestive tract.
- These input and output neurons of the enteric nervous system are linked by interneurons.
- Some of the output neurons are excitatory and some are inhibitory.
- Neurons that release acetylcholine as a neurotransmitter promote contraction of digestive tract smooth muscle.
- The neurotransmitters nitric oxide and vasoactive intestinal peptide act in concert to cause relaxation.
- The extrinsic nerves are from both branches of the autonomic system.
- The sympathetic system tends to inhibit or slow down digestive tract contraction and secretion.
- The parasympathetic nervous system dominates in quiet, relaxed situations when general maintenance types of activities such as digestion can proceed optimally.
- Parasympathetic nerve fibers, which arrive primarily by way of the vagus nerve, tend to increase smooth muscle motility and promote secretion of digestive enzymes and hormones.
- Tucked within the mucosa of certain regions of the digestive tract are endocrine gland cells that release hormones into the blood upon appropriate stimulation.
- Gastrointestinal hormones are carried to other areas of the digestive tract where they exert either excitatory or inhibitory influences on smooth muscle and exocrine gland cells.
- Gastrointestinal hormones are released primarily in response to specific local changes in the luminal contents, acting either directly on the endocrine gland cells or indirectly through the intrinsic plexuses or extrinsic autonomic nerves.

*Receptor activation alters digestive activity through neural reflexes and hormonal pathways.*

- The wall of the digestive tract contains three different types of sensory receptors that respond to local chemical or mechanical changes in the digestive tract: (1) chemoreceptors, (2) mechanoreceptors, and (3) osmoreceptors.
- The effector cells include smooth muscle, exocrine gland, and endocrine gland cells.

## MOUTH

*The oral cavity is the entrance to the digestive tract.*

- Entry to the digestive tract is through the mouth or oral cavity.
- The palate separates the mouth from the nasal passages.
- Toward the front of the mouth, the palate is made of bone, forming what is known as the hard palate.
- The palate toward the rear of the mouth is called the soft palate.
- The tongue is composed of voluntarily controlled skeletal muscle.
- The pharynx acts as a common passageway for both the digestive system and the respiratory system.

*The teeth are responsible for chewing.*

- The first step in the digestive process is mastication.
- Occlusion allows food to be ground and crushed between the tooth surfaces.
- The purposes of chewing are: (1) to grind and break food up into smaller pieces to facilitate swallowing, (2) to mix food with saliva, and (3) to stimulate taste buds.

*Saliva begins carbohydrate digestion, is important in oral hygiene, and facilitates speech.*

- Saliva is produced by three major pairs of salivary glands, the: (1) sublingual, (2) submandibular and (3) parotid glands.
- Saliva is composed of water, proteins and electrolytes.
- The process of salivation occurs in two stages.
- First, the glandular portion of the salivary glands, the acini, produces a primary secretion with an electrolyte composition similar to that of plasma.
- Second, as the primary secretion flows through the salivary ducts, the duct cells reabsorb sodium and chloride ions from the saliva.

- The salivary salt concentration is considerably below that of plasma.
- The most important salivary proteins—amylase, mucus, and lysozyme—contribute to the functions of saliva as follows: (1) salivary amylase breaks polysaccharides down into disaccharides; (2) saliva facilitates swallowing by providing mucus; (3) saliva exerts some antibacterial action by means of lysozyme, an enzyme that lyses or destroys certain bacteria; (4) saliva serves as a solvent for molecules that stimulate the taste buds; (5) saliva aids speech by facilitating movements of the lips and tongue; (6) saliva plays an important role in oral hygiene; and (7) saliva buffers the acids in food as well as acids produced by bacteria in the mouth, thereby helping to prevent dental caries.

*Salivary secretion is continuous and can be reflexly increased.*

- The simple, or unconditioned, salivary reflex occurs when chemoreceptors and pressure receptors within the oral cavity respond to the presence of food.
- These receptors initiate impulses in afferent nerve fibers that carry the information to the salivary center located in the medulla of the brain stem.
- The salivary center sends impulses via the extrinsic autonomic nerves to the salivary glands to promote increased salivation.
- With the acquired, or conditioned, salivary reflex, salivation occurs without oral stimulation.
- Just thinking about, seeing, smelling, or hearing the preparations of pleasant food initiates salivation through this reflex.
- Inputs that arise outside the mouth and are mentally associated with the pleasure of eating act through the cerebral cortex to stimulate the medullary salivary center.
- Parasympathetic stimulation, which exerts the dominant role in salivary secretion, produces a prompt and abundant flow of watery saliva that is rich in enzymes.
- Sympathetic stimulation produces a much smaller volume of thick saliva that is rich in mucus.

*Digestion in the mouth is minimal; no absorption of nutrients occurs.*

- Importantly, some therapeutic agents can be absorbed by the oral mucosa, a prime example being a vasodilator drug, nitroglycerin, which is used by certain cardiac patients to relieve anginal attacks associated with myocardial ischemia.

## PHARYNX AND ESOPHAGUS

*Swallowing is a sequentially programmed all-or-none reflex.*

- The pressure of the bolus in the pharynx stimulates pharyngeal pressure receptors, which send afferent impulses to the swallowing center located in the medulla.
- Swallowing is an example of a sequentially programmed all-or-none reflex in which multiple responses are triggered in a specific timed sequence.
- Swallowing is initiated voluntarily, but once it is initiated, it cannot be stopped.

*During the oropharyngeal stage of swallowing, food is prevented from entering the wrong passageways.*

- Food is prevented from reentering the mouth during swallowing by the position of the tongue against the hard palate.
- The uvula is elevated and lodges against the back of the throat, sealing off the nasal passage.
- Contraction of laryngeal muscles aligns the vocal folds in tight apposition to each other, thus sealing the glottis entrance.
- The bolus tilts the epiglottis backward down over the closed glottis.
- Because the respiratory passages are temporarily sealed off during swallowing, respiration is briefly inhibited.
- Pharyngeal muscles contract to force the bolus into the esophagus.

*The pharnygoesophageal sphincter prevents air from entering the digestive tract during breathing.*

- The upper esophageal sphincter, is the pharyngoesophageal sphincter and the lower sphincter is the gastroesophageal sphincter.

*Peristaltic waves push food through the esophagus.*

- The swallowing center initiates a primary peristaltic wave that sweeps from the beginning to the end of the esophagus, forcing the bolus ahead of if through the esophagus to the stomach.
- Peristalsis refers to the ringlike contractions of the circular smooth muscle that move progressively forward with a stripping motion.
- Liquids fall quickly to the lower esophageal sphincter by gravity and then must wait about five seconds until the primary peristaltic wave finally arrives before they can pass through the gastroesophageal sphincter.

*The gastroesophageal sphincter prevents reflux of gastric contents.*

- If gastric contents do flow back into the esophagus, the acidity of these contents irritates

the esophagus, causing the esophageal discomfort known as heartburn.
- In the condition known as achalasia, the lower esophageal sphincter contracts vigorously; food accumulates in the esophagus.

*Esophageal secretion is entirely protective.*
- Esophageal secretion is entirely mucus.
- Mucus provides lubrication for passage of food and protects the esophageal wall from acid and enzymes in the gastric juice if gastric reflux should occur.

## STOMACH

*The stomach stores food and begins protein digestion.*
- The fundus is the portion of the stomach that lies above the esophageal opening.
- The main part of the stomach is the body.
- The lower portion is called the antrum.
- The terminal portion of the stomach consists of the pyloric sphincter.
- It is important that the stomach store food and meter it into the duodenum at a rate that does not exceed the small intestine's capacities.
- A second function of the stomach is to secrete hydrochloric acid and enzymes that begin protein digestion.

*Gastric filling involves receptive relaxation.*
- There are four aspects to gastric motility: (1) gastric filling, (2) gastric storage, (3) gastric mixing, and (4) gastric emptying.
- Gastric filling relies on the basic plasticity of the stomach smooth muscle and receptive relaxation of the stomach as it fills.
- Plasticity refers to the ability of smooth muscle to maintain a constant tension over a wide range of lengths.

*Gastric storage takes place in the body of the stomach.*
- The plasticity of smooth muscle is augmented by reflex relaxation of the stomach as it fills.
- Food emptied into the stomach from the esophagus is stored in the relatively quiet body without being mixed.
- Food is gradually fed from the body into the antrum, where mixing does take place.

*Gastric mixing takes place in the antrum of the stomach.*
- The strong antral peristaltic contractions are responsible for the mixing of food with gastric secretions to produce chyme.
- The antral peristaltic contractions provide the driving force for gastric emptying.
- The main gastric factor that influences the strength of contraction is the amount of chyme in the stomach.
- Distention of the stomach triggers increased gastric motility through a direct effect of stretch on the smooth muscle as well as through involvement of the intrinsic plexuses, the vagus nerve, and the stomach hormone gastrin.

*Gastric emptying is largely controlled by factors in the duodenum.*
- Factors in the duodenum are of primary importance in controlling the rate of gastric emptying.
- The four most important duodenal factors that influence gastric emptying are: (1) fat, (2) acid, (3) hypertonicity, and (4) distention.
- The presence of one or more of these stimuli in the duodenum activates duodenal receptors, thereby triggering either a neural or hormonal response.
- The subsequent reduction in antral peristaltic activity slows down the rate of gastric emptying.
- The neural response is mediated through the intrinsic nerve plexuses and the autonomic nerves.
- Collectively, these reflexes are called the enterogastric reflex.
- The hormonal response involves the release from the duodenal mucosa of several hormones collectively known as enterogastrones: secretin, cholecystokinin, and gastric inhibitory peptide.
- Peristaltic contractions occur in the empty stomach before the next meal.
- In conjunction with the sensation of hunger before the next meal, peristaltic contractions begin again.
- Both the sensation of hunger and the increased peristaltic activity are triggered simultaneously by the reduced amount of glucose being metabolized by the brain.

*Emotions can influence gastric motility.*
- Sadness and fear generally tend to decrease motility, whereas anger and aggression tend to increase it.
- Intense pain from any part of the body tends to inhibit motility.

*The stomach does not actively participate in vomiting.*
- Vomiting, or emesis, is the forceful expulsion of gastric contents out through the mouth.
- The major force for expulsion comes from contraction of the respiratory muscles namely the diaphragm and the abdominal muscles.
- The complex act of vomiting is coordinated by a vomiting center in the medulla.

- Nausea, retching, and vomiting can be initiated by afferent input to the vomiting center from a number of receptors throughout the body.
- The causes of vomiting include the following: (1) tactile stimulation of the back of the throat; (2) irritation or distension of the stomach or duodenum; (3) elevated intracranial pressure; (4) rotation or acceleration of the head producing dizziness; (5) intense pain arising from a variety of organs; (6) chemical agents, emetics; and (7) psychogenic vomiting induced by emotional factors.

*Gastric digestive juice is secreted by glands located at the base of gastric pits.*

- The cells responsible for gastric secretion are located in the lining of the stomach, the gastric mucosa, which is divided into two distinct areas: (1) the oxyntic mucosa, which lines the body and fundus, and (2) the pyloric gland area, which lines the antrum.
- The mucosal gland cells are found in gastric pits.
- Three types of secretory cells are found in the walls of the pits in the oxyntic mucosa.
- The mucous neck cells secrete a thin, watery mucus.
- Chief cells secrete the enzyme precursor pepsinogen and parietal cells secrete HCl and intrinsic factor.
- The gastric pits of the pyloric gland area secrete mucus and pepsinogen.
- No acid is secreted in the pyloric gland area.
- More importantly, endocrine cells in the pyloric gland area secrete the hormone gastrin into the blood.

*Hydrochloric acid activates pepsinogen.*

- Parietal cells actively secrete HCl.
- The pH of the luminal contents falls as low as two as a result of this HCl secretion.

*Pepsinogen, once activated, begins protein digestion.*

- Hydrochloric acid: (1) activates the enzyme precursor pepsinogen to an active enzyme, pepsin, and provides an acid medium that is optimal for pepsin activity; (2) aids in the breakdown of connective tissue and muscle fibers; and (3) along with salivary lysozyme, kills most of the microorganisms ingested with the food.
- The major digestive constituent of gastric secretion is pepsinogen.
- Pepsinogen is stored in the chief cell's cytoplasm within secretory vesicles known as zymogen granules.
- Hydrochloric acid cleaves off a small fragment of the pepsinogen molecule, converting it to the active form of the enzyme, pepsin.
- Pepsin initiates protein digestion by splitting proteins to yield peptide fragments.

*Mucus is protective.*

- The surface of the gastric mucosa is covered by a layer of mucus.
- Mucus protects the gastric mucosa against mechanical injury.
- Mucus helps protect the stomach wall from self-digestion.
- Being alkaline, mucus helps protect against acid injury by neutralizing HCl in the vicinity of the gastric lining.

*Intrinsic factor is essential for absorption of vitamin $B_{12}$.*

- Intrinsic factor is important in the absorption of certain vitamins.
- In the absence of intrinsic factor, pernicious anemia results.
- Pernicious anemia is caused by an autoimmune attack against the parietal cells.
- Occasionally, the oxyntic mucosa atrophies.
- The most detrimental consequence of gastric mucosal atrophy is loss of intrinsic factor and the subsequent development of pernicious anemia.
- Gastrin stimulates the parietal and chief cells.

*Multiple regulatory pathways influence the parietal and chief cells.*

- Four chemical messengers primarily influence the secretion of gastric juices. These include acetylcholine, histamine, gastrin, and somatostatin.

*Control of gastric secretion involves three phases.*

- Gastric secretion is divided into three phases, the: (1) cephalic, (2) gastric, and (3) intestinal phase.
- The cephalic phase of gastric secretion refers to the increased secretion of HCl and pepsinogen even before the food reaches the stomach.
- Thinking about, tasting, smelling, chewing, and swallowing food increases gastric secretion by means of vagal nerve activity.
- Vagal stimulation of the intrinsic plexuses promotes increased secretion of HCl and pepsinogen by the secretory cells.
- Vagal stimulation of the pyloric gland cells causes a release of gastrin.
- The gastric phase of gastric secretion occurs when food actually reaches the stomach.
- Stimuli acting in the stomach—namely, proteins, distention, caffeine, or alcohol—increase gastric

secretion by means of overlapping efferent pathways.
- Protein initiates short local reflexes in the intrinsic nerve plexuses to stimulate the secretory cells.
- Protein stimulates vagal fibers to the stomach.
- Vagal activity further enhances intrinsic nerve stimulation of the secretory cells and triggers the release of gastrin.
- Protein directly stimulates the release of gastrin.
- The intestinal phase of gastric secretion encompasses the factors originating in the small intestine that influence gastric secretion.
- The presence of the products of protein digestion in the duodenum stimulates further gastric secretion by triggering the release of an intestinal gastrin that is carried by the blood to the stomach.
- This is known as the excitatory component.
- The inhibitory component of the intestinal phase of gastric secretion is dominant over the excitatory component.
- The inhibitory component is important in helping to shut off the flow of gastric juices as chyme begins to be emptied into the small intestine.

*Gastric secretion gradually decreases as food empties from the stomach into the intestine.*
- As the meal is gradually emptied into the duodenum, the major stimulus for enhanced gastric secretion is withdrawn.
- After food leaves the stomach the pH falls very low.
- Gastric secretion is inhibited.
- The same stimuli that inhibit gastric motility inhibit gastric secretion.

*The gastric mucosal barrier protects the stomach lining from gastric secretions.*
- The properties of the gastric mucosa that enable the stomach to contain acid without injuring itself constitute the gastric mucosal barrier.
- The barrier is occasionally broken so that the gastric wall is injured by its acidic and enzymatic contents.
- When this occurs, an erosion, or peptic ulcer, of the stomach wall exists.
- In the early 1990s, the bacterium *Helicobacter pylori* was pinpointed as the cause of more than 80 percent of all peptic ulcers.
- Being equipped with four to six flagella, the organisms tunnel through and take up residence under the stomach's thick mucus layer.
- *H. pylori* contributes to ulcer formation by secreting toxins that cause a persistent inflammation.
- Frequent exposure to some chemicals can break the gastric mucosal barrier; the most important of these are ethyl alcohol and nonsteroid anti-inflammatory drugs.
- The persistence of stressful situations is frequently associated with ulcer formation.
- Two of the most serious consequences of ulcers are: (1) hemorrhage resulting from damage of submucosal capillaries, and (2) perforation of the stomach wall due to complete erosion through the wall.

*Carbohydrate digestion continues in the body of the stomach; protein digestion begins in the antrum.*
- Carbohydrate digestion continues under the influence of salivary amylase.
- Because food is not mixed with gastric secretions in the body of the stomach, very little protein digestion occurs here.

*The stomach absorbs alcohol and aspirin but no food.*
- No food or water is absorbed into the blood from the stomach mucosa.
- Even though none of the ingested food is absorbed from the stomach, two noteworthy non-nutrient substances are absorbed directly by the stomach: (1) ethyl alcohol, and (2) aspirin.

## PANCREATIC AND BILIARY SECRETIONS

*The pancreas is a mixture of exocrine and endocrine tissue.*
- The pancreas is a mixed gland that contains both exocrine and endocrine tissue.
- The predominant exocrine portion consists of grapelike clusters of secretory cells that form sacs known as acini.
- The smaller endocrine portion consists of isolated islands of endocrine tissue, the islets of Langerhans.

*The exocrine pancreas secretes digestive enzymes and an aqueous alkaline fluid.*
- The exocrine pancreas secretes a pancreatic juice consisting of two components: a potent enzymatic secretion and an aqueous alkaline secretion that is rich in sodium bicarbonate.
- The three types of pancreatic enzymes are: (1) proteolytic enzymes, (2) pancreatic amylase, and (3) pancreatic lipase.
- The three major proteolytic enzymes secreted by the pancreas are trypsinogen, chymotrypsinogen, and procarboxypeptidase.
- When trypsinogen is secreted into the duodenal lumen, it is activated to its active enzyme form, trypsin, by enterokinase.

- Chymotrypsinogen and procarboxypeptidase are converted to their active forms, chymotrypsin and carboxypeptidase within the duodenal lumen.
- Each of the proteolytic enzymes attacks different peptide linkages.
- Pancreatic amylase plays an important role in carbohydrate digestion by converting polysaccharides into disaccharides.
- Pancreatic lipase is extremely important because it is the only enzyme secreted throughout the entire digestive system that can accomplish digestion of fat.
- Pancreatic enzymes function best in a neutral or slightly alkaline environment.
- The alkaline fluid secreted by the pancreas into the duodenal lumen serves the important function of neutralizing the acidic chyme.

*Pancreatic exocrine secretion is regulated by secretin and CCK.*

- A small amount of parasympathetically induced pancreatic secretion occurs during the cephalic phase of digestion.
- The release of two major enterogastrones, secretin and cholecystokinin, in response to chyme in the duodenum plays the central role in the control of pancreatic secretion.
- The primary stimulus specifically for secretin release is acid in the duodenum.
- Secretin is carried by the blood to the pancreas, where it stimulates the duct cells to markedly increase their secretion of a sodium bicarbonate—rich aqueous fluid into the duodenum.
- This mechanism provides a control system for maintaining neutrality of the chyme in the intestine.
- Cholecystokinin is important in the regulation of pancreatic enzyme secretion.
- The circulatory system transports cholecystokinin to the pancreas where it stimulates the pancreatic acinar cells to increase secretion.

*The liver performs various important functions including bile production.*

- The biliary system includes the liver, the gallbladder, and associated ducts.
- The liver's importance to the digestive system is its secretion of bile salts, but it performs a wide variety of other functions, including the following: (1) metabolic processing of the major categories of nutrients; (2) detoxification or degradation of body wastes and hormones as well as drugs and other foreign compounds; (3) synthesis of plasma proteins; (4) storage of glycogen, fats, copper, iron, and many vitamins; (5) activation of vitamin D; (6) removal of bacteria and worn-out red blood cells; and (7) excretion of cholesterol and bilirubin.
- Venous blood enters the liver by means of a unique and complex vascular connection between the digestive tract and the liver that is known as the hepatic portal system.
- The veins draining the digestive tract do not directly join the inferior vena cava.
- The veins from the stomach and intestine enter the hepatic portal vein, which carries the products absorbed from the digestive tract directly to the liver for processing, storage, or detoxification.

*The liver lobules are delineated by vascular and bile channels.*

- The liver is organized into functional units known as lobules.
- At the outer edge of the lobule are three vessels: a branch of the hepatic artery, a branch of the portal vein, and a bile duct.
- Blood from the branches of both the hepatic artery and the portal vein flows from the periphery of the lobule in large sinusoids, which run between rows of liver cells to the central vein.
- The central vein of all liver lobules converge to form the hepatic vein.
- Hepatocytes continuously secrete bile into thin channels, the bile canaliculi, which carry the bile to a bile duct at the periphery of the lobule.

*Bile is continuously secreted by the liver and is diverted to the gallbladder between meals.*

- The opening of the bile duct into the duodenum is guarded by the sphincter of Oddi.
- When the sphincter is closed, the bile is diverted back up into the gallbladder.

*Bile salts are recycled through the enterohepatic circulation.*

- Bile consists of an aqueous alkaline fluid as well as several organic constituents, including bile salts, cholesterol, lecithin, and bilirubin.
- Bile salts are derivatives of cholesterol.
- Most bile salts are reabsorbed back into the blood by special active transport mechanisms located in the terminal ileum.
- This recycling of bile salts between the small intestine and the liver is referred to as the enterohepatic circulation.

*Bile salts aid fat digestion and absorption.*

- Detergent actions refers to the bile salt's ability to convert large fat globules into a liquid emulsion that consists of many small fat droplets suspended in the aqueous chyme, thus increasing the surface area available for attack by pancreatic lipase.

- Bile salts, along with cholesterol and lecithin, play an important role in facilitating fat absorption through micellar formation.
- Micelles, being water soluble by virtue of their hydrophilic shells, can dissolve water-insoluble substances in their lipid-soluble cores.

*Bilirubin is a waste product excreted in the bile.*

- Bilirubin does not play a role in digestion but is one of the few waste products excreted in the bile.
- Bilirubin is the end product resulting from the degradation of the heme portion of hemoglobin.
- Bilirubin is a yellow pigment that gives bile its yellow color.
- If bilirubin is formed more rapidly than it is excreted, it accumulates in the body and causes jaundice.
- Jaundice can be brought about in three different ways: (1) prehepatic or hemolytic jaundice, (2) hepatic jaundice, and (3) posthepatic or obstructive jaundice.

*Bile salts are the most potent stimulus for increased bile secretion.*

- Any substance that increases bile secretion is called a choleretic.
- The most potent choleretic is bile salts themselves.
- Secretin stimulates an aqueous alkaline bile secretion by the liver.
- Vagal stimulation of the liver plays a minor role in bile secretion during the cephalic phase of digestion.

*The gallbladder stores and concentrates bile between meals and empties during meals.*

- The presence of food in the duodenal lumen triggers the release of cholecystokinin.
- Cholecystokinin stimulates contraction of the gallbladder and relaxation of the sphincter of Oddi, so bile is discharged into the duodenum.

*Hepatitis and cirrhosis are the most common liver disorders.*

- Hepatitis is an inflammatory disease of the liver that results from a variety of causes, including viral infection or exposure to toxic agents.
- Repeated or prolonged hepatic inflammation, usually in association with chronic alcoholism, can lead to cirrhosis, a condition in which damaged hepatocytes are permanently replaced by connective tissue.

## SMALL INTESTINE

- The small intestine is the site where most digestion and absorption take place.
- The small intestine is arbitrarily divided into three segments: (1) the duodenum, (2) the jejunum, and (3) the ileum.

*Segmentation contractions mix and slowly propel the chyme.*

- Segmentation consists of oscillating, ringlike contractions of the circular smooth muscle along the length of the small intestine.
- The contractile rings occur every few centimeters, dividing the small intestine into segments like a chain of sausages.
- After a brief period of time, the contracted segments relax, and ringlike contractions appear in the previously relaxed areas.
- Shortly thereafter, the areas of contraction and relaxation alternate again.
- In this way, the chyme is chopped, churned, and thoroughly mixed.
- Segmentation contractions are initiated by the smaller intestine's pacesetter cells, which produce a basic electrical rhythm.
- Parasympathetic stimulation enhances segmentation.
- Sympathetic stimulation depresses segmental activity.

*The migrating motility complex sweeps the intestine clean between meals.*

- When most of the meal has been absorbed, segmentation contractions cease and are replaced between meals by the migrating motility complex.
- The waves start at the stomach and migrate down the intestine, "sweeping" any remnants of the preceding meal plus mucosal debris and bacteria forward toward the colon.

*The ileocecal juncture prevents contamination of the small intestine by colonic bacteria.*

- The anatomical arrangement is such that valvelike folds of tissue protrude from the ileum into the lumen of the cecum.
- This ileocecal valve is easily pushed open, but the folds of tissue are forcibly closed when the cecal contents attempt to move backward.
- The smooth muscle within the last several centimeters of the ileal wall is thickened, forming a sphincter, the ileocecal sphincter, which is under neural and hormonal control.

*Small intestine secretions do not contain any digestive enzymes.*

- The exocrine glands located in the small intestine mucosa secrete into the lumen an aqueous salt and mucus solution known as the succus entericus.

- The mucus in the secretion provides protection and lubrication.

*The small intestine enzymes complete digestion intracellularly.*

- Digestion within the small intestine lumen is accomplished by the pancreatic enzymes, with fat digestion being enhanced by bile secretion.
- Fat digestion is completed within the small intestine lumen.
- Hairlike projections on the luminal surface of the small intestine epithelial cells form the brush border, which contains three different categories of enzymes: (1) enterokinase, (2) the disaccharidases, and (3) the amino peptidases.

*The small intestine is remarkably well adapted for its primary role in absorption.*

- Most absorption occurs in the duodenum and jejunum.
- The small intestine has an abundant reserve absorptive capacity.
- The following special modifications of the small intestine mucosa greatly increase the surface area available for absorption: (1) the inner surface of the small intestine is thrown into circular folds; (2) projecting from this folded surface are microscopic fingerlike projections known as villi; and (3) even smaller hairlike projections known as microvilli arise from the luminal surface of the epithelial cells.
- Each villus has the following four major components: (1) epithelial cells that cover the surface of the villus—these cells possess carriers for absorption of specific nutrients and electrolytes from the lumen as well as the intracellular digestive enzymes that complete carbohydrate and protein digestion. (2) A connective tissue core formed by the lamina propria. (3) A capillary network. (4) A terminal lymphatic vessel known as a central lacteal.
- Like renal transport, intestinal absorption may be an active or passive process, with active absorption involving energy expenditure during at least one of the steps in the transepithelial transport process.

*The mucosal lining experiences rapid turnover.*

- The epithelial cells lining the small intestine slough off and are replaced at a rapid rate as a result of high mitotic activity in the crypts of Lieberkühn.
- The old cells that are sloughed off into the lumen are digested, with the cell constituents being absorbed into the blood and reclaimed for synthesis of new cells, among other things.

*Energy-dependent sodium absorption drives passive water absorption.*

- Sodium may be absorbed both passively and actively.
- Most water absorption in the digestive tract depends on the active carrier that pumps sodium into the lateral spaces.
- Water entering the space reduces the osmotic pressure but raises the hydrostatic pressure.

*Carbohydrate and protein are both absorbed by secondary active transport and enter the blood.*

- Dietary carbohydrate is presented to the small intestine for absorption mainly in the forms of the disaccharides, maltose, sucrose, and lactose.
- Glucose and galactose are both absorbed by secondary active transport.
- The operation of these cotransport carriers depends on the sodium concentration gradient established by the energy-consuming basolateral sodium-potassium pump.
- Glucose leaves the cell down its own concentration gradient to enter the blood.
- Not only are ingested proteins digested and absorbed, but endogenous proteins are digested and absorbed as well.
- Amino acids are absorbed across the intestinal cells by secondary active transport.

*Digested fat is absorbed passively and enters the lymph.*

- Fat absorption is quite different from carbohydrate and protein absorption because of the insolubility of fat in water presents a special problem.
- Biliary components facilitate absorption of these fatty end products through formation of micelles.
- Once the micelles reach the luminal membranes of the epithelial cells, the monoglycerides and free fatty acids passively diffuse from the micelles through the lipid component of the epithelial cell membranes to enter the interior of these cells.
- Once within the interior of epithelial cells, the monoglycerides and free fatty acids are resynthesized into triglycerides.
- The large, coated fat droplets, known as chylomicrons, enter the central lacteals rather than the capillaries because of the structural differences between these two vessels.
- Fat can be absorbed into the lymphatics but not directly into the blood.
- Fat absorption is a passive process.

*Vitamin absorption is largely passive.*

- Water-soluble vitamins are primarily absorbed with water, whereas fat-soluble vitamins are carried in the micelles and absorbed passively with the end products of fat digestion.

*Iron and calcium absorption is regulated.*

- Two main steps are involved in the absorption of iron into the blood: (1) absorption of iron from the lumen into the intestinal epithelial cells, and (2) absorption of iron from the epithelial cells into the blood.
- Iron is transported in the blood by a plasma protein carrier known as transferrin.
- Iron that is not immediately needed remains stored within the epithelial cells in a granular form called ferritin.
- Iron stored as ferritin is lost in the feces.
- Absorption of calcium is accomplished partly by passive diffusion but mostly by active transport.
- Vitamin D can stimulate this active transport only after it has been activated in the liver and kidneys, a process that is enhanced by parathyroid hormone.

*Most absorbed nutrients immediately pass through the liver for processing.*

- Anything absorbed into the digestive capillaries first must pass through the hepatic biochemical factory before entering the general circulation.
- Fat is picked up by the central lacteal and enters the lymphatic system instead.
- The absorbed fat is carried by systemic circulation to the liver and to other tissues of the body.

*Extensive absorption by the small intestine keeps pace with secretion.*

- The digestive juices are not lost from the body.
- After the constituents of the juices are secreted into the digestive tract lumen and perform their function, they are returned to the plasma.
- The only secretory product that escapes from the body is bilirubin, a waste product that must be eliminated.

*Biochemical balance among the stomach, pancreas, and small intestine is normally maintained.*

- The acid-base balance of the body is not altered by digestive processes.

*Diarrhea results in loss of fluid and electrolytes.*

- The other common digestive tract disturbance that can lead to loss of fluid and an acid-base imbalance is diarrhea.
- The following are the causes of diarrhea: (1) the most common cause of diarrhea is excessive intestinal motility, which arises either from local irritation of the gut wall by bacteria or viral infection of the intestine or from emotional stress; (2) diarrhea also occurs when excess osmotically active particles, such as those found in lactase deficiency, are present in the digestive tract lumen; (3) toxins of the bacterium Vibrio cholerae and certain other microorganisms promote the secretion of excessive amounts of fluid by the small intestine mucosa, resulting in profuse diarrhea.

## Large Intestine

*The large intestine is primarily a drying and storage organ.*

- The large intestine consists of the colon, cecum, appendix, and rectum.
- Since most digestion and absorption have been accomplished in the small intestine, the contents delivered to the colon consist of indigestible food residues, unabsorbed biliary components, and the remaining fluid.
- The colon extracts more water and salt from the contents.
- What remains to be eliminated is known as feces.

*Haustral contractions slowly shuffle the colonic contents back and forth.*

- The colon's primary method of motility is haustral contractions initiated by the autonomous rhythmicity of colonic smooth muscle cells.

*Mass movements propel feces long distances.*

- Haustral contractions are largely controlled by locally mediated reflexes involving the intrinsic plexuses.
- Slow colonic movement allows bacteria time to grow and accumulate in the large intestine.
- The gastrocolic reflex is mediated from the stomach to the colon by gastrin and by the extrinsic autonomic nerves.
- The gastroileal reflex moves the remaining small intestine contents into the large intestine, and the gastrocolic reflex pushes the colonic contents into the rectum, triggering the defecation reflex.

*Feces are eliminated by the defecation reflex.*

- When mass movements of the colon move fecal material into the rectum, the resultant distention of the rectum stimulates stretching receptors in the rectal wall, thus initiating the defecation reflex.
- The defecation reflex causes the internal anal sphincter to relax and the rectum and sigmoid colon to contract more vigorously.
- If the external anal sphincter is also relaxed, defecation occurs.

*Constipation occurs when the feces become too dry.*
- If defecation is delayed too long, constipation may result.
- When the colonic contents are retained for longer periods of time than normal, more than the usual amount of water is absorbed.

*Large intestine secretion is entirely protective.*
- Colonic secretion consists of an alkaline mucus solution, whose function is to protect the large intestine mucosa from mechanical and chemical injury.
- No digestion takes place in the large intestine because there are no digestive enzymes.

*The colon contains myriads of beneficial bacteria.*
- About 500 species of bacteria reside in the colon.

*The large intestine absorbs salt and water, converting the luminal contents into feces.*
- The colon normally absorbs some salt and water.
- The main waste product excreted in the feces is bilirubin.

*Intestinal gases are absorbed or expelled.*
- Flatus is derived primarily from two sources: (1) swallowed air; and (2) gas produced by bacterial fermentation in the colon.
- Some foods, such as beans, contain types of carbohydrates for which humans lack digestive enzymes.
- These fermentable carbohydrates enter the colon, where they are attacked by gas producing bacteria.

### Overview of Gastrointestinal Hormones
- Chyme in the stomach stimulates the release of gastrin, which acts on parietal and chief cells to increase secretion of HCl and pepsinogen.
- Gastrin enhances gastric motility, stimulates ileal motility, relaxes the ileocecal sphincter, and induces mass movements in the colon.
- Secretion performs four major interrelated functions: (1) it inhibits gastric emptying to prevent further acid from entering the duodenum until the acid that is present is neutralized; (2) it inhibits gastric secretion to reduce the amount of acid being produced; (3) it stimulates the pancreatic duct cells to produce a large volume of aqueous sodium bicarbonate secretion, which is emptied into the duodenum to neutralize the acid; and (4) it stimulates secretion by the liver of a sodium bicarbonate-rich bile, which likewise is emptied into the duodenum to assist in the neutralization process.
- Cholecystokinin also performs several important interrelated functions: (1) it inhibits gastric motility and secretion, thereby allowing adequate time for the nutrients already in the duodenum to be digested and absorbed; (2) it stimulates the pancreatic acinar cells to increase secretion of pancreatic enzymes, which continue the digestion of these nutrients in the duodenum; and (3) it causes contraction of the gallbladder and relaxation of the sphincter of Oddi so that bile is emptied into the duodenum to aid fat digestion and absorption.
- Cholecystokinin is an important regulator of food intake.
- Cholecystokinin plays a key role in satiety, the sensation of having had enough to eat.
- Gastric inhibitory peptide plays an important role in inhibiting gastric motility and secretion in response to fat, acid, hypertonicity, and distention in the duodenum.
- Gastric inhibitory peptide initiates the release of insulin in anticipation of the absorption of the meal.

## Key Terms

absorption
accessory digestive organs
bilirubin
bile salts
bolus
carboxypeptidase
cholecystokinin
choleretic
chymotrypsin
chymotrypsinogen
cubilin
dental carries
digestion
digestive tract
emetics
enteric amylase
enteric nervous system
enterokinase
enterogastric reflex
gallbladder
gallstones
gastric inhibitory peptide
gastrin
gastric filling
gastroileal reflex
haustra
heartburn
hepatic portal system
hepatocyte
hernias
hiatal hernia
hydrolysis
islets of langerhans
intrinsic factor
long reflex
malabsorption

malocclusion
mass movements
mesentery
micelle
microvilli
migrating motility complex
motility
mucosa
muscularis externa
myenteric plexus (Auerbach's)
pancreatic lipase
pacesetter potential
parotid glands
peptic ulcer
pepsin
pepsinogen
peristalsis
plasticity
procarboxypeptidase
retropulsion
salivary amylase
secretin
secretion
segmentation
serosa
steatorrhea
short reflexes
sublingual glands
submucosa
succus entericus
swallowing
taeniae coli
villi
vomiting center
xerostomia
zymogen granules

## Review Exercises

*Answers are in the appendix.*

**True/False**

_____ 1. The simplest form of carbohydrate is the simple sugars or disaccharides.

_____ 2. Starch, glycogen, and cellulose are all disaccharides.

_____ 3. The end products of fat digestion are triglycerides and free fatty acids.

_____ 4. Digestion is accomplished by enzymatic hydrolysis.

_____ 5. The lamina propria is a layer in the submucosa of the digestive tract.

_____ 6. Auerbach's plexus is located in the submucosa.

_____ 7. Meissner's and Haustra's plexuses together are termed the enteric nervous system.

_____ 8. The palate separates the mouth from the nasal passages.

_____ 9. The tongue is composed of voluntarily controlled smooth muscle.

_____ 10. Buccal glands are found in the rugae of the stomach.

_____ 11. Occlusion is when you cannot swallow.

_____ 12. Bicarbonate buffers are found in saliva.

_____ 13. Salivary amylase breaks down polysaccharides.

_____ 14. The hard outer covering, or the crown, of a tooth is enamel.

_____ 15. Swallowing is known as deglutition.

_____ 16. Burping is called borborygmia.

_____ 17. The antrum is that portion of the stomach that lies above the esophageal opening.

_____ 18. The interior of the stomach is thrown into deep folds known as rugae.

_____ 19. Gastrin influences the rate of gastric emptying.

_____ 20. Cholecystokinin is responsible for the contraction of the gallbladder.

_____ 21. Malocclusion is the forceful expulsion of gastric contents out through the mouth.

_____ 22. Chief cells secrete mucus.

_____ 23. Oxyntic cells is another name for mucous neck cells.

_____ 24. Hydrochloric acid activates pepsinogen into the active enzyme pepsin.

_____ 25. The major digestive constituent of gastric secretion is trypsinogen.

_____ 26. Peptic ulcers frequently occur when the gastric mucosal barrier is severely eroded or broken.

_____ 27. The stomach absorbs aspirin.

_____ 28. Water is absorbed from the stomach contents.

_____ 29. The exocrine portion of the pancreas is known as the islets of Langerhans.

_____ 30. Trypsinogen is activated to its active enzyme form trypsin by HCl.

_____ 31. Pancreatic juice is an aqueous acidic solution.

_____ 32. Cholecystokinin is important in the regulation of pancreatic enzyme secretion.

_____ 33. The principal clinical manifestation of pancreatic exocrine insufficiency is steatorrhea.

_____ 34. Macrophages of the liver are called Kupffer cells.

_____ 35. Bile is produced in the gallbladder.

_____ 36. Most absorption occurs in the duodenum and jejunum.

_____ 37. Small intestine secretions do not contain any digestive enzymes.

_____ 38. Villi, the fingerlike projections in the small intestine, are also known as the brush border.

_____ 39. The acid-base balance of the body is not altered by the digestive processes.

_____ 40. The colon's primary method of motility is Haustral contractions.

_____ 41. Cholecystokinin inhibits gastric motility.

_____ 42. Secretin inhibits gastric secretion.

**Fill in the Blank**

43. The exocrine glands located in the small intestine mucosa secrete into the lumen an aqueous salt and mucus solution known as the ____________________________.

44. ___________________ activates the pancreatic enzyme trypsinogen.

45. ______________________ hydrolyze the small peptide fragments into their amino acids components, completing protein digestion.

46. Projecting from the small intestines' folded surface are microscopic fingerlike projections known as __________________________.

47. _________________ may be caused by damage to or reduction of the surface area of the small intestine.

48. Dipping down into the mucosal surface between the villi are shallow invaginations known as the ________________________.

49. Iron is transported in the blood by a plasma-protein carrier known as _____________________.

50. The first step in the digestive process is _________________.

51. Saliva is composed of about _______ $H_2O$ and _______ protein and electrolytes.

52. ________________ is a condition characterized by salivary secretions.

53. With ____________________________________ salivation occurs without oral stimulation.

54. The __________ separates the mouth from the nasal passages.

55. Some therapeutic agents can be absorbed by the oral mucosa. An example is _______________.

56. A(n) ________________ is a ball of food.

57. The ______________________ consists of moving the bolus from the mouth through the pharynx and into the esophagus.

58. The lower esophageal sphincter is the _____________________________.

59. _______________ refers to ringlike contractions of the circular smooth muscle that move progressively forward with a stripping motion.

60. ______________________________ do not involve the swallowing center.

61. _______________ is a condition in which the lower esophageal sphincter fails to relax during swallowing but instead contracts more vigorously.

62. The small fingerlike projection at the bottom of the cecum is the _______________.

63. The colon's primary method of motility is ____________________________.

64. The large intestine consists of the _________, _________, ___________, and _________.

65. The presence of gas percolating through the luminal contents gives rise to gurgling sounds known as ______________.

66. _____________________________________ plays an important role in inhibiting gastric motility and secretion in response to fat, acid, hypertonicity, and distention in the duodenum.

67. The _______ is the portion of the stomach that lies above the esophageal opening.

68. The lower portion of the stomach is called the _____________.

69. The mucosal gland cells are found in ____________________.

70. Endocrine cells in the PGA secrete the hormone _____________ into the blood.

71. Pepsinogen is stored in the chief cell's cytoplasm within secretory vesicles known as _________________________.

72. ___________ phase of gastric secretion occurs when food actually reaches the stomach.

73. A(n) _________________ occurs when the gastric wall is injured by its acidic and enzymatic contents.

74. ___________________ is a condition characteristic of excessive undigested fat in the feces.

75. The ________ is the largest and most important metabolic organ in the body.

76. The opening of the bile duct into the duodenum is guarded by the ________________.

77. ______________ are derivatives of cholesterol.

78. Any substance that increases bile secretion by the liver is called a(n) ________________.

79. The presence of food in the duodenal lumen triggers the release of ________.

80. ______________ is a condition in which damaged hepatocytes are permanently replaced by connective tissue.

81. Digestion is accomplished by enzymatic _________________.

**Matching**
*Match the following characteristics to the conditions listed below.*

a. xerostomia
b. heartburn
c. achalasia
d. emesis
e. gastric mucosal atrophy
f. peptic ulcer
g. steatorrhea
h. jaundice
i. hepatitis
j. cirrhosis
k. lactose intolerance
l. malabsorption
m. gluten enteropathy
n. borborygmi

_____ 82. A condition in which damaged hepatocytes are permanently replaced by connective tissue.

_____ 83. This condition is characterized by diminished salivary secretion.

_____ 84. This is the forceful expulsion of gastric contents out through the mouth.

_____ 85. This clinical manifestation is excessive undigested fat in the feces.

_____ 86. This condition is impaired absorption in the intestines.

_____ 87. This is the gurgling sound of intestinal gas through the luminal contents.

_____ 88. This condition occurs when the gastric contents flow back into the esophagus.

_____ 89. In this condition the small intestine is abnormally sensitive to gluten.

_____ 90. In this condition there is a loss of intrinsic factor and the subsequent development of pernicious anemia.

_____ 91. This condition occurs when bilirubin accumulates in the body.

_____ 92. This fairly common disorder involves a deficiency of lactose.

_____ 93. This condition is an inflammatory disease of the liver.

_____ 94. This condition is characterized by an erosion of the stomach wall.

_____ 95. This disorder is characterized by an accumulation of food in the esophagus.

**Multiple Choice**

96. Which category of foodstuffs is characterized by having peptide bonds?
   a. starches
   b. carbohydrates
   c. proteins
   d. fats
   e. triglycerides

97. Where is the salivary center located within the brain?
   a. pons
   b. medulla
   c. cerebellum
   d. cerebrum
   e. hypothalamus

98. Which of the following is absorbed in the mouth?
   a. starch
   b. amino acids
   c. lipids
   d. ethyl alcohol
   e. nitroglycerin

99. Which of the hormones trigger gastric motility?
   a. gastrin
   b. pepsin
   c. trypsin
   d. gastrokinin
   e. villikinin

100. Which of the following pancreatic proteolytic enzymes are converted to their active form by trypsin?
   a. trypsinogen
   b. pepsinogen
   c. enterokinase
   d. pancreatic lipase
   e. procarboxypeptidase

101. Which of the following has specialized cells known as Kupffer cells?
   a. Islets of Langerhans
   b. gallbladder
   c. liver
   d. salivary glands
   e. small intestines

102. The opening of the bile duct into the duodenum is guarded by this sphincter?
   a. esophageal sphincter
   b. pyloric sphincter
   c. ileocecal sphincter
   d. sphincter of Acinus
   e. sphincter of Oddi

103. Which of the following play an important role in facilitating fat absorption through micellar formation?
   a. apoenzymes
   b. bile salts
   c. cholexystokinin
   d. disaccharidases
   e. enterokinase

104. Approximately how much fluid is absorbed by the small intestine per day?
   a. 500 ml
   b. 1,000 ml
   c. 1,200 ml
   d. 9,000 ml
   e. 9,500 ml

105. Which of the following organisms cause peptic ulcers?
   a. *Vibrio cholerae*
   b. *Enterogastro antrumi*
   c. *Clostridium pepti*
   d. *Helicobacter pylori*

**Modified Multiple Choice**

*Indicate which hormone(s) is (are) being described by writing the appropriate letter in the blank using the following answer code.*

A = both secretin and CCK
B = secretin
C = CCK
D = gastrin

_____ 106. Secreted by the pyloric gland area.
_____ 107. Stimulated primarily by the presence of fat (fatty acids).
_____ 108. Stimulated primarily by the presence of protein.
_____ 109. Secretion is inhibited by a low pH.
_____ 110. Inhibit(s) gastric secretion.
_____ 111. Stimulate(s) gastric secretion.
_____ 112. Inhibits gastric motility.
_____ 113. Stimulate(s) pancreatic digestive enzyme secretion.
_____ 114. Stimulate(s) pancreatic aqueous $NaHCO_3$ secretion.
_____ 115. Stimulate(s) gallbladder contraction.
_____ 116. Known as enterogastrone.
_____ 117. Secreted by the small intestine mucosa.
_____ 118. Stimulate(s) the pancreatic acinar cells.
_____ 119. Stimulate(s) the pancreatic duct cells.
_____ 120. Stimulated primarily by the presence of acid.

*Identify which part of the digestive system is associated with each item in question by writing the appropriate letter in the blank using the following answer code.*

A = small intestine
B = large intestine
C = mouth
D = esophagus
E = stomach

_____ 121. Absorbs weak acids such as aspirin.
_____ 122. Primarily motility is haustrations.
_____ 123. Receptive relaxation occurs.
_____ 124. Secondary peristaltic waves occur.
_____ 125. Primary motility is segmentation.
_____ 126. BER occurs at a rate of 3 contractions/minute.
_____ 127. Absorbs nitroglycerin.
_____ 128. Involved with secondary stage of deglutition.
_____ 129. Involved with mastication.
_____ 130. Food is converted into chyme.
_____ 131. Chyme is converted into feces.
_____ 132. Main site of digestion and absorption.
_____ 133. Secretion is entirely mucus.
_____ 134. Protein digestion commences.
_____ 135. Carbohydrate digestion commences.
_____ 136. Fat digestion commences.

## Points to Ponder

1. How selective is the digestive tract in its absorption process? How would you explain the absorption of indigestible compounds such as inulin? What about the absorption of known carcinogens?

2. What happens to the pH in your mouth when you sleep with your mouth open?

3. After studying this chapter how do you feel about the disorder discussed in television advertisements known as irregularity?

4. In terms of digestion what is happening when food is cooked?

5. Why do dental procedures usually stimulate salivation?

6. What are the different types of muscular action that occur during peristalsis?

7. What is the purpose of pepsin in the stomach?

8. What are "hunger pangs"?

9. What pacemakers occur in the digestive tract?

10. Why is the absorption of fats more complex than the absorption of proteins and carbohydrates?

## Clinical Perspectives

1. Are there serious consequences when a person strains to have a bowl movement while constipated?
2. How do you suppose the bile flows through the ducts of the biliary systems?
3. What causes the dryness of "morning mouth"?
4. How would you explain heartburn to your roommate?
5. Your friend suffers from lactose intolerance. She wants to know why she has a problem. After reading this chapter on the digestive system, how would you explain lactose intolerance to her?
6. Your father has had his gall bladder removed (cholecystectomy). He knows that you are taking a physiology course and asks you if the gallbladder is essential for the digestive process. What would you tell him?
7. Your friend wants to know what gastroesophageal reflux disease is and how it can be treated. How would you explain this problem to him?
8. What are some factors that can cause cirrhosis of the liver? Jaundice?
9. What causes gallstones?
10. Why is the enteric nervous system sometimes called the "enteric brain"?

## PhysioEdge Activities

**Related to Text:**
Media Exercise 15.1: The Stomach.
Media Exercise 15.2: Stomach: Cellular Level.
Media Exercise 15.3: The Intestines and Associated Organs.

**Related to Figures:**
*Figure 16.2.* For an interaction related to this figure, see Media Exercise 16.3: The Intestine and Associated Organs.
*Figure 16.5.* For an animation of this figure, click the Gastrointestinal Motility tab in the Gastrointestinal Physiology tutorial.
*Figure 16.6.* For an animation of this figure, click the Gastrointestinal Motility tab in the Gastrointestinal Physiology tutorial.
*Figure 16.7.* For an animation of this figure, click the Gastrointestinal Motility tab in the Gastrointestinal Physiology tutorial. For an interaction related to this figure, see Media Exercise 16.1: The Stomach.
*Figure 16.12.* For an animation of this figure, click the Gastrointestinal Secretions tab in the Gastrointestinal tutorial. For an interaction related to this figure, see Media Exercise 16.3: The Intestine and Associated Organs.
*Figure 16.18.* For an animation of this figure, click the Gastrointestinal Secretions tab in the Gastrointestinal tutorial. For an interaction related to this figure, see Media Exercise 16.3: The Intestine and Associated Organs.
*Figure 16.19.* For an animation of this figure, click the Gastrointestinal Motility tab in the Gastrointestinal Physiology tutorial. For an interaction related to this figure, see Media Exercise 16.3: The Intestine and Associated Organs.
*Figure 16.21.* For an animation of this figure, click the Digestion and Absorption tab (p.1) in the Gastrointestinal tutorial.
*Figure 16.26.* For an animation of this figure, click the Digestion and Absorption tab (p.1) in the Gastrointestinal tutorial.

## Media Resources

**PhysioEdge CD-ROM**
For a visual review of concepts in this chapter, check out:

Tutorial:Gastrointestinal Physiology
Media Exercise 16.1: The Stomach
Media Exercise 16.2: Stomach: Cellular Level
Media Exercise 16.3: The Intestine and Associated Organs

**Book Companion Website**
The website for this book contains a wealth of helpful study aids, as well as many ideas for further reading and research. Log on to:

**http://www.brookscole.com/sherwoodhp6**

Select Chapter 16 from the drop-down menu, or click on one of the many resources areas, including **Case Histories**, which introduce clinical aspects of human physiology. For this chapter check out Case History 6: Just Stress.

For Suggested Readings, consult **InfoTrac® College Edition**, your online research library, at:

**http://infotrac.thomsonlearning.com**

# 17

# Energy Balance and Temperature Regulation

## Chapter Overview

Food intake is essential to power cell activities. Energy balance and thus body weight are maintained by controlling food intake. The thousands of chemical reactions within the body are frequently referred to as the metabolic processes of the organism. Metabolism requires energy and has an optimum temperature range. The survival of cells depends on certain intracellular chemical reactions. In maintaining homeostasis the body systems utilize chemical processes for the benefit of all cells. In this chapter, the balance of energy and the regulation of temperature necessary for homeostasis are considered.

The first law of thermodynamics states that energy cannot by created or destroyed but can be converted to another form. The second law of thermodynamics states that energy will be lost as energy is converted from one form to another. Thus, energy balance is interrelated to temperature production. Inasmuch as energy is not supplied in a form that can be utilized directly, heat is produced as the body converts the energy from one form to another. To meet the energy needs of the organism, and at the same time maintain an optimum temperature, various body systems interact to provide an energy balance and adequate methods of thermoregulation. The controls are the same; i.e., nervous and endocrine. The phenomenon is homeostasis.

## Chapter Outline

ENERGY BALANCE

*Most food energy is ultimately converted into heat in the body.*

- According to the first law of thermodynamics, energy can neither be created nor destroyed, but it can be converted from one form to another.
- The energy in nutrient molecules that is not used to energize work is either transformed into thermal energy or heat.
- The energy in ingested foodstuffs constitutes energy input to the body.
- Energy output or expenditure falls into two categories: external work and internal work.

*The metabolic rate is the rate of energy use.*

- The rate at which energy is expended by the body during both external and internal work is known as the metabolic rate.
- The so-called basal metabolic rate is a reflection of the body's "idling speed," or the minimal waking rate of internal energy expenditure.

*Energy input must equal output to maintain a neutral energy balance.*

- Since energy cannot be created or destroyed, energy input must equal energy output.
- There are three possible states of energy balance: (1) neutral energy balance, (2) positive energy balance, and (3) negative energy balance.
- To maintain a constant body weight, energy acquired through food intake must equal energy expenditure by the body.

*Food intake is controlled primarily by the hypothalamus.*

- Control of food intake is primarily a function of the hypothalamus.
- The hypothalamus is considered to house a pair of appetite centers located in the lateral regions of the hypothalamus, one on each side, and another pair of satiety centers located in the ventromedial area.
- The appetite centers tell us to eat, whereas the satiety centers tell us when we have had enough.
- The following factors are among those that have been hypothesized as contributing to the control of food intake: (1) the size of fat stores, (2) the extent of gastrointestinal distention, (3) the extent of glucose utilization, (4) the intensity of cell power production, (5) the level of cholecystokinin, (6) the influences of neurotransmitters, and (7) psychosocial influences.

*Obesity occurs when more kilocalories are consumed than are burned up.*

- Obesity is defined as excessive fat content in the adipose tissue stores.

- The causes of obesity are many and obscure. Some factors that may be involved with obesity include: (1) emotional disturbances where overeating replaces other gratifications; (2) development of an excessive number of fat cells as a result of overfeeding; (3) certain endocrine disorders such as hypothyroidism; (4) disturbances of the satiety-appetite regulatory centers in the hypothalamus; (5) hereditary tendencies; (6) the palatability of available food; and (7) lack of physical exercise.
- It is known that obesity, especially of the android type, can predispose an individual to illness and premature death from a multitude of diseases.
- The gene that produces leptin, the obese (ob) gene, was first identified in a strain of genetically obese mice.
- Their obesity arises from a failure to produce leptin as a result of a defective ob gene.
- Most obese people do not have a problem producing leptin.
- Scientists suspect that the problem lies with faulty leptin receptors in the brain that do not respond appropriately to high levels of circulating leptin.
- Other genetic errors may be at fault.
- The pattern of development of obesity in mice that possess the tubby gene more closely parallels that of obese humans than does the obesity of mice with a mutated ob gene.
- Investigators have identified a gene similar to the tubby gene in humans.
- High-fat diets are especially obesity-producing for several reasons: fewer calories are used to convert ingested fat into stored fat than to convert ingested carbohydrate or protein into stored fat; as fat intake increases at the expense of reduced carbohydrate consumption, glucose utilization decreases.
- Low glucose utilization promotes increased food intake.

*People suffering from anorexia nervosa have a pathological fear of gaining weight.*

- Chronic diseases such as renal failure, cancer, and tuberculosis are commonly accompanied by a lack of appetite.
- Researchers have identified a protein called cachetin, which is released by the immune system and is believed to contribute to the development of emaciation.
- Patients with anorexia nervosa, most commonly adolescent girls and young women, have a morbid fear of becoming fat.
- Because they have an aversion to food, they eat very little and consequently lose considerable weight, perhaps even starving themselves to death.
- Other characteristics of the condition include altered secretion of many hormones, absence of menstrual periods, and low body temperature.

## Temperature Regulation

*Internal core temperature is homeostatically maintained at 100° F.*

- Humans are usually in environments cooler than their bodies, so they must constantly generate heat internally to maintain their body temperatures.
- Most people suffer convulsion when the internal body temperature reaches about 106° F.
- The upper limit is considered to be 110° F.
- Most of the body's tissues can transiently withstand substantial cooling.
- The cooled tissues need less nourishment than they do at normal body temperature because of their pronounced reduction in metabolic activity.
- Normal body temperature has traditionally been considered 98.6° F (37° C).
- A recent study indicates that normal body temperature varies among individuals and varies throughout the day, ranging from 96.0° F in the morning to 99.9° F in the evening.
- From a thermoregulatory viewpoint, the body may be conveniently viewed as a central core surrounded by an outer shell.
- Skin temperatures may fluctuate between 68° and 104° F without damage.
- We are accustomed to thinking in terms of oral, rectal, or axillary temperature because these are easy sites for monitoring body temperature.
- Recently available is a temperature-monitoring instrument that scans the heat generated by the eardrum and converts this temperature into an oral equivalent.
- Most people's core temperature normally varies during the day, with the lowest level occurring early in the morning before rising and the highest point occurring in late afternoon.
- Women also experience a monthly rhythm in core temperature connected with their menstrual cycle.
- The core temperature increases during exercise.
- The core temperature may vary slightly with exposure to extremes of temperatures.

*Heat input must balance heat output to maintain a stable core temperature.*

- Balance between heat input and output is frequently disturbed by: (1) changes in internal heat production by exercise, and (2) changes in the external environmental temperature.
- If the core temperature starts to fall, heat production is increased and heat loss is minimized so that normal temperature can be restored.

- If the temperature starts to rise above normal, it can be corrected by increasing heat loss and reducing heat production.

*Heat exchange takes place by radiation, conduction, convection, and evaporation.*

- Heat always moves down its own concentration gradient, that is, down a thermal gradient from a warmer to a cooler region.
- Radiation is the emission of heat energy from the surface of a warm body in the form of electromagnetic waves.
- Conduction is the transfer of heat between objects of differing temperatures that are in direct contact with each other.
- Convection refers to the transfer of heat energy by air currents.
- When water evaporates from the skin surface, the heat required to transform water from a liquid to a gaseous state is absorbed from the skin, thereby cooling the body.
- Sweating is an active evaporative heat-loss process under sympathetic nervous control.
- Sweat is a dilute salt solution that is actively extruded to the surface of the skin by sweat glands dispersed all over the body.
- The most important factor determining the extent of evaporation of sweat is the relative humidity of the surrounding air.

*The hypothalamus integrates a multitude of thermosensory inputs.*

- The hypothalamus serves as the body's thermostat.
- The hypothalamus must be continuously apprized of both the skin temperature and the core temperature by means of specialized temperature-sensitive receptors called thermoreceptors.
- Peripheral thermoreceptors monitor skin temperature.
- The core temperature is monitored by central thermoreceptors.

*Shivering is the primary involuntary means of increasing heat production.*

- The hypothalamus takes advantage of the fact that increased skeletal muscle activity generates more heat.
- Shivering consists of rhythmic, oscillating skeletal muscle contractions that occur at a rapid rate of 10 to 20 per second.
- Shivering is a very effective mechanism for increasing heat production.
- Nonshivering thermogenesis also plays a role in thermoregulation.
- Nonshivering thermogenesis is mediated by the hormones epinephrine and thyroid hormone.

*The magnitude of heat loss can be adjusted by varying the flow of blood through the skin.*

- Most of the flow of blood through the skin is for purposes of temperature regulation.
- Skin vasomotor responses are coordinated by the hypothalamus.
- Increased sympathetic activity to the skin vessels produces heat-conserving vasoconstriction in response to cold exposure, whereas decreased sympathetic activity produces heat-losing vasodilation of the skin vessels in response to heat exposure.

*The hypothalamus simultaneously coordinates heat-production and heat-loss mechanisms.*

- In response to cold exposure, the posterior region of the hypothalamus directs increased heat production such as by shivering, while simultaneously decreasing heat loss by skin vasoconstriction and other measures.
- During heat exposure the anterior part of the hypothalamus reduces heat production by decreasing skeletal muscle activity and promotes increased heat loss by inducing skin vasodilation.
- When even the maximal skin vasodilation is inadequate to rid the body of excess heat, sweating is brought into play to accomplish further heat loss through evaporation.

*During a fever, the hypothalamic thermostat is reset at an elevated temperature.*

- Fever refers to an elevation in body temperature as a result of infection or inflammation.
- In response to microbial invasion, white blood cells release endogenous pyrogen, which acts on the hypothalamic thermoregulatory center to raise the setting of the thermostat.
- The hypothalamus now maintains the temperature at the new set level instead of maintaining normal body temperature.
- Endogenous pyrogen raises the set point of the hypothalamic thermostat during fever production by triggering the local release of prostaglandins.
- Aspirin reduces a fever by inhibiting the synthesis of prostaglandins.

*Hyperthermia can occur unrelated to infection.*

- The most common cause of hyperthermia is sustained exercise A completely different way in which hyperthermia can be brought about is by excessive heat production in connection with abnormally high circulating levels of epinephrine or thyroid hormone.

## Key Terms

anorexia nervosa
appetite centers
arcuate nucleus
basal metabolic rate (BMR)
body mass index (BMI)
calorie
cholecystokinin (CCK)
conduction
convection
core temperature
diet-induced thermogenesis
direct calorimetry
electromagnetic waves (heat waves)
endogenous pyrogen
energy balance
energy equivalent of $O_2$
evaporation
external work
feeding (appetite) centers
glucostatic theory
hyperthermia
hypothermia
indirect calorimetry
internal work
ischymetric theory
kilocalorie (calorie)
leptin
lipostatic theory
melanocortins
metabolic rate
neuropeptide y (NPY)
non-exercise activity thermogenesis (NEAT)
nonshivering (chemical) thermogenesis
nucleus tractus solitarius
obesity
orexins
radiation
satiety
satiety centers
shivering
sweating
temperature regulation
thermal energy (heat)
thermal gradient
thermoneutral zone
thermoreceptors
tubby gene
uncoupling proteins

## Review Exercises

*Answers are in the appendix.*

**True/False**

_____ 1. Internal work refers to the energy expended when skeletal muscles are contracted to move external objects.

_____ 2. Energy cannot be created or destroyed or converted from one form to another form.

_____ 3. The rate at which energy is expended by the body is known as the metabolic rate.

_____ 4. The appetite centers are located in the lateral regions of the thalamus.

_____ 5. The glucostatic theory proposes that satiety is signaled by increased glucose utilization.

_____ 6. Cachetin is released by the immune system.

_____ 7. Obesity is defined as excessive fat content in the adipose-tissue stores.

_____ 8. Skin temperature may fluctuate between 68° and 104° F without damage.

_____ 9. Rectal temperatures are the same as oral temperatures.

_____ 10. Heat moves down a thermal gradient from a warmer to a cooler region.

_____ 11. Conduction is the transfer of heat using air currents.

_____ 12. The hypothalamus serves as the body's thermostat.

_____ 13. Sweating is an active evaporative heat-loss process.

_____ 14. Most of the blood flow through the skin is for purposes of temperature regulation.

**Fill in the Blank**

15. ______________ sympathetic activity to the skin vessels produces heat-conserving vasoconstriction in response to cold exposure.

16. Nonshivering thermogenesis is mediated by the hormones ________________ and ______________________________.

17. In response to microbial invasion, certain white blood cells release a chemical known as __________________________.

18. In response to heat exposure the _____________ part of the hypothalamus reduces heat production by __________________ skeletal muscle activity and promotes increased heat loss by inducing skin _________________________.

19. Aspirin reduces a fever by inhibiting the synthesis of ________________________.

20. The most common cause of hyperthermia is _____________________________.

21. Sweating is under ______________________ nervous control.

22. ________________ refers to the transfer of heat energy by air currents.

23. The prominent feature in anorexia nervosa is a(n) _________________________.

24. According to the _______________ theory, increased fat storage in adipose tissue signals satiety.

25. Energy output or expenditure falls into two categories: _____________________ and _____________________.

26. ____________ energy balance exists if the amount of energy in food intake is greater than the amount of energy expended.

27. Chronic diseases such as _________________, __________, and ____________________ are commonly accompanied by a lack of appetite.

28. ______________ is the emission of heat energy from the surface of a warm body.

**Matching**

*Match the situation or condition to the approximate temperature in degrees Fahrenheit.*

| | | |
|---|---|---|
| _____ 29. | Upper limit compatible with life. | a. 100° F |
| _____ 30. | Central core temperature. | b. 68° to 104° F |
| _____ 31. | Oral temperature. | c. 99.7° F |
| _____ 32. | Skin temperature. | d. 110° F |
| _____ 33. | Rectal temperature. | e. 106° F |
| _____ 34. | Convulsions at this temperature and higher. | f. 98.6° F |
| _____ 35. | Core temperature during hard exercise. | g. 104° F |

## Multiple Choice

36. Which of the following are released in response to endogenous pyrogens?
    a. cachetin
    b. cholecystokinin
    c. prostaglandins
    d. brown fat
    e. electromagnetic waves

37. Which of the following nervous activities is utilized to control the skin vasomotor responses when the body is subjected to heat exposure?
    a. decreased parasympathetic activity
    b. increased sympathetic activity
    c. increased parasympathetic activity
    d. increased thermoreceptor activity
    e. decreased sympathetic activity

38. Which of the following nervous activities is utilized to control the skin vasomotor responses when the body is subjected to cold exposure?
    a. decreased parasympathetic activity
    b. increased sympathetic activity
    c. increased parasympathetic activity
    d. increased thermoreceptor activity
    e. decreased sympathetic activity

39. Which part of the brain houses the body's thermostat?
    a. cerebral cortex
    b. cerebellum
    c. pons
    d. hypothalamus
    e. medulla

40. Which part of the brain houses the satiety centers?
    a. pons
    b. thalamus
    c. hypothalamus
    d. medulla
    e. pars nervosa

41. Which of the following activates shivering in response to cold exposure?
    a. pons
    b. thalamus
    c. hypothalamus
    d. medulla
    e. cerebellum

42. Which of the following is mediated by epinephrine?
    a. shivering
    b. nonshivering thermogenesis
    c. increased glucose utilization
    d. decreased thyroid hormone production
    e. decreased in physical exercise

43. Which of the following increase carbohydrate consumption?
    a. cachetin
    b. serotonin
    c. dopamine
    d. neuropeptide
    e. insulin

44. Which of the following decrease carbohydrate consumption?
    a. glucagon
    b. neuropeptide
    c. dopamine
    d. epinephrine
    e. cachetin

45. Which of the following is characterized by a low body temperature?
    a. hard exercise
    b. microbial invasion
    c. obesity
    d. anorexia nervosa
    e. cancer

46. Which mouse gene causes mice to develop obesity in a pattern that closely parallels the pattern in obese humans?
    a. ob gene
    b. tubby gene
    c. cachetin gene
    d. fat gene
    e. lipostatic gene

47. Which type of obesity is characterized by abdominal fat distribution around the waist?
    a. glucostatic obesity
    b. android obesity
    c. ischymetric obesity
    d. psychosocial obesity
    e. gynoid obesity

48. Which of the following represents the normal variation in most people's core temperature during the day?
    a. 0° C
    b. 5° C
    c. 3° C
    d. 1° C
    e. 7° C

**Modified Multiple Choice**

*Indicate which form of heat loss is being described by writing the appropriate letter in the blank using the following answer code.*

A = convection
B = evaporation
C = radiation
D = conduction

_____ 49. emission of heat energy from the surface of a warm body in the form of electromagnetic waves
_____ 50. heat loss by means of conversion of water from a liquid to a gaseous state
_____ 51. transfer of heat between objects of differing temperatures that are in direct contact with each other
_____ 52. transfer of heat by air or water currents

*Indicate which form of thermal descriptor is being described by writing the appropriate letter in the blank by using the appropriate code.*

A = energy equivalent
B = external work
C = calorie
D = basal metabolic rate
E = internal work
F = specific dynamic action

_____ 53. energy expended by contraction of skeletal muscles to move external objects to move the body
_____ 54. all forms of biological energy expenditure that do not accomplish work outside the body
_____ 55. obligatory increase in metabolic rate that occurs as a consequence of food intake
_____ 56. amount of heat required to raise the temperature of 1 gram of water 1 degree C
_____ 57. quantity of heat produced per liter of oxygen consumed
_____ 58. minimal waking rate of internal energy expenditure

*Indicate which of the following physiological conditions is being described by writing the appropriate letter in the blank using the appropriate code.*

A = hypothermia
B = obesity
C = frostbite
D = hyperthermia
E = anorexia nervosa
F = heat stroke
G = heat exhaustion

_____ 59. Excessive fat content in adipose tissue stores
_____ 60. Morbid fear of becoming overweight
_____ 61. State of collapse resulting from reduced blood pressure brought about as a consequence of over taxing the heat loss mechanism
_____ 62. Heat loss mechanisms are absent despite the fact that the body temperature is rapidly rising
_____ 63. Freezing of tissues as a result of cold exposure
_____ 64. Elevation in body temperature due to reasons other than an infection
_____ 65. Slowing of all body processes as a result of generalized cooling

## *Points to Ponder*

1. Using common sense and simple logic how do you know that the hypothalamus raises the thermostat setting during fever?

2. Should you take aspirin at the onset of fever? Why do you have fever in response to microbial invasions?

3. What would you do if your roommate or member of your family developed anorexia nervosa? Could you help their situation?

4. Should we begin to treat obesity as a disorder or continue with the present social attitude?

5. In the body, what do anabolic reactions accomplish? Catabolic reactions?

6. Under what major condition is gluconeogenesis employed?

7. In what form do lipids enter the lymphatic system?

8. How is the brain's glucose supply kept constant?

9. Why are some amino acids called essential?

10. Contrast the metabolic rate with the basal metabolic rate.

## *Clinical Perspectives*

1. What is the relationship between diet and cancer?

2. How does dieting affect one's metabolic rate?

3. Why do you wake up cold when you fall asleep uncovered?

4. Your friend asks you why she feels warmer on hot, humid days than on hot, dry days. How would you answer her question?

5. What are some of the ways in which the normal body temperature can vary?

6. Your friend has stopped smoking and has gained weight. How would you explain this physiological weight gain to your friend?

7. Why do surgeons put a patient into a state of hypothermia during open heart or brain surgery?

8. How would you explain the fact that ketoacidosis and dehydration are produced in a person with type I diabetes?

9. How can a high fat diet in childhood lead to increased numbers of adipocytes? Explain how this process is related to the ability of adipocytes to regulate the insulin sensitivity of skeletal muscles in adults.

10. Explain the relationship between insulin resistance, obesity, exercise, and non-insulin diabetes mellitus.

## *Experiment of the Day*

1. Using your automobile as an "experimental animal," list as many analogies as you can concerning energy balance and thermoregulation.

## *PhysioEdge Activities*

**Related to Text:**
Media Exercise 17.1: Basics of Energy Balance.
Media Exercise 17.2: Body Temperature Regulation.

**Related to Figures:**
No tutorials or media exercises directly related to figures.

## Media Resources

**PhysioEdge CD-ROM**
For a visual review of concepts in this chapter, check out:

Media Exercise 17.1: Basics of Energy Balance
Media Exercie 17.2: Body Temperature Regulation

**Book Companion Website**
The website for this book contains a wealth of helpful study aids, as well as many ideas for further reading and research. Log on to:

**http://www.brookscole.com/sherwoodhp6**

Select Chapter 17 from the drop-down menu, or click on one of the many resources areas.

For Suggested Readings, consult **InfoTrac® College Edition**, your online research library, at:

**http://infotrac.thomsonlearning.com**

# 18

# Principles of Endocrinology: The Central Endocrine Glands

## Chapter Overview

The two major control systems within the body are the nervous system and the endocrine system. The nervous system usually deals with rapid responses while the endocrine system generally provides prolonged, sustained control. Endocrine glands release hormones, blood-borne chemical messengers that act on target cells located long distances from the endocrine glands. The nervous system controls the endocrine system primarily through the hypothalamus. The hypothalamus secretes tropic hormones, which act on endocrine glands, and nontropic hormones, which act on nonendocrine target cells.

The hypothalamus and endocrine glands secrete hormones, or chemical messengers, which evoke responses in target cells elsewhere in the body. Synthesis, storage, secretion, transportation, and mechanism of action all depend on the chemical properties of the hormones. There are chemically three types of hormones. Hormone secretion is controlled by negative feedback. Hormones, regardless of their source, utilize the blood to reach their targets. Hypothalamic neurons secrete two nontropic hormones into capillaries in the posterior pituitary.

The hypothalamus also secretes seven tropic hormones, which are transported to their target, the anterior pituitary, via the hypothalamic-hypophyseal portal system. Growth, which is under endocrine control, is a typical example of how the nervous system ultimately controls prolonged events. Growth also demonstrates the interaction of hormones that brings about attainment of genetic potential. Homeostasis is a very complex phenomenon.

## Chapter Outline

GENERAL PRINCIPLES OF ENDOCRINOLOGY

*Hormones exert a variety of regulatory effects throughout the body.*

- Endocrinology is the study of the homeostatic chemical adjustments and other activities accomplished by hormones.
- The endocrine system is one of the body's two major control systems.
- The endocrine system primarily controls activities that require duration rather than speed including the following: (1) regulating organic metabolism and water and electrolyte balance; (2) inducing adaptive changes to help the body cope with stressful situations; (3) promoting smooth, sequential growth and development; (4) controlling reproduction; (5) regulating red cell production; and (6) along with the autonomic nervous system, controlling and integrating both circulation and the digestion and absorption of food.
- A hormone that has as its primary function the regulation of hormone secretion by another endocrine gland is classified functionally as a tropic hormone.
- A nontropic hormone primarily exerts its effects on nonendocrine target tissues.
- The following points add to the complexity of the endocrine system: (1) a single endocrine gland may produce multiple hormones; (2) a single hormone may be secreted by more than one endocrine gland; (3) frequently, a single hormone has more than one type of target cell and therefore induce more than one type of effect; (4) a single target cell may be influenced by more than one hormone; (5) the same chemical messenger may be either a hormone or a neurotransmitter; and (6) some organs are exclusively endocrine in function whereas other organs of the endocrine system perform nonendocrine functions in addition to secreting hormones.

*The effective plasma concentration of a hormone is normally regulated by changes in its rate of secretion.*

- The plasma concentration of free, biologically active hormone depends on several factors: (1) the hormone's rate of secretion into the blood by the

endocrine gland; (2) its rate of removal from the blood by metabolic inactivation and excretion in the urine; (3) for lipophilic hormones, its extent of binding to plasma proteins; and (4) for a few hormones, its rate of metabolic activation.
- Negative feedback is a prominent feature of hormonal control systems.
- Many endocrine control systems involve neuroendocrine reflexes, which include neural as well as hormonal components.
- The purpose of such reflexes is to produce a sudden increase in hormone secretion in response to a specific stimulus, frequently a stimulus external to the body.
- The secretion rates of all hormones rhythmically fluctuates up and down as a function of time.
- The most common endocrine rhythm is the diurnal or circadian rhythm, which is characterized by repetitive oscillations in hormone levels that are very regular and have a frequency of one cycle every 24 hours.
- Endocrine rhythms are locked on, or "entrained," to external cues, called zeitgebers, such as the light-dark cycle or the activity cycle.
- Some endocrine cycles operate on time scales other than a circadian rhythm—some much shorter than a day and some much longer.

*The effective plasma concentration of a hormone is influenced by its transport, metabolism, and excretion.*
- Eventually, all hormones are metabolized by enzyme-mediated reactions that modify the hormonal structure in some way.
- Hormone metabolism is not always a mechanism for removal of used hormones.
- In some cases, a hormone is activated by metabolism.
- The liver is the most common site for metabolic hormonal inactivation or activation.
- The primary means of eliminating hormones and their metabolites from the blood is by urinary excretion.

*Endocrine disorders result from hormone excess or deficiency or decreased target-cell responsiveness.*
- Endocrine disorders most commonly result from abnormal plasma concentrations of a hormone due to inappropriate secretion rates; that is, hyposecretion or hypersecretion.
- If an endocrine gland is secreting too little of its hormone because of an abnormality within that gland, the condition is referred to as primary hyposecretion.
- If the endocrine gland is normal but is secreting too little hormone because of a deficiency of its tropic hormone, the condition is known as secondary hyposecretion.
- The following are among the many different factors that may be responsible for hormone deficiency: (1) genetic, (2) dietary, (3) chemical or toxic, (4) immunologic, (5) other disease processes, (6) iatrogenic, and (7) idiopathic.
- The most common method of treating hormone hyposecretion is administration of a hormone that is the same as the one that is deficient.
- Hypersecretion by a particular endocrine gland is designated as primary or secondary depending on whether the defect lies in that gland or is due to excessive stimulation from the outside.
- Hypersecretion may be caused by (1) tumors that ignore the normal regulatory input and continuously secrete excess hormone and (2) immunologic factors.
- Endocrine dysfunction can also occur as a result of inadequate responsiveness of the target cells to the hormone.

*The responsiveness of a target cell can be varied by regulating the number of its hormone-specific receptors.*
- A target cell's response to a hormone is correlated with the number of the cell's receptors occupied by molecules of that hormone.
- A hormone can influence the activity of another hormone at a given target cell in one of three ways: permissiveness, synergism, and antagonism.

## HYPOTHALAMUS AND PITUITARY

*The pituitary gland consists of anterior and posterior lobes.*
- The pituitary gland or hypophysis, is a small endocrine gland located in a bony cavity at the base of the brain just below the hypothalamus.
- The pituitary has two anatomically and functionally distinct lobes, the posterior pituitary and the anterior pituitary.
- The posterior pituitary is also termed the neurohypophysis.
- The anterior pituitary is also known as the adenohypophysis.

*The hypothalamus and posterior pituitary act as a unit to secrete vasopressin and oxytocin.*
- The hypothalamus and posterior pituitary form a neuroendocrine system that consists of a population of neurosecretory neurons whose cell bodies lie in two well-defined clusters in the hypothalamus and whose axons pass down through the connecting stalk to terminate on capillaries in the posterior pituitary.

- The posterior pituitary does not actually produce any hormones.
- It simply stores and releases into the blood two small peptide hormones, vasopressin and oxytocin, which are synthesized by the neuronal cell bodies in the hypothalamus.
- Vasopressin has two major effects that correspond to its two names: (1) it enhances the retention of water by the kidneys, and (2) it causes contraction of arteriolar smooth muscle.
- Oxytocin stimulates contraction of the uterine smooth muscle to aid in expulsion of the baby during childbirth, and it promotes ejection of milk from the mammary glands during breast-feeding.

*Most anterior pituitary hormones are tropic.*

- Unlike the posterior pituitary, the anterior pituitary itself synthesizes the hormones that it releases into the blood.
- Different cell populations within the anterior pituitary produce and secrete six established peptide hormones: (1) growth hormone, (2) thyroid-stimulating hormone, (3) adrenocorticotropic hormone, (4) follicle-stimulating hormone, (5) luteinizing hormone, and (6) prolactin.

*Hypothalamic releasing and inhibiting hormones help regulate anterior pituitary hormone secretion.*

- The two most important factors that regulate anterior pituitary hormone secretion are: (1) hypothalamic hormones, and (2) feedback from target-gland hormones.
- The secretion of each of the anterior pituitary hormones is stimulated or inhibited by one or more of the seven generally accepted hypophysiotropic hormones.
- Depending on their actions, these small peptides are called releasing hormones or inhibiting hormones.
- The anatomical and functional link between the hypothalamus and the anterior pituitary is an unusual capillary-to-capillary connection, the hypothalamic-hypophyseal portal system.

*Target-gland hormones inhibit hypothalamic and anterior pituitary hormone secretion via negative feedback.*

- In most cases, hypophysiotropic hormones initiate a three-hormone sequence: (1) hypothalamic releasing hormone, (2) anterior pituitary tropic hormone, and (3) peripheral target hormone.
- The target-gland hormone also acts to suppress secretion of the tropic hormone that is driving it.

## ENDOCRINE CONTROL OF GROWTH

*Growth depends on growth hormone but is influenced by other factors as well.*

- Growth requires net synthesis of proteins and includes lengthening of long bones as well as increases in the size and number of cells in the soft tissues throughout the body.
- An individual's maximum growth capacity is genetically determined.
- Attainment of this full growth potential further depends on: (1) an adequate diet, (2) freedom from chronic disease and stressful environmental conditions, and (3) a normal milieu of growth-influencing hormones.

*Growth hormone is essential for growth, but it also exerts metabolic effects not related to growth.*

- The overall metabolic effect of growth hormone is to mobilize fat stores as a major energy source while conserving glucose for glucose-dependent tissues such as the brain.
- Growth of soft tissues is accomplished by: (1) increasing the number of cells (hyperplasia), and (2) increasing the size of cells (hypertrophy).
- Growth of the long bones resulting in increased height is the most dramatic effect of growth hormone.
- Bones grow in length as a result of proliferation of the cartilage cells in the epiphyseal plates.
- Growth in thickness of bone is achieved by the addition of new bone on top of the already existing bone on the outer surface.

*Growth hormone is essential for growth, but it also exerts metabolic effects not related to growth.*

- Growth hormone does not act directly on its target cells to bring about its growth-producing actions.
- These effects are directly brought about by peptide mediators known as somatomedins, whose synthesis, in turn, is stimulated by growth hormone.
- These peptides are also referred to as insulinlike growth factors.
- The major site of somatomedin production is the liver.
- Growth hormone secretion is regulated by two hypophysiotropic hormones.
- Two antagonistic regulatory hormones from the hypothalamus are involved in the control of growth hormone secretion: growth hormone-releasing hormone and growth hormone-inhibiting hormone.
- Negative-feedback loops participate in the regulation of growth hormone secretion.
- Growth hormone secretion displays a well-characterized diurnal rhythm.

- Approximately one hour after the onset of deep sleep, growth hormone secretion increases up to five times the daytime value.
- Bursts in secretion occur in response to exercise, stress, and hypoglycemia, the major stimuli for increased secretion.
- Growth hormone uses up fat stores and promotes synthesis of body proteins.

*Bone grows in thickness and in length by different mechanisms, both stimulated by growth hormone.*

- Growth hormone deficiency may be caused by a pituitary defect or occur secondarily to hypothalamic dysfunctions.
- Hyposecretion of growth hormone in a child results in dwarfism.
- Growth may be thwarted because the tissues fail to respond normally to growth hormone (Laron dwarfism).
- The symptoms of this condition resemble those of severe growth hormone deficiency even though blood levels of growth hormone are actually high.
- In some instances, growth hormone levels are adequate and target cell responsiveness is normal, but somatomedins are deficient.
- The short stature of African pygmies is attributable to a genetic deficit of the most potent of the somatomedins.
- If overproduction of growth hormone begins in childhood the condition is known as gigantism.
- If growth hormone hypersecretion occurs after adolescence, a disproportionate growth pattern produces a disfiguring condition known as acromegaly.

*Abnormal growth hormone secretion results in aberrant growth patterns.*

- Diseases related to both deficiencies and excesses of growth hormone can occur.

*Other hormones besides growth hormones are essential for normal growth.*

- Several other hormones in addition to growth hormone contribute in special ways to overall growth.
- These hormones include thyroid hormone, insulin, androgens, and estrogens.

## Key Terms

acromegaly
active hormone
adenohypophysis
adrenocorticotropic hormone
amine hormones
androgens
antagonism
anterior pituitary gland
antioxidant
chrondrocyte
circadian rhythm
clock proteins
diurnal (circadian) rhythm
down regulation
dwarfism
endocrine system
endocrinology
entrained
follicle-stimulating hormone
giantism
growth hormone
hormone response element (HRE)
hormones
hyperplasia
hypersecretion
hypertrophy
hypophysiotropic hormone
hypophysis
hyposecretion
hypothalamic-hypophyseal portal system
hypothalamus
insulin-like growth factor
jet lag
Laron dwarfism
luteinizing hormone
melanocyte-stimulating hormone
melanopsin
melatonin
molecular chaperone machinery
neuroendocrine reflex
neurohormones
neurohypophysis
neurosecretory neurons
nontropic hormones
orphan receptors
ossification
osteoblasts
paracrines
permissiveness
peptide hormones
pineal gland
pituicytes
posterior pituitary gland
preprohormones
prohormones
prolactin
protein hormones
short-loop negative feedback
somatomedins
somatostatin
steroid hormones
suprachiasmatic nucleus
synergism
thyroid stimulating hormone
tropic hormone
zeitgebers

## *Review Exercises*

*Answers are in the appendix.*

**True/False**

_____ 1. A postnatal growth spurt occurs during a child's first year of life.

_____ 2. Bone is a living tissue.

_____ 3. Bones grow in length as a result of proliferation of the cartilage cells in the epiphyseal plates.

_____ 4. The major site of somatomedin production is the liver.

_____ 5. Growth hormone deficiency may be caused by a pituitary defect.

_____ 6. Thyroid hormone is believed to be an important growth-promoting factor.

_____ 7. Androgens ultimately terminate linear growth by stimulating complete conversion of the epiphyseal plates to bone.

_____ 8. The posterior pituitary does not actually produce any hormones.

_____ 9. The posterior pituitary is also known as the adenohypophysis.

_____ 10. Oxytocin causes contraction of arteriolar smooth muscle.

_____ 11. Prolactin enhances breast development and milk production in females.

_____ 12. The most intimate means of intracellular communication is by gap junctions.

_____ 13. Neurotransmitters are local chemical messengers whose effect is exerted only on neighboring cells.

_____ 14. A tropic hormone primarily exerts its effects on nonendocrine target tissues.

_____ 15. A single hormone may be secreted by only one endocrine gland.

_____ 16. All steroid and thyroid hormones are hydrophilic.

_____ 17. Cholesterol is the common precursor for all steroid hormones.

_____ 18. The majority of the hydrophilic hormones circulate in the blood to their target cells reversibly bound to plasma proteins.

_____ 19. Negative feedback exists when the output of a system opposes a change in input.

_____ 20. The liver is the most common site for metabolic hormonal inactivation or activation.

**Fill in the Blank**

21. The posterior pituitary consists of neuronal terminals plus glial-like supporting cells called ________________.

22. FSH and LH are collectively referred to as ______________________.

23. The __________________ is a small endocrine gland located in a bony cavity at the base of the brain just below the hypothalamus.

24. The pituitary is connected to the hypothalamus by a thin stalk, the _______________.

25. ________________________________ refers to a pituitary hormone's action at the hypothalamus to inhibit the release of its stimulatory neurohormone.

26. ______________ contributes to the pubertal growth spurt by promoting protein synthesis and bone growth.

27. The bone cells that produce the organic matrix are known as __________________.

28. ____________ is similar to bone, except that it is not calcified.

29. A long bone basically consists of a fairly uniform cylindrical shaft, the _____________ with a flared articulating knob at either end, a(n) _____________.

30. ______________ is a connective-tissue sheath that covers the outer bone surface.

31. Hyposecretion of growth hormone in a child results in _________________.

32. Overproduction of growth hormone beginning in childhood results in ____________________.

33. ______________ are long-range chemical messengers that are specifically secreted into the blood by endocrine glands in response to an appropriate signal.

34. ____________________________________ stimulates growth hormone secretion.

35. ____________________________________ regulates skin coloration by controlling the dispersion of granules containing the pigment _____________.

36. Growth of soft tissues is accomplished by: (1) increasing the number of cells (________________) by stimulating cell division, and (2) increasing the size of cells (________________) by favoring synthesis of proteins.

37. The _____________________, which secretes blood-borne chemical messengers known as hormones, specializes in intracellular communications.

38. Once the hormone is bound to the receptor, the hormone-receptor complex binds with DNA at a specific attachment site on DNA known as the _____________________.

39. ________________ are local chemical messengers whose effect is exerted only on neighboring cells in the immediate environment of their site of secretion.

40. A hormone that has as its primary function the regulation of hormone secretion by another endocrine gland is classified functionally as a(n) _____________ hormone.

41. ______________ which include the hormones secreted by the adrenal cortex, gonads, and most placental hormones, are neutral lipids derived from cholesterol.

42. Preprohormones are synthesized by _________________ on the ______________ ____________________.

43. Endocrine rhythms are locked on or "entrained" to external cues referred to as _________________.

44. The secretion rates of all hormones rhythmically fluctuate up and down as a function of time. The most common endocrine rhythm is the ________________, or _________________, which is characterized by repetitive oscillations in hormone levels that are very regular and have a frequency of one cycle every twenty-four hours.

45. With _______________________, one hormone must be present in adequate amounts for the full exertion of another hormone's effect.

46. __________________ occurs when the actions of several hormones are complementary and their combined effect is greater than the sum of their separate effects.

47. __________________ occurs when one hormone causes the loss of another hormone's receptors, reducing the effectiveness of the second hormone.

48. The secretion of each of the anterior pituitary hormones is stimulated or inhibited by one or more of the seven generally accepted hypothalamic __________________________________________.

49. The anatomical and functional link between the ____________________ and ____________________________ is an unusual capillary-to-capillary connection called the hypothalamic-hypophyseal portal system.

50. The ___________—thyroid hormone and adrenomedullary catecholamines—have unique synthetic and secretory pathways.

51. New cartilage cells are commonly called ____________________.

52. If growth hormone hypersecretion occurs after adolescence the disproportionate growth pattern produces a disfiguring condition known as _________________.

**Matching**

*Match the endocrine gland to its hormone.*

| | | |
|---|---|---|
| _____ 53. cortisol | | a. adrenal medulla |
| _____ 54. oxytocin | | b. posterior pituitary |
| _____ 55. vitamin D | | c. islets of Langerhans |
| _____ 56. estrogen | | d. pineal gland |
| _____ 57. melatonin | | e. hypothalamus |
| _____ 58. glucagon | | f. thyroid follicular cells |
| _____ 59. inhibin | | g. kidneys |
| _____ 60. testosterone | | h. duodenum |
| _____ 61. thymosin | | i. heart |
| _____ 62. prolactin | | j. placenta |
| _____ 63. TRH | | k. skin |
| _____ 64. growth hormone | | l. thymus |
| _____ 65. calcitonin | | m. liver |
| _____ 66. chorionic gonadotropin | | n. anterior pituitary |
| _____ 67. gastrin | | o. stomach |
| _____ 68. atrial natriuretic peptide | | p. parathyroid |
| _____ 69. FSH | | q. thyroid C cells |
| _____ 70. CRH | | r. gonads |
| _____ 71. thyroxine | | s. adrenal cortex |
| _____ 72. aldosterone | | |
| _____ 73. progesterone | | |
| _____ 74. PTH | | |
| _____ 75. insulin | | |

_____ 76. somatomedins
_____ 77. LH
_____ 78. GHIH
_____ 79. epinephrine
_____ 80. TSH
_____ 81. renin
_____ 82. somatostatin
_____ 83. androgens
_____ 84. ACTH
_____ 85. vasopressin
_____ 86. PIH
_____ 87. norepinephrine
_____ 88. cholecystokinin
_____ 89. erythropoietin
_____ 90. PRH
_____ 91. GHRH
_____ 92. secretin

**Multiple Choice**

93. Which of the following is the hormone of the pineal gland?
 a. melanocyte
 b. melanin
 c. melatonin
 d. MSH
 e. PGH

94. Which of the following hormones has as its target an endocrine gland?
 a. paracrines
 b. tropic hormones
 c. nontropic hormones
 d. neurotransmitters
 e. neuromodulators

95. Which of the following promotes breast development and stimulates milk secretion?
 a. leutinizing hormone
 b. somatomedins
 c. renin
 d. oxytocin
 e. prolactin

96. Which of the following decreases plasma calcium concentration?
 a. parathyroid hormone
 b. glucagon
 c. cortisol
 d. calcitonin
 e. oxytocin

97. Which of the following are external cues on which endocrine rhythms lock or entrain?
 a. cicadas
 b. synergisms
 c. Zwitte cues
 d. pituicytes
 e. zeitgebers

98. Which of the following is a condition resulting from hypersecretion of growth hormone after adolescence?
    a. Laron dwarfism
    b. achondroplastic dwarfism
    c. gigantism
    d. acromegaly
    e. exophthalmic goiter

99. Which of the following hormones increase uterine contractility?
    a. oxytocin
    b. vasopressin
    c. PRH
    d. GHRH
    e. none of the above

100. Which of the following hormones produces vasoconstriction in arterioles?
    a. oxytocin
    b. aldosterone
    c. cortisol
    d. ACTH
    e. vasopressin

101. Which of the following conditions occurs when the tissues fail to respond normally to growth hormone?
    a. dwarfism
    b. achondroplastic dwarfism
    c. gigantism
    d. Laron dwarfism
    e. acromegaly

102. Which of the following condition occurs when hypersecretion of growth hormone begins in early childhood?
    a. dwarfism
    b. gigantism
    c. acromegaly
    d. Laron gigantism
    e. achondroplastic gigantism

**Modified Multiple Choice**

*Indicate whether the following characteristics apply to the endocrine system or the nervous system by writing the appropriate letter in the blank using the answer code below.*

A = applies to the endocrine system
B = applies to the nervous system
C = applies to both the endocrine and nervous systems

_____ 103. Structural continuity in the system.
_____ 104. Releases hormones into the blood.
_____ 105. Has an influence on other major control systems.
_____ 106. Secretes chemical messengers that affect target cells.
_____ 107. Chemical messengers act at a long distance from their site of secretion.
_____ 108. Specificity is dependent on specificity of target cell responsiveness.
_____ 109. Controls activities that require longer duration rather than speed.
_____ 110. Duration of action is brief (milliseconds).
_____ 111. Speed of response is long (minutes to days) or longer.

*Indicate the major hormone category that is being described by writing the appropriate letter in the blank using the answer code below.*

A = steroids
B = catecholamines
C = peptides
D = thyroid hormones

_____ 112. Synthesized by the endoplasmic reticulum-Golgi-complex system.
_____ 113. Synthesized by enzymatic modification of cholesterol.
_____ 114. Synthesized within colloid.
_____ 115. Once synthesized, actively transported into preformed vesicles for storage.

*Indicate the type of attraction that is being described by writing the appropriate letter in the blank using the answer code below.*

A = lipophilic (hydrophobic)
B = hydrophilic (lipophobic)

_____ 116. steroids
_____ 117. thyroid hormone
_____ 118. peptides
_____ 119. catecholamines

## Points to Ponder

1. Why do hypothalamic neurons extend into the posterior pituitary but not into the anterior pituitary?

2. Why must there be releasing and inhibiting hormones secreted by the hypothalamus? Why not evolve a system whereby the endocrine gland responds in the negative feedback loop?

3. How does growth fit in homeostasis when homeostasis means a state of physiological equilibrium produced by a balance of functions and chemical composition within an organism?

4. What is the function of the receptor site on a plasma membrane?

5. How can LH and FSH affect both male and female organs?

6. How can the adenohypophysis be specialized to produce so many hormones?

7. How do glucagon and insulin work together to regulate the concentration of blood glucose?

8. Explain why the pituitary gland is sometimes referred to as the "master gland" in the body and why this reference is misleading.

9. With respect to the endocrine system, how can stress lead to certain diseases?

10. What relationship does the adrenal medulla have to the autonomic nervous system?

## *Clinical Perspectives*

1. How would you determine that the negative feedback system has failed for a particular hormone. What would you test for?

2. Based on function, why is ADH also called vasopressin?

3. Does milk production cease once the menstrual cycle resumes after pregnancy? Explain your answer.

4. Anabolic steroids or anabolic-androgenic steroids are synthetic derivatives of testosterone, the male hormone responsible for the anabolic (tissue building) and androgenic (masculinizing) effects seen in normal postpubertal males. What are the pros and cons for athletes using anabolic steroids?

5. Why do diabetics have to inject insulin instead of taking it orally?

6. What is the relationship between prostaglandins and aspirin?

7. Why would oxytocin be given to a pregnant woman? How does it function?

8. How do glucocorticoids inhibit the immune system?

9. Why do individuals take melatonin pills? Do you think they work? Explain your answers.

10. Suppose your body's immune system made antibodies against your insulin receptors. What affect might this condition have on your carbohydrate and fat metabolism.

## *PhysioEdge Activities*

**Related to Text:**
Media Exercise 18.1: Overview of the Classic Endocrine Glands.
Media Exercise 18.2: Anterior Pituitary Gland.
Media Exercise 18.3: Posterior Pituitary Gland.
Media Exercise 18.4: Endocrine Roles by Organs of Mixed or Uncertain Functions.

**Related to Figures:**
No tutorials or media exercises directly related to figures.

## *Media Resources*

**PhysioEdge CD-ROM**
For a visual review of concepts in this chapter, check out:

Media Exercise 18.1: Overview of the Classic Endocrine Glands
Media Exercie 18.2: Anterior Pituitary Gland
Media Exercise: 18.3: Posterior Pituitary Gland
Media Exercise: 18.4: Endocrine Roles by Organs of Mixed or Uncertain Functions

**Book Companion Website**
The website for this book contains a wealth of helpful study aids, as well as many ideas for further reading and research. Log on to:

**http://www.brookscole.com/sherwoodhp6**

Select Chapter 18 from the drop-down menu, or click on one of the many resources areas.

For Suggested Readings, consult **InfoTrac® College Edition**, your online research library, at:

**http://infotrac.thomsonlearning.com**

# 19

# The Peripheral Endocrine Glands

## Chapter Overview

The endocrine system, by means of blood-borne hormones it secretes, generally regulates activities that require duration rather than speed. The role of each body system is equally important in maintaining homeostasis. The peripheral endocrine glands provide an efficient control of nutrient metabolism and mineral balance. The glands in this part of the endocrine system include the thyroid, adrenal, endocrine, pancreas, and parathyroid. These glands respond indirectly to the central nervous system and through their various hormones control the metabolic rate, fuel metabolism, and calcium and phosphate balances. In most instances there are dual controls for increased and decreased activity.

Maintaining adequate nutrient levels to meet the cellular requirements of the organism involves a complex set of hormonal controls. The peripheral endocrine glands not only meet the needs but respond to excesses and shortages by stimulating the various body systems involved with the storage, release, conversion, and excretion of nutrients and minerals. Homeostasis is maintained and the cells survive.

## Chapter Outline

THYROID GLAND

*The major cells that secrete thyroid hormones are organized into colloid-filled follicles.*

- The chief constituent of the colloid is a large, complex molecule known as thyroglobulin, within which are incorporated the thyroid hormones in their various stages of synthesis.
- The follicular cells produce two iodine-containing hormones derived from the amino acid tyrosine: tetraiodothyronine (thyroxine) and triiodothyronine.
- These two hormones, collectively referred to as thyroid hormone, are important regulators of overall basal metabolic rate.
- Interspersed in the interstitial spaces between the follicles is another secretory type cell, the C cells, so called because they secrete the peptide hormone calcitonin, which plays a role in calcium metabolism.

*Thyroid hormone is synthesized and stored on the thyroglobulin molecule.*

- The synthesis, storage, and secretion of thyroid hormone involve the following steps: (1) Tyrosine becomes incorporated in the much larger thyroglobulin molecules, which are exported from the follicular cells into the colloid by exocytosis. (2) The thyroid captures iodine from the blood and transfers it into the colloid by means of a very active "iodine pump." (3) Within the colloid iodine is quickly attached to a tyrosine within the thyroglobulin molecule. Attachment of one iodine to tyrosine yields monoiodotyrosine (MIT). Attachment of two iodines to tyrosine yields diiodotyrosine (DIT). (4) Coupling of two DITs yields tetraiodothyronine (thyroxine). Coupling of one MIT and one DIT yields triiodothyronine.

*To secrete thyroid hormone, the follicular cells phagocytize thyroglobulin-laden colloid.*

- Upon appropriate stimulus for thyroid hormone secretion, the follicular cells internalize a portion of the thyroglobulin-hormone complex by phagocytizing a piece of colloid.
- The biologically active thyroid hormones are split off.
- The MIT and DIT are of no endocrine value.

*Most of the secreted $T_4$ is converted into $T_3$ outside the thyroid.*

- Triiodothyronine is the major biologically active form of thyroid hormone at the cellular level, even though the thyroid secretes mostly tetraiodothyronine.
- Three different plasma proteins are important in thyroid hormone binding: thyroxine-binding globulin, albumin, and thyroxine-binding prealbumin.

*Thyroid hormone is the main determinant of the basal metabolic rate and exerts other effects as well.*

- Virtually every tissue in the body is affected either directly or indirectly by thyroid hormone.
- Thyroid hormone increases the body's overall basal metabolic rate.
- Closely related to thyroid hormone's overall metabolic effect is its calorigenic effect.
- Thyroid hormone modulates the rates of many specific reactions involved in fuel metabolism.
- The effects of thyroid hormone on metabolic fuels are multifaceted; not only can it influence the synthesis and degradation of carbohydrate, fat, and protein, but small or large amounts of the hormone may induce opposite effects.
- Thyroid hormone increases target cell responsiveness to catecholamines, the chemical messengers used by the sympathetic nervous system and its hormonal reinforcements from the adrenal medulla.
- Thyroid hormone increases heart rate and force of contraction, thus increasing cardiac output.
- In response to the heat load generated by the calorigenic effect of thyroid hormone, peripheral vasodilation occurs to carry the extra heat to the body surface for elimination to the environment.
- Thyroid hormone not only stimulates growth hormone secretion but also promotes the effects of growth hormone on the synthesis of new structural proteins and on skeletal growth.
- Thyroid hormone plays a crucial role in the normal development of the nervous system.

*Thyroid hormone is regulated by the hypothalamus-pituitary-thyroid axis.*

- Thyroid-stimulating hormone, the thyroid tropic hormone from the anterior pituitary, is the most important physiological regulator of thyroid hormone secretion.
- Thyroid hormone, in negative-feedback fashion, "turns off" TSH secretion, whereas the hypothalamic thyrotropin-releasing hormone (TRH), in tropic fashion, "turns on" TSH secretion by the anterior pituitary.

*Abnormalities of thyroid function include both hypothyroidism and hyperthyroidism.*

- Hypothyroidism can result: (1) from primary failure of the thyroid gland itself; (2) secondary to a deficiency of TRH, TSH, or both; or (3) from an inadequate dietary supply of iodine.
- The most common cause of hyperthyroidism is Grave's disease, an autoimmune disease in which the body erroneously produces thyroid-stimulating immunoglobulin (TSI), an antibody whose target is the TSH receptors on the thyroid cells.
- Less frequently, hyperthyroidism occurs secondary to excess TRH or TSH or in association with a hypersecreting thyroid tumor.

*A goiter develops when the thyroid gland is over stimulated.*

- A goiter refers to an enlarged thyroid gland.
- Hypothyroidism secondary to hypothalamic or anterior pituitary failure will not be accompanied by a goiter because the thyroid is not being adequately stimulated.
- With hypothyroidism caused by thyroid gland failure or lack of iodine, a goiter does develop because the circulating level of thyroid hormone is so low that there is little negative-feedback inhibition on the anterior pituitary, and TSH secretion is therefore elevated.
- Excessive TSH secretion resulting from hypothalamic or anterior pituitary defect would obviously be accompanied by a goiter because of over stimulation of thyroid growth.
- In Grave's disease, a hypersecreting goiter occurs because TSI promotes growth of the thyroid as well as enhances secretion of thyroid hormone.
- Hyperthyroidism resulting from overactivity of the thyroid in the absence of over stimulation, such as that caused by an uncontrolled thyroid tumor, is not accompanied by a goiter.

## ADRENAL GLANDS

*Each adrenal gland consists of a steroid-secreting cortex and a catecholamine-secreting medulla.*

- Each adrenal gland is actually composed of two endocrine organs: (1) the adrenal cortex, and (2) the adrenal medulla.

*The adrenal cortex secretes mineralocorticoids, glucocorticoids, and sex hormones.*

- The adrenal cortex produces a number of different adrenocortical hormones, all of which are steroids derived from the common precursor molecule, cholesterol.
- Mineralocorticoids influence mineral balance.
- Glucocorticoids, primarily cortisol, play a major role in glucose metabolism as well as in protein and lipid metabolism.
- The sex hormones are identical or similar to those produced by the gonads.

*Mineralocorticoids' major effects are on sodium and potassium balance and blood pressure homeostasis.*

- Mineralocorticoids are essential for life.
- Without aldosterone, a person rapidly dies from circulatory shock because of a marked fall in plasma volume.

- Aldosterone secretion is increased by: (1) activation of the renin-angiotensin-aldosterone system by factors related to a reduction in sodium and a fall in blood pressure, and (2) direct stimulation of the adrenal cortex by a rise in plasma potassium concentration.

*Glucocorticoids exert metabolic effects and play a key role in adaptation to stress.*

- Cortisol stimulates hepatic gluconeogenesis.
- Cortisol inhibits glucose uptake and use by many tissues, but not the brain, thus sparing glucose for use by the brain, which absolutely requires it as metabolic fuel.
- Cortisol stimulates protein degradation in many tissues, especially muscle.
- Cortisol facilitates lipolysis, the breakdown of lipid stores in adipose tissue, thus releasing free fatty acids into the blood.
- Cortisol must be present in adequate amounts to permit catecholamines to induce vasoconstriction.
- Cortisol plays a key role in adaptation to stress.
- Cortisol is also known to alter mood and behavior.
- Administration of large amounts of glucocorticoids inhibits almost every step of the inflammatory response.

*Cortisol secretion is regulated by the hypothalamus-pituitary-adrenal axis.*

- Cortisol secretion by the adrenal cortex is regulated by a long-loop negative-feedback system involving the hypothalamus and anterior pituitary.
- Adrenocorticotropic hormone from the anterior pituitary stimulates the adrenal cortex to secrete cortisol.
- ACTH enhances many steps in the synthesis of cortisol.
- The ACTH-producing cells in turn secrete only at the command of corticotropin-releasing hormone from the hypothalamus.
- The other major factor that is independent of, and in fact can override, the stabilizing negative-feedback control is stress.

*The adrenal cortex secretes both male and female sex hormones in both sexes.*

- No hormones are unique to either males or females (except those from the placenta during pregnancy) because small amounts of the sex hormone of the opposite sex are produced by the adrenal cortex in both sexes.
- The only adrenal sex hormone that has any biological importance is the androgen dehydroepiandrosterone (DHEA).
- This androgen is responsible for androgen-dependent processes in the female, such as growth of pubic and axillary hair, enhancement of the pubertal growth spurt, and development and maintenance of the female sex drive.
- ACTH controls adrenal androgen secretion.
- DHEA inhibits gonadotropin-releasing hormone, just as testicular androgens do.
- Some scientists suspect that the age-related decline of DHEA and other hormones, such as human growth hormone and melatonin, plays a role in some of the problems associated with getting older.
- Early studies with DHEA replacement therapy demonstrated some physical improvement, but the most pronounced effect was a marked increase in psychological well-being and an improved ability to cope with stress.

*The adrenal gland may secrete too much or too little of any of its hormones.*

- Excess mineralocorticoid secretion may be caused by: (1) a hypersecreting adrenal tumor made up of aldosterone-secreting cells, or (2) inappropriately high activity of the renin-angiotensin system.
- Excessive cortisol secretion can be caused by: (1) overstimulation of the adrenal cortex by excessive amounts of CRH and/or ACTH; (2) adrenal tumors that uncontrollably secrete cortisol independent of ACTH; or (3) ACTH-secreting tumors located in places other than the pituitary, most commonly in the lung.
- Excess adrenal androgen secretion, a masculinizing condition, is more common than the extremely rare feminizing condition of excess adrenal estrogen secretion.
- The adrenogenital syndrome is most commonly caused by an inherited enzymatic defect in the cortisol steroidogenic pathway.
- In primary adrenocortical insufficiency, also known as Addison's disease, all layers of the adrenal cortex are undersecreting.
- Secondary adrenocortical insufficiency may occur because of a pituitary or hypothalamic abnormality, resulting in insufficient ACTH secretion.

*The adrenal medulla is a modified sympathetic postganglionic neuron.*

- The adrenal medulla is actually a modified part of the sympathetic nervous system.
- The adrenal medulla is composed of modified postganglionic sympathetic neurons.
- Postganglionic sympathetic neurons in the adrenal medulla do not possess axonal fibers that terminate on effector organs.

- The ganglionic cell bodies within the adrenal medulla release their chemical transmitter directly into the circulation upon stimulation by the preganglionic fiber.

*Epinephrine and norepinephrine vary their affinities for the different adrenergic receptor types.*
- Like sympathetic fibers, the adrenal medulla does release norepinephrine, but its most abundant secretory output is epinephrine.
- Epinephrine and norepinephrine both accelerate the heart rate.
- Both induce generalized arteriolar vasoconstriction of the skin, digestive tract, and kidneys.
- Epinephrine brings about vasodilation of the blood vessels that supply skeletal muscles and the heart.
- Epinephrine is able to exert some unique effects, such as its metabolic effects, because it reaches places not supplied by sympathetic fibers.
- Epinephrine secretion always accompanies a generalized sympathetic discharge, so sympathetic activity indirectly exerts control over actions performed by epinephrine.

*Epinephrine reinforces the sympathetic nervous system and exerts additional metabolic effects.*
- The sympathetic system and epinephrine both exert widespread effects on organ systems that are ideally suited for fight-or-flight responses.
- Under the influence of epinephrine and the sympathetic system, the rate and strength of cardiac contraction increase, resulting in increased cardiac output, and their vasoconstrictor effects increase total peripheral resistance.
- Vasodilation of coronary and skeletal muscle blood vessels induced by epinephrine shifts blood to the heart and skeletal muscles from other vasoconstricted regions of the body.
- Epinephrine dilates the respiratory airways.
- Epinephrine prompts the mobilization of stored carbohydrate and fat to provide immediately available energy for use as needed to fuel muscular work.
- Epinephrine stimulates both hepatic gluconeogenesis and glycogenolysis.
- Epinephrine and the sympathetic system may further add to this hyperglycemic effect by inhibiting the secretion of insulin.
- Epinephrine affects the central nervous system to promote a state of arousal and increased CNS alertness.
- Both epinephrine and norepinephrine cause sweating.
- Epinephrine acts on smooth muscles within the eyes to dilate the pupil and flatten the lens.

*Sympathetic stimulation of the adrenal medulla is solely responsible for epinephrine release.*
- Catecholamine secretion by the adrenal medulla is controlled entirely by sympathetic input to the gland.
- Among the major factors that stimulate increased adreno-medullary output are a variety of stressful conditions such as physical or psychological trauma, hemorrhage, illness, exercise, hypoxia, cold exposure, and hypoglycemia.
- Adrenomedullary dysfunction is very rare.
- No adverse effects have been attributed to a deficiency of epinephrine.
- The only catecholamine disorder is pheochromocytoma, a rarely occurring catecholamine secreting tumor.

## Integrated Stress Response

*The stress response is a generalized pattern of reactions to any situation that threatens homeostasis.*
- Besides epinephrine, a number of other hormones are involved in the overall stress response.
- The predominant hormonal response is activation of the CRH-ACTH-cortisol system.
- The sympathetic nervous system and the epinephrine secreted at its bidding both inhibit insulin and stimulate glucagon.

*The multifaceted stress response is coordinated by the hypothalamus.*
- The hypothalamus receives input concerning physical and emotional stressors from virtually all areas of the body and from many receptors throughout the body.
- In response, the hypothalamus directly activates the sympathetic nervous system, secretes CRH to stimulate ACTH and cortisol release, and triggers the release of vasopressin.
- Sympathetic stimulation brings about the secretion of epinephrine.

*Activation of the stress response by chronic psychosocial stressors may be harmful.*
- There is strong circumstantial evidence for a link between chronic exposure to psychosocial stressors and the development of pathological conditions such as atherosclerosis and high blood pressure, although no definitive cause-and-effect relationship has been ascertained.

## Endocrine Control of Fuel Metabolism

*Fuel metabolism includes anabolism, catabolism, and interconversions among energy-rich organic molecules.*

- Metabolism refers to all the chemical reactions that occur within the cells of the body
- Anabolism refers to the buildup or synthesis of larger organic macromolecules from the small organic molecular subunits.
- Catabolism refers to the breakdown or degradation of large, energy-rich organic molecules within cells.
- Mobilized glucose, fatty acid, and amino acid molecules can be used as needed for energy production or cellular synthesis elsewhere in the body.
- Essential nutrients such as the essential amino acids and vitamins cannot be formed in the body by conversion from another type of organic molecule.

*Because food intake is intermittent, nutrients must be stored between meals.*
- Dietary intake is intermittent, not continuous.

*The brain must be continuously supplied with glucose.*
- Excess circulating glucose is stored as glycogen.
- Additional glucose is transformed into fatty acids and glycerol, which are used to synthesize triglycerides.
- Excess circulating fatty acids derived from dietary intake also become incorporated into triglycerides.
- Excess circulating amino acids not needed for protein synthesis are converted to glucose and fatty acids.
- The major site of energy storage for excess nutrients of all three classes is adipose tissue.
- Glycogen and triglyceride reservoirs serve solely as energy depots.
- A substantial amount of energy is stored as structural protein, primarily in muscle, the most abundant protein mass in the body.
- The brain normally depends on the delivery of adequate blood glucose as its sole source of energy.
- When new dietary glucose is not entering the blood, tissues not obligated to use glucose shift their metabolic gears to burn fatty acids instead, thus sparing glucose for the brain.

*Metabolic fuels are stored during the absorptive state and are mobilized during the post-absorptive state.*
- Ingested nutrients are being absorbed and are entering the blood during the absorptive or fed, state.
- During this time, glucose is plentiful and serves as the major energy source.
- The average meal is completely absorbed in about four hours.
- On a typical three-meals-a-day diet, no nutrients are being absorbed from the digestive tract during late morning, late afternoon, and throughout the night.
- These times constitute the postabsorptive state, or fasting state.
- During the absorptive state, the glut of absorbed nutrients is swiftly removed from the blood and placed into storage; during the postabsorptive state, these stores are catabolized to maintain the blood concentrations at levels necessary to sustain tissue energy demands.

*Lesser energy sources are tapped as needed.*
- Several other organic molecules play a lesser role as energy sources—namely, glycerol, lactic acid, and ketone bodies.

*The pancreatic hormones, insulin and glucagon, are the most important in regulating fuel metabolism.*
- Scattered throughout the pancreas are clusters of endocrine cells known as the islets of Langerhans.
- The most abundant pancreatic endocrine cell type is the beta cell, the site of insulin synthesis and secretion.
- Next most important are the alpha cells, which produce glucagon.
- The delta cells are the pancreatic site of somatostatin synthesis, whereas the least common islet cells, the PP cells, secrete pancreatic polypeptide.
- The overall effect of pancreatic somatostatin is to inhibit digestion of nutrients and to decrease nutrient absorption.
- Pancreatic polypeptide seems to be concerned primarily with inhibiting gastrointestinal function.

*Insulin lowers blood glucose, amino acid, and fatty acid levels and promotes their storage.*
- Insulin lowers the blood levels of glucose, fatty acids, and amino acids and promotes their storage.
- Insulin facilitates glucose transport into most cells.
- Insulin stimulates glycogenesis and inhibits glycogenolysis.
- Insulin further decreases hepatic glucose output by inhibiting gluconeogenesis, the conversion of amino acids into glucose in the liver.
- Insulin increases the transport of glucose into adipose tissue cells by means of GLUT-4 recruitment.
- Glucose serves as a precursor for the formation of fatty acids and glycerol.
- Insulin activates enzymes that catalyze the production of fatty acids from glucose derivatives.
- Insulin promotes the entry of fatty acids from the blood into adipose tissue cells.

- Insulin inhibits lipolysis, thus reducing the release of fatty acids from adipose tissue into the blood.
- Insulin promotes the active transport of amino acids from the blood into muscles and other tissues.
- Insulin increases the rate of amino acid incorporation into protein by stimulating the cells' protein synthesizing machinery.
- Insulin inhibits protein degradation.

*The primary stimulus for increased insulin secretion is an increase in blood glucose concentration.*
- The primary control of insulin secretion is a direct negative-feedback system between the pancreatic beta cells and the concentration of glucose in the blood flowing to them.
- An elevated plasma amino acid level directly stimulates the beta cells to increase insulin secretion.
- The major gastrointestinal hormones, especially gastric inhibitory peptide, stimulate pancreatic insulin secretion.
- The increase in parasympathetic activity that occurs in response to food in the digestive tract stimulates insulin release.
- Sympathetic stimulation and the concurrent increase in epinephrine both inhibit insulin secretion.

*The symptoms of diabetes mellitus are characteristic of an exaggerated post-absorptive state.*
- Diabetes mellitus is by far the most common of all endocrine disorders.
- The most prominent feature of diabetes mellitus is hyperglycemia.
- A large urine volume occurs both in diabetes mellitus due to insulin insufficiency and in diabetes insipidus due to vasopressin deficiency.
- Type I diabetes is characterized by a lack of insulin secretion.
- Type II diabetes is characterized by normal or even increased insulin secretion but reduced sensitivity of insulin's target cells to its presence.

*Insulin excess causes brain-starving hypoglycemia.*
- Insulin excess can occur in a diabetic patient when too much insulin has been injected for the person's caloric intake and exercise level, resulting in so-called insulin shock.
- An abnormally high blood insulin level may occur in a nondiabetic individual with a beta cell tumor or in whom the beta cells are over-responsive to glucose.
- The consequences of insulin excess are primarily manifestations of the effects of hypoglycemia on the brain.

*Glucagon in general opposes the actions of insulin.*
- Many physiologists view the insulin-secreting beta cells and the glucagon-secreting alpha cells as a coupled endocrine system whose combined secretory output is a major factor in the regulation of fuel metabolism.
- Glucagon affects many of the same metabolic processes that are influenced by insulin, but in most cases glucagon's actions are opposite to those of insulin.

*Glucagon secretion is increased during the postabsorptive state.*
- Glucagon secretion is increased during the postabsorptive state and decreased during the absorptive state.
- The major factor regulating glucagon secretion is a direct effect of the blood glucose concentration on the endocrine pancreas.
- There is a direct negative-feedback relationship between blood glucose concentration and the alpha cells' rate of secretion.
- An elevated blood glucose level inhibits glucagon secretion.

*Glucagon and insulin work together to maintain blood glucose and fatty acid levels.*
- A negative-feedback relationship exists between blood glucose concentration and both the beta and alph cells'rates of secretion.

*Glucagon excess can aggravate the hyperglycemia of diabetes mellitus.*
- No known clinical abnormalities are attributable to glucagon deficiency or excess.
- Since glucagon is a hormone that raises blood glucose, its excess intensifies the hyperglycemia of diabetes mellitus.

*Epinephrine, cortisol, growth hormone, and thyroid hormone also exert direct metabolic effects.*
- The stress hormonesepinephrine and cortisol both increase blood levels of glucose and fatty acids.
- Cortisol appears to contribute to the maintenance of blood glucose concentration during long-term starvation.
- Growth hormone, like cortisol, appears to contribute to the maintenance of blood glucose concentrations during starvation.

## ENDOCRINE CONTROL OF CALCIUM METABOLISM

*Plasma calcium must be closely regulated to prevent changes in neuromuscular excitability.*

- Three hormones: (1) parathyroid hormone, (2) calcitonin, and (3) vitamin D control calcium and phosphate metabolism.
- Extracellular fluid calcium plays a vital role in a number of essential activities including the following: (1) neuromuscular excitability, (2) excitation-contraction coupling in cardiac and smooth muscle, (3) stimulus-secretion coupling, (4) maintenance of tight junctions between cells, and (5) clotting of blood.

*Control of calcium metabolism includes regulation of both calcium homeostasis and calcium balance.*

- Regulation of calcium metabolism depends on hormonal control of exchanges between the ECF and three other compartments: bone, kidneys, and intestine.

*Parathyroid hormone raises free plasma calcium levels by its effects on bone, kidneys, and intestine.*

- Parathyroid hormone (PTH) is a peptide hormone secreted by the parathyroid glands.
- The overall effect of PTH is to increase the calcium concentration of plasma.
- Parathyroid hormone also acts to lower plasma phosphate concentration.
- Parathyroid hormone uses bone as a "bank" from which it withdraws calcium as needed to maintain the plasma calcium level.
- Parathyroid hormone stimulates calcium conservation and promotes phosphate elimination by the kidneys during the formation of urine.
- The primary regulator of PTH secretion is the plasma concentration of free calcium.
- Parathyroid hormone secretion is increased in response to a fall in plasma calcium concentration and decreased by a rise in plasma calcium levels.
- Calcitonin lowers the plasma calcium concentration but is not important in the normal control of calcium metabolism.
- Calcitonin, the hormone produced by the C cells of the thyroid gland, also exerts an influence on plasma calcium levels.
- Since calcitonin reduces plasma calcium levels, this system constitutes a second simple negative-feedback control over plasma calcium concentration, one that is opposed to the PTH system.
- Vitamin D is actually a hormone that increases calcium absorption in the intestine.
- The final factor involved in the regulation of calcium metabolism is cholecalciferol, or vitamin D, a steroidlike compound that is essential for calcium absorption in the intestine.
- The skin is actually an endocrine gland, and vitamin D is a hormone.
- The most dramatic and biologically important effect of activated vitamin D is to increase calcium absorption in the intestine.

*PTH's immediate effect is to promote the transfer of calcium from bone fluid into plasma.*

- Hypoparathyroidism leads to hypocalcemia and hyperphosphatemia.
- The major consequence associated with vitamin D deficiency is impaired intestinal absorption of calcium.
- The rates of bone formation and reabsorption are about equal throughout most of adult life.
- A reduction in bone mass is known as osteoporosis.

*PTH acts on the kidneys to conserve calcium and eliminate phosphate.*

- PTH stimulates calcium conservation and promotes phosphate elimination by the kidneys during the formation of urine.

*Calcitonin lowers plasma calcium concentration but is not important in the normal control of calcium metabolism*

- Calcitonin exerts its effect on plasma calcium levels.

*PTH indirectly promotes absorption of calcium and phosphate by the intestine.*

- PTH indirectly increases both calcium and phosphate absorption from the small intestine by means of its role in vitamin D activation.

*Vitamin D is actually a hormone that increases calcium absorption in the intestine.*

- The final factor involved in the regulation of calcium metabolism is cholecalciferal or vitamin D.

*Phosphate metabolism is controlled by the same mechanisms that regulate calcium metabolism.*

- Phosphate is regulated directly by vitamin D and indirectly by the plasma calcium-PTH feedback loop.
- Excess PTH secretion (hyperparathyroidism) is due to a tumor in one of the parathyroid glands.
- Most cases of hypoparathyroidism are due to the removal of the parathyroid glands.

## Key Terms

absorptive (fed) state
Addison's disease (primary adrenocortical insufficiency)
adrenal cortex
adrenal genital syndrome
adrenal medulla
aldosterone
alpha cells
anabolism
beta cells
c cells
calcitonin
calcitriol
calcium balance
calcium homeostasis
calorigenic effect
catabolism
cholecalciferal (vitamin D)
colloid
Conn's syndrome
cretinism
Cushings syndrome
dehydroepiandrosterone (DHEA)
delta cells
diabetes mellitus
epinephrine
essential nutrients
exophthalmos
follicle
general adaptation syndrome
glucagon
glucocorticoids
gluconeogenesis
glucose transporters (GLUT)
glycogenesis
glycogenolysis
goiter
grave's disease (hyperthyroidism)
hirsutism
hydroxyapatite crystals
hypercalcemia
hyperglycemia
hyperparathyroidism
hyperthyroidism
hypocalcemia
hypoglycemia
hypoparathyroidism
hypothyroidism
insulin
insulin antagonists
insulin shock
intermediary (fuel) metabolism
ketogenesis
metabolism
mineralocorticoids
myxedema
osteoporosis
osteoprotegerin
pheochromocytoma
postabsorptive (fasting) state
pp cells
precocious pseudopuberty
primary hyperaldosteronism (Conn's disease)
pseudohermaphroditism
reactive hypoglycemia
somatostatin
stress
sympathomimetic effect
tetraiodothyronine
thyroglobulin
thyrotropic-releasing hormone
thyroxine
transporter recruitment
triiodothyronine
vitamin D

## Review Exercises

*Answers are in the appendix.*

**True/False**

_____ 1. The MIT and DIT are of no endocrine value.

_____ 2. T4 is the major biologically active form of thyroid hormone at the cellular level.

_____ 3. There are four different plasma proteins that are important in thyroid hormone binding.

_____ 4. Thyroid hormone increases the body's overall basal metabolic rate.

_____ 5. Thyroid hormone decreases target-cell responsiveness to catecholamines.

_____ 6. If an individual has hypothyroidism from birth, he has a condition known as myxedema.

_____ 7. A hypersecreting goiter occurs because TSI promotes growth of the thyroid as well as enhancing secretion of thyroid hormone.

_____ 8. Adrenal means next to the kidney.

_____ 9. About 20 percent of the adrenal gland is composed of the cortex.

_____ 10. Without aldosterone, a person rapidly dies from circulatory shock because of the marked fall in plasma volume.

_____ 11. Cortisol stimulates protein degradation in many tissues, except muscle.

_____ 12. Sodium retention is also known as hypokalemia.

_____ 13. A true hermaphrodite has the gonads of both sexes.

_____ 14. Addison's disease is also known as secondary adrenocortical insufficiency.

_____ 15. Epinephrine dilates the respiratory airways.

_____ 16. Both epinephrine and norepinephrine cause sweating.

_____ 17. Excess circulating glucose is stored as glycogen.

_____ 18. Structural protein is the most abundant protein mass in the body.

_____ 19. The average meal is completely absorbed in about 10 hours.

_____ 20. The most abundant pancreatic endocrine-cell type is the alpha cell.

_____ 21. The beta cell is the site of insulin synthesis and secretion.

_____ 22. The delta cells produce glucagon.

_____ 23. Diabetes mellitus is by far the most common of all endocrine disorders.

_____ 24. Insulin is the only hormone capable of lowering blood glucose levels.

_____ 25. Type II diabetes is characterized by a lack of insulin secretion.

_____ 26. Glucagon secretion is increased during the postabsorptive state and decreased during the absorptive state.

_____ 27. An elevated blood glucose level inhibits glucagon secretion.

_____ 28. The overall effect of PTH is to decrease the calcium concentration of plasma.

_____ 29. Parathyroid hormone uses bone as a "bank."

_____ 30. PTH secretion is increased in response to a fall in plasma calcium concentration and decreased by a rise in plasma calcium levels.

_____ 31. Calcitonin increases plasma calcium levels.

_____ 32. Vitamin D is a hormone.

_____ 33. Phosphate is regulated directly by vitamin D.

_____ 34. A fall in plasma phosphate causes a decrease in plasma calcium.

_____ 35. Hypercalcemia reduces the excitability of muscle and nervous tissue.

**Fill in the Blank**

36. ________________ refers to all of the chemical reactions that occur within the cells of the body.

37. ________________ refers to the build-up or synthesis of larger organic macromolecules from the small organic molecular subunits.

38. ________________ refers to the breakdown or degradation of large, energy-rich organic molecules within cells.

39. Scattered throughout the pancreas are clusters of endocrine cells known as the ________________________.

40. ________________ are a group of compounds produced by the liver during glucose sparing.

41. ________________ is the production of glycogen.

42. Hypoglycemia refers to _______ blood glucose.

43. ______________ are layers of osteocytes entombed within the bone that they have deposited around themselves.

44. ____________________________________________ separates the mineralized bone itself from the plasma within the central canals.

45. ______________, the hormone produced by the C-cells of the thyroid gland, exerts an influence on plasma calcium levels.

46. The final factor involved in the regulation of calcium metabolism is ______________.

47. When demineralized bones become soft and deformed, bowing to the pressures of weight bearing, is known as ____________ in children and ____________________ in adults.

48. A reduction in bone mass is known as ________________________.

49. The chief constituent of the colloid is a large, complex molecule known as _____________________.

50. Normal thyroid function is referred to as ______________________.

51. ______________________ is an autoimmune disease in which the body erroneously produces thyroid-stimulating immunoglobulin (TSI).

52. A(n) _________ refers to an enlarged thyroid gland.

53. The ___________________ secrete hormones important in the metabolism of nutrient molecules, adaptation to stress, and maintenance of salt balance.

54. ____________ are so called because they secrete the peptide hormone calcitonin.

55. MIT is short for ______________________________________.

56. $T_4$ is short for _________________.

57. The three layers of the cortex are: (1) ________________________ –the outermost layer, (2) ________________________ –middle and largest layer, and (3) __________________________ –the innermost layer.

58. ______________ plays an important role in carbohydrate, protein, and fat metabolism.

59. _______________ syndrome is an excessive cortisol secretion.

60. ____________________ refers to the condition where a woman develops a male pattern of body hair.

61. ________________ disease is where all layers of the adrenal cortex are under-secreting.

62. Epinephrine and norepinephrine are stored in _________________________________.

63. The only catecholamine disorder is a(n) ______________________________, a rarely occurring catecholamine-secreting tumor.

**Matching**

*Match the various causes of thyroid dysfunction to the plasma concentrations of hormones and presence of goiters.*

a. increased thyroid hormone, decreased TSH and no goiter
b. decreased thyroid hormone, increased TSH with goiter
c. increased thyroid hormone, decreased TSH with goiter
d. decreased thyroid hormone, decreased TRH and/or decreased TSH and no goiter
e. increased thyroid hormone, increased TRH and/or increased TSH with goiter

_____ 64. primary failure of thyroid gland
_____ 65. anterior pituitary failure or hypothalamic failure
_____ 66. excess hypothalamic secretion or excess pituitary secretion
_____ 67. lack of dietary iodine
_____ 68. hypersecreting thyroid tumor
_____ 69. grave's disease

*Match the condition to the disorder.*

_____ 70. excess aldosterone
_____ 71. excess androgen
_____ 72. deficient cortisol and aldosterone
_____ 73. excess cortisol
_____ 74. deficient cortisol

a. Cushing syndrome
b. Addison's disease
c. Conn's syndrome
d. Secondary adrenocortical insufficiency
e. Adrenogenital syndrome

**Multiple Choice**

75. Which of the following refers to a potassium depletion?
a. hypocalcemia
b. osteomalacia
c. hypokalemia
d. osteoporosis
e. pheochromocytoma

76. Which of the following refers to a reduction in bone mass?
a. hirsutism
b. osteoporosis
c. osteomalacia
d. hypocalcemia
e. hypokalemia

77. Which of the following refers to the prominent bulging eyes characteristic of Grave's disease?
   a. goiter
   b. hirsutism
   c. osteomalacia
   d. optic protrusia
   e. exophthalmos

78. Which disorder is characterized by a lack of insulin?
   a. type I diabetes
   b. type II diabetes
   c. diabetes insipidus
   d. pheochromocytoma
   e. insulemia

79. Which of the following is characterized by the development of a male pattern of body hair in women?
   a. Pheochromocytoma
   b. Ostomalacia
   c. Goiter
   d. Exophthalmos
   e. Hirsutism

80. Which of the following disorders is characterized by the demineralization of bones in adults?
   a. rickets
   b. osteoporosis
   c. osteomalacia
   d. hirsutism
   e. pheochromocytoma

81. Which of the following disorders is characterized by a catecholamine-secreting tumor?
   a. pheochromocytoma
   b. catechofibroma
   c. catecholemia
   d. exophthalmos
   e. osteomalacia

82. Which disorder is characterized by low blood calcium?
   a. hypokalemia
   b. hypercalcemia
   c. hypocalcemia
   d. osteoporosis
   e. osteomalacia

83. Which disorder refers to an enlarged thyroid gland?
   a. hypothyroidism
   b. hyperthyroidism
   c. pheochromocytoma
   d. exophthalmos
   e. goiter

84. Which disorder is characterized by reduced sensitivity of insulin's target cells?
   a. insulemia
   b. type I diabetes
   c. type III diabetes
   d. diabetes insipidus
   e. type II diabetes

**Modified Multiple Choice**

*Indicate the effect the item on the left has on the secretion rate of the hormone on the right by circling the appropriate letter using the answer code below.*

A = increases
B = decreases
C = has little or no effect on

85. Increased TSH _____ thyroid hormone.
86. Increased TRH _____ TSH.
87. Decreased thyroid hormone _____ TSH.
88. Decreased iodine in the diet _____ thyroid hormone.
89. Increased ACTH _____ aldosterone.
90. Increased ACTH _____ cortisol.
91. Stress _____ cortisol.
92. Stress _____ epinephrine.
93. Increased renin-angiotensin _____ aldosterone.
94. Increased plasma potassium ion _____ aldosterone.
95. Increased blood glucose _____ insulin.
96. Increased blood glucose _____ glucagon.
97. Increased plasma calcium _____ parathyroid hormone.
98. Increased plasma calcium _____ calcitonin.
99. Increased TSH _____ calcitonin.
100. Increased plasma phosphate ion _____ parathyroid hormone.

*Indicate the metabolic effects each on each of the following hormones by writing the appropriate letter using the answer code below.*

A = increases
B = decreases
C = has no effect on

101. Insulin _____ blood glucose.
102. Insulin _____ fatty acids.
103. Insulin _____ blood amino acids.
104. Insulin _____ muscle protein.
105. Glucagon _____ blood glucose.
106. Glucagon _____ blood fatty acids.
107. Glucagon _____ blood amino acids.
108. Glucagon _____ muscle protein.
109. Epinephrine _____ blood glucose.
110. Epinephrine _____ blood fatty acids.
111. Epinephrine _____ blood amino acids.
112. Epinephrine _____ muscle protein.
113. Cortisol _____ blood glucose.
114. Cortisol _____ blood fatty acids.
115. Cortisol _____ blood amino acids.
116. Cortisol _____ muscle protein.
117. Growth hormone _____ blood glucose.
118. Growth hormone _____ blood fatty acids.
119. Growth hormone _____ blood amino acids.
120. Growth hormone _____ muscle protein.

## Points to Ponder

1. Recent research indicates that diabetes mellitus is genetic. If this is true then the use of insulin only increases the frequency of the disease in the human population. How do we reduce the frequency of the disease and still help those who have the disorder?

2. How much of your behavior can you attribute to the hormones discussed in this and the previous chapter?

3. What do you think causes osteoporosis? Is it an estrogen deficiency? Do men have estrogen levels?

4. List the steps and describe the process of biosynthesis of thyroid hormones.

5. Explain the statement that glucocorticoids have a catabolic effect on protein metabolism.

6. The hypothalamic-pituitary axis may be activated under stress. Explain the process by which it occurs.

7. What is the normal effect of insulin on potassium?

8. What does an elevated blood glucose indicate?

9. What are two possible complications of diabetic ketoacidosis or HHNK?

10. What is the most common cause of Addison's disease?

## Clinical Perspectives

1. After a total thyroidectomy, what is the significance of elevated serum thyroglobulin in patients with metastatic disease?

2. What effects do abnormally high levels of glucocortocoids have on glucose metabolism?

3. Explain the way aldosteronism develops in response to congestive heart failure.

4. What are the factors that contribute to hypoglycemia in Addison's disease?

5. What is the purpose of measuring fasting blood glucose levels?

6. What are the advantages of the second generation oral sulfonylureas?

7. Why is it important to identify individuals who are at risk for developing diabetes mellitus?

8. Why is edema absent in chronic aldosteronism?

9. Why is ketoacidosis harmful to a diabetic?

10. What is the reason for osmotic diuresis in ketoacidosis?

## *PhysioEdge Activities*

**Related to Text:**
Media Exercise 19.1: Thyroid Functions.
Media Exercise 19.2:Adrenal Gland Functions.
Media Exercise 19.3: Endocrine Pancreas and Fuel Metabolism.
Media Exercise 19.4: Endocrine Control of Calcium Metabolism.

**Related to Figures:**
*Figure 19.1.* For an interaction related to this figure, see Media Exercise 19.1: Thyroid Functions.
*Figure 19.3.* For an interaction related to this figure, see Media Exercise 19.1: Thyroid Functions.
*Figure 19.7.* For an interaction related to this figure, see Media Exercise 19.2: Adrenal Gland Functions.

## *Media Resources*

**PhysioEdge CD-ROM**
For a visual review of concepts in this chapter, check out:

Media Exercise 19.1: Thyroid Functions
Media Exercie 19.2: Adrenal Gland Functions
Media Exercise: 19.3: Endocrine Pancreas and Fuel Metabolism
Media Exercise: 19.4: Endocrine Control of Calcium Metabolism

**Book Companion Website**
The website for this book contains a wealth of helpful study aids, as well as many ideas for further reading and research. Log on to:

**http://www.brookscole.com/sherwoodhp6**

Select Chapter 19 from the drop-down menu, or click on one of the many resources areas, including **Case Histories**, which introduce clinical aspects of human physiology. For this chapter check out Case History 4: Starvation in the Midst of Plenty; Case History 5: Eight Years Later.

For Suggested Readings, consult **InfoTrac® College Edition**, your online research library, at:

**http://infotrac.thomsonlearning.com**

# The Reproductive System

## Chapter Overview

The reproductive system contributes very little to the maintenance of homeostasis of the body. An individual can survive quite normally without reproducing. However, the survival of the species depends on the reproductive system. This is the major function of the system.

While the reproductive system is not a primary factor in overall homeostasis of the body, there are unique situations in the reproductive process that involve homeostasis. These situations include Sertoli cells, endometrial cells, and the hormonal controls within the reproductive system.

The testes and ovaries produce the gametes with reduced chromosome numbers necessary for zygote formation through fusion of the egg and sperm. The reproductive system also produces hormones, which play a role in initiation and completion of this reproductive function of the species. This body system is controlled by the central nervous system and the endocrine system. The various body systems maintain homeostasis in the body, and cells of the body survive. But, if it were not for a pair of reproductive systems there would be no body.

## Chapter Outline

INTRODUCTION

*The reproductive system is not essential for survival of the individual, but it is necessary for survival of the species and has a profound impact on a person's life.*

- Reproductive capability depends on an intricate relationship among the hypothalamus, anterior pituitary, reproductive organs, and target cells of sex hormones.

*The reproductive system includes the gonad and reproductive tract.*

- The primary reproductive organs, or gonads, perform the dual function of: (1) gametogenesis, and (2) secreting sex hormones.
- Testosterone in the male and estrogen in the female are responsible for the secondary sex characteristics.
- The essential reproductive functions of the male are: (1) spermatogenesis, and (2) delivery of sperm to the female.
- The essential female reproductive functions include: (1) oogenesis, (2) reception of sperm, (3) fertilization, (4) gestation, (5) parturition, and (6) lactation.

*Reproductive cells each contain a half set of chromosomes.*

- Somatic cells contain 46 chromosomes (the diploid number).
- Gametes contain only one of each homologous pair for a total of 23 chromosomes (the haploid number).

*Gametogenesis is accomplished by meiosis.*

- During meiosis, a specialized diploid germ cell undergoes one chromosome replication followed by two nuclear divisions.
- Thus, sperm and ova each have a unique haploid number of chromosomes.
- When fertilization takes place, a sperm and ovum fuse to form the start of a new individual with 46 chromosomes.

*The sex of an individual is determined by the combination of sex chromosomes.*

- Twenty-two of the chromosome pairs are autosomal chromosomes.
- The remaining pair of chromosomes contain the sex chromosomes.

*Sex differentiation along male or female lines depends on the presence or absence of masculinizing determinants.*

- Gonadal specificity appears during the seventh week of intrauterine life.
- The sex-determining region of the Y chromosome "masculinizes" the gonads by stimulating production of H-Y antigen by primitive gonadal cells.

- By 10 to 12 weeks of gestation, the sexes can be easily distinguished by the anatomical appearance of the external genitalia.
- In males, the reproductive tract develops from the Wolffian ducts.
- In females, the Mullerian ducts differentiate into the reproductive tract.
- Development of the reproductive tract along male or female lines is determined by the presence or absence of two hormones secreted by the fetal testes—testosterone and Mullerian-inhibiting factor.

## MALE REPRODUCTIVE PHYSIOLOGY

*The scrotal location of the testes provides a cooler environment, essential for spermatogenesis.*

- In the last months of fetal life, the testes begin a slow descent, passing out of the abdominal cavity through the inguinal canal into the scrotum.
- Descent of the testes into this cooler environment is essential because spermatogenesis is temperature sensitive and cannot occur at normal body temperature.

*The testicular Leydig cells secrete masculinizing testosterone.*

- About 80 percent of the testicular mass consists of highly coiled seminiferous tubules, within which spermatogenesis takes place.
- The endocrine cells that produce testosterone—the Leydig cells or interstitial cells—are located in the connective tissue between the seminiferous tubules.
- Before birth, testosterone secretion by the fetal testes is responsible for masculinizing the reproductive tract and external genitalia and for promoting descent of the testes into the scrotum.
- At puberty, the Leydig cells start secreting testosterone, and spermatogenesis is initiated in the seminiferous tubules.
- Testosterone is responsible for growth and maturation of the entire male reproductive system.
- Ongoing testosterone secretion is essential for spermatogenesis and for maintaining a mature male reproductive tract throughout adulthood.
- Testosterone is responsible for development of sexual libido at puberty and helps to maintain the sex drive in the adult male.
- All male secondary sexual characteristics depend on testosterone for their development and maintenance.
- Testosterone has a general protein anabolic effect and promotes bone growth.
- Testosterone secretion and spermatogenesis occur continuously throughout the male's life.

*Spermatogenesis yields an abundance of highly specialized, mobile sperm.*

- Two functionally important cell types are present in the seminiferous tubules: germ cells and Sertoli cells.
- During meiosis, each primary spermatocyte forms two secondary spermatocytes during the first meiotic division, finally yielding four spermatids as a result of the second meiotic division.
- Sperm are essentially "stripped-down" cells from which most of the cytosol and the organelles not needed for the task of delivering the sperm's genetic information to an ovum have been extruded.
- A spermatozoon has four parts: a head, an acrosome, a midpiece, and a tail.
- The developing germ cells arising from a single primary spermatocyte remained joined by cytoplasmic bridges.
- If it were not for the sharing of cytoplasm so that all the haploid cells are provided with the products coded for by X chromosomes until sperm development is complete, the Y-bearing, male-producing sperm would not be able to develop and survive.

*Throughout their development, sperm remain intimately associated with Sertoli cells.*

- The Sertoli cells form a ring that extends from the outer basement membrane to the lumen of the tubule.
- The supportive Sertoli cells perform the following functions essential for spermatogenesis: (1) The tight junctions between adjacent Sertoli cells form a blood-testes barrier. (2) Sertoli cells provide nourishment for sperm cells. (3) Sertoli cells engulf cytoplasm extruded from spermatids during their remodeling and destroy defective germ cells that fail to successfully complete all stages of spermatogenesis. The Sertoli cells secrete into the lumen seminiferous tubule fluid, which "flushes" sperm into the epididymis. Sertoli cells secrete androgen-binding protein. The Sertoli cells are the site of action for control of spermatogenesis by both testosterone and FSH.
- The Sertoli cells release inhibin, which acts in negative-feedback fashion to regulate FSH secretion.

*LH and FSH from the anterior pituitary control testosterone secretion and spermatogenesis.*

- The testes are controlled by two gonadotropic hormones secreted by the anterior pituitary: (1) luteinizing hormone (LH), and (2) follicle-stimulating hormone (FSH).

- Luteinizing hormone acts on the Leydig cells to regulate testosterone secretion.
- Follicle stimulating hormone acts on the seminiferous tubules, especially the Sertoli cells, to enhance spermatogenesis.
- Secretion of both LH and FSH from the anterior pituitary is stimulated by a single hypo-thalamic hormone, gonadotropin-releasing hormone (GnRH).
- Follicle-stimulating hormone is required for spermatid remodeling.

*Gonadotropin-releasing hormone activity increases at puberty.*
- The pubertal process is initiated by an increase in GnRH activity sometime between eight and twelve years of age.
- At puberty, the hypothalamus becomes less sensitive to feedback inhibition by testosterone.

*The reproductive tract stores and concentrates sperm and increases their fertility.*
- The remainder of the male reproductive system consists of: (1) a tortuous pathway of tubes that transport sperm from the testes to the outside of the body; (2) several glands, which contribute secretions that are important to the viability and motility of the sperm; and (3) the penis, which is designed to penetrate and deposit the sperm within the vagina of the female.
- As they leave the testes, the sperm are incapable of either movement or fertilization.
- Sperm gain both capabilities during their passage through the epididymis.
- The epididymis concentrates the sperm by absorbing most of the fluid that enters from the seminiferous tubules.
- The ductus deferens serves as an important site for sperm storage.

*The accessory sex glands contribute the bulk of the semen.*
- The seminal vesicles: (1) supply fructose, (2) secrete prostaglandins, (3) provide more than half of the semen, and (4) secrete fibrinogen.
- The prostate gland: (1) secretes an alkaline fluid that neutralizes the acidic vaginal secretions, and (2) provides clotting enzymes and fibrinolysin.
- The bulbourethral glands secrete a mucus-like substance that provides lubrication for sexual intercourse.

*Prostaglandins are ubiquitous, locally acting chemical messengers.*
- Prostaglandins are produced in virtually all tissues from arachidonic acid, a fatty acid constituent of the phospholipids within the plasma membrane.
- Prostaglandins are designated as belonging to one of three groups—PGA, PGE and PGF—according to structural variations in the five-carbon ring that they contain at one end.
- Besides enhancing sperm transport in semen, prostaglandins are known or suspected to exert other actions in the female reproductive system and in the respiratory, urinary, digestive, nervous and endocrine systems, in addition to having effects on platelet aggregation, fat metabolism, and inflammation.
- Specific prostaglandins have been medically administered in such diverse situations as inducing labor, treating asthma, and treating gastric ulcers.

## Sexual Intercourse between Males and Females

*The male sex act is characterized by erection and ejaculation.*
- The male sex act involves two components: (1) erection, and (2) ejaculation.
- The sexual response cycle encompasses broader physiological responses that can be divided into four phases: (1) the excitement phase, (2) the plateau phase, (3) the orgasmic phase, and (4) the resolution phase.

*Erection is accomplished by penis vasocongestion.*
- Erection is caused by engorgement of the penis with blood.
- During sexual arousal, the arterioles reflexly dilate and the erectile tissue fills with blood, causing the penis to enlarge both in length and width and to become more rigid.
- A larger buildup of blood and further enhancement of erection is achieved by a reduction in venous outflow.
- Tactile stimulation of the glans reflexly triggers increased parasympathetic vasodilator activity and decreased sympathetic vasoconstrictor activity to the penile arterioles.
- This response is the major instance of direct parasympathetic control over blood vessel caliber in the body.
- Parasympathetic impulses promote secretion of lubricating mucus from the bulbourethral glands and the urethral glands in preparation for coitus.
- The erection reflex is a spinal reflex that can be either facilitated or inhibited by higher brain centers through descending pathways that also terminate on the autonomic nerves supplying the penile arterioles.

- Psychic stimuli can induce an erection.

*Ejaculation includes emission and expulsion.*
- Ejaculation is accomplished by a spinal reflex.
- Sympathetic impulses cause sequential contraction of smooth muscles in the prostate, reproductive ducts, and seminal vesicles.
- This contractile activity delivers prostatic fluid, then sperm, and finally seminal vesicle fluid into the urethra.
- The filling of the urethra with semen triggers nerve impulses that activate a series of skeletal muscles at the base of the penis.
- Rhythmic contractions of these muscles forcibly expel the semen through the urethra to the exterior.
- During the resolution phase following orgasm, sympathetic vasoconstrictor impulses slow the inflow of blood into the penis so that the erection subsides.

*The female sexual cycle is very similar to the male cycle.*
- Both sexes experience the same four phases of sexual cycle.
- The excitement phase in females can be initiated by either physical or psychological stimuli.
- Tactile stimulation of the clitoris and surrounding perineal area is an especially powerful stimulus.
- Spinal reflexes bring about parasympathetically induced vasodilation of arterioles throughout the vagina and external genitalia.
- Vasocongestion of the vaginal capillaries forces fluid out of the vessels into the vaginal lumen.
- This fluid serves as the primary lubricant for intercourse.
- Further vasocongestion of the lower third of the vagina reduces its inner capacity so that it tightens around the thrusting penis.
- Simultaneously, the uterus raises upward, lifting the cervix and enlarging the upper two-thirds of the vagina to create a space for ejaculate deposition.
- The sexual response culminates in orgasm as sympathetic impulses trigger rhythmic contractions of the pelvic musculature.

## Female Reproductive Physiology

*Complex cycling characterizes female reproductive physiology.*
- The release of ova is intermittent, and secretion of female sex hormones displays wide cyclical swings.
- During each menstrual cycle, the female reproductive tract is prepared for the fertilization and implantation of an ovum released from the ovary at ovulation.
- The ovaries perform the dual function of oogenesis and secreting the female sex hormones, estrogen and progesterone.
- Estrogen in the female is responsible for maturation and maintenance of the entire female reproduction system and establishment of female secondary sexual characteristics.
- Estrogen is essential for ova maturation and release, development of physical characteristics that are sexually attractive to males, and transport of sperm from the vagina to the site of fertilization in the oviduct.
- Estrogen contributes to breast development in anticipation of lactation.
- Progesterone is important in preparing a suitable environment for nourishing a developing embryo/fetus and for contributing to the breasts' ability to produce milk.

*The steps of gametogenesis are the same in both sexes but the timing and outcome sharply differ.*
- The undifferentiated primordial germ cells in the fetal ovaries divide mitotically to give rise to millions of oogonia by the fifth month of gestation, at which time mitotic proliferation ceases.
- Known now as primary oocytes, they contain 46 replicated chromosomes.
- Before birth, each primary oocyte is surrounded by a single layer of granulosa cells to form the primary follicle.
- No new oocytes or follicles appear after birth.
- A follicle is destined for one of two fates once it starts to develop: it will reach maturity and ovulate, or it will degenerate to form scar tissue.
- By menopause, few, if any, primary follicles remain.
- The primary oocyte within a primary follicle is still a diploid cell.
- Just before ovulation, the primary oocyte, whose nucleus has been in meiotic arrest for years, completes its first meiotic division.
- This division yields two daughter cells.
- Almost all of the cytoplasm remains with one of the daughter cells, now called the secondary oocyte, which is destined to become the ovum.
- The chromosomes of the other daughter cell together with a small share of cytoplasm form the first polar body.
- The nutrient-poor polar body soon degenerates.
- Sperm entry into the secondary oocyte is needed to trigger the second meiotic division.
- Twenty-three unpaired chromosomes remain in what is now the mature ovum.

- These 23 maternal chromosomes unite with the 23 paternal chromosomes of the penetrating sperm to complete fertilization.

*The ovarian cycle consists of alternating follicular and luteal phases.*

- After the onset of puberty, the ovary constantly alternates between two phases: (1) the follicular phase, which is dominated by the presence of maturing follicles; and (2) the luteal phase, which is characterized by the presence of the corpus luteum.
- The released ovum is quickly drawn into the oviduct.
- Rupture of the follicle at ovulation signals the end of the follicular phase.

*The ovarian cycle is regulated by complex hormonal interactions.*

- Gonadal function in the female is directly controlled by the anterior pituitary gonadotropic hormones, FSH and LH.
- These hormones are regulated by pulsatile episodes of hypothalamic gonadotropin-releasing hormone and feedback actions of gonadal hormones.
- Antrum formation is induced by FSH.
- Both FSH and estrogen stimulate proliferation of granulosa cells.
- Both LH and FSH are required for synthesis and secretion of estrogen by the follicle.
- Thecal cells readily produce androgens.
- Granulosa cells are readily able to convert androgens into estrogens.
- Luteinizing hormone acts on the thecal cells to stimulate androgen production, whereas FSH acts on the granulosa cells to promote the conversion of thecal androgens into estrogen.
- Part of the estrogen produced by the growing follicle is secreted into the blood and is responsible for steadily increasing plasma estrogen levels during the follicular phase.
- The secreted estrogen inhibits the hypothalamus and anterior pituitary in negative-feedback fashion.
- The plasma FSH level declines during the follicular phase as the estrogen level rises.
- The LH secretion continues to rise slowly during the follicular phase despite inhibition of GnRH secretion.
- Ovulation and subsequent luteinization of the ruptured follicle are triggered by an abrupt, massive increase in LH secretion.
- This LH surge brings about four major changes in the follicle: (1) It halts estrogen synthesis by the follicular cells. (2) It reinitiates meiosis in the oocyte of the developing follicle. (3) It triggers production of locally acting prostaglandins. (4) It causes differentiation of follicular cells into the luteal cells.
- The LH surge is triggered by a positive-feedback effect.
- Thus, LH enhances estrogen production by the follicle, and the resultant peak estrogen concentration stimulates LH secretion.
- Luteinizing hormone "maintains" the corpus luteum.
- No progesterone is secreted during the follicle phase.
- The follicular phase is dominated by estrogen and the luteal phase by progesterone.
- Even though a high level estrogen stimulates LH secretion, progesterone, which dominates the luteal phase, powerfully inhibits LH secretion as well as FSH secretion.
- Inhibition of FSH and LH by progesterone prevents new follicular maturation and ovulation during the luteal phase.
- Under progesterone's influence, the reproductive system is gearing up to support the just-released ovum, should it be fertilized.

*Cyclic uterine changes are caused by hormonal changes during the ovarian cycle.*

- The menstrual cycle consists of three phases: the menstrual phase, the proliferative phase, and the secretory or progestational phase.
- The menstrual phase is characterized by discharge of blood and endometrial debris from the vagina.
- The first day of menstruation is considered to be the start of a new cycle and coincides with termination of the ovarian luteal phase and onset of the follicular phase.
- As the corpus luteum degenerates, circulating levels of estrogen and progesterone drop precipitously.
- The fall in ovarian hormone levels also stimulates release of a uterine prostaglandin that causes vasoconstriction of the endometrial vessels.
- The entire uterine lining sloughs during each menstrual period except for a deep, thin layer of epithelial cells and glands from which the endometrium will regenerate.
- Local uterine prostaglandin stimulates mild rhythmic contractions of the uterine myometrium to help expel the blood and endometrial debris from the uterine cavity.
- Excessive uterine contractions caused by overproduction of prostaglandin are responsible for menstrual cramps (dysmenorrhea).
- Withdrawal of estrogen and progesterone upon degeneration of the corpus luteum leads

simultaneously to sloughing of the endometrium and development of new follicles in the ovary under the influence of rising gonadotropic hormone levels.
- The proliferative phase of the uterine cycle begins concurrent with the last portion of the ovarian follicular phase as the endometrium starts to repair.
- The estrogen-dominant proliferative phase lasts from the end of menstruation to ovulation.
- Following ovulation, a new corpus luteum is formed.
- Progesterone acts on the thickened, estrogen-primed endometrium to convert it to a richly vascularized, glycogen-filled tissue.
- This period is called the secretory phase, or the progestational phase.
- If fertilization and implantation do not occur, the corpus luteum degenerates and a new follicular phase and menstruation phase begin once again.

*Fluctuating estrogen and progesterone levels produce cyclical changes in cervical mucus.*
- Under the influence of estrogen during the follicular phase, the mucus secreted by the cervix is abundant, clear, and thin.
- After ovulation, under the influence of progesterone from the corpus luteum, the mucus becomes thick and sticky, essentially forming a plug across the cervical opening.

*Pubertal changes in females are similar to those in males.*
- Unlike the fetal testes, the fetal ovaries do not need to be functional, because feminization of the female reproductive system automatically takes place in the absence of fetal testosterone secretion without the presence of female sex hormones.
- The resultant secretion of estrogen by the activated ovaries induces growth and maturation of the female reproductive tract as well as development of the female secondary sexual characteristics.
- Three other pubertal changes in females--growth of axillary and pubic hair, the pubertal growth spurt, and development of libido--are attributed to a spurt in adrenal androgen secretion at puberty, not to estrogen.
- The pubertal rise in estrogen closes the epiphyseal plates.
- The cessation of a woman's menstrual cycles at menopause is attributable to the limited supply of ovarian follicles present at birth.
- The termination of reproductive potential in a middle-aged woman is "preprogrammed" at her own birth.
- Menopause is preceded by a period of progressive ovarian failure characterized by increasingly irregular cycles and dwindling estrogen levels.
- Postmenopausal women are not completely devoid of estrogen, however, because adipose tissue, the liver, and the adrenal cortex continue to produce estrogen.
- The absence of ovarian estrogen is responsible for the physical postmenopausal changes that occur in the reproductive tract, including vaginal dryness and gradual atrophy of the genital organs.
- Postmenopausal women still have a sex drive because of their adrenal androgens.

*Menopause is unique to females.*
- The dramatic loss of ovarian estrogen that accompanies menopause has impacts on other body systems.
- Estrogen helps build strong bones, thus shielding premenopausal women from the bone-thinning condition of osteoporosis.
- Estrogen exerts cardioprotective effects that give females a fifteen- to twenty-year advantage over males in evading atherosclerotic coronary disease.
- Coronary artery disease is the leading cause of death in women.
- Postmenopausal estrogen deficiency leads to a decrease in the production of HDL-cholesterol and an increase in LDL-cholesterol.
- Estrogen helps modulate the actions of epinephrine and norepinephrine on the arteriolar walls.
- Transient increases in the flow of warm blood through the superficial vessels are responsible for "hot flashes."
- About 20 percent of menopausal and postmenopausal women in the United States take supplemental estrogen to counter the effects of their natural decline in estrogen.
- Those on hormone replacement therapy enjoy stronger bones, a reduced risk of heart attacks, fewer hot flashes, and fewer mood swings than those not using estrogen.
- Estrogen therapy increases the risk of breast cancer, a detrimental side effect that can be diminished to some extent by taking estrogen in combination with progesterone.

*The oviduct is the site of fertilization.*
- Fertilization, the union of male and female gametes, normally occurs in the ampulla, the upper third of the oviduct.
- The dilated end of the oviduct cups around the ovary and contains fimbriae, fingerlike projections that contract in a sweeping motion to guide the released ovum into the oviduct.

- Fertilization must occur within 24 hours after ovulation.
- Sperm can survive up to five days in the female reproductive tract.
- Once sperm are deposited in the vagina at ejaculation, they must travel through the cervical canal, through the uterus, and then up to the egg in the upper third of the oviduct.
- Once the sperm have entered the uterus, contractions of the myometrium churn them around in "washing-machine" fashion.
- Sperm transport in the oviduct is facilitated by antiperistaltic contractions of the oviduct smooth muscle.
- These myometrial and oviduct contractions that facilitate sperm transport are induced by the high estrogen level that exists prior to ovulation, perhaps aided by seminal prostaglandins.
- To fertilize an ovum, a sperm must first pass through the corona radiata and the zona pellucida surrounding it.
- The acrosomal enzymes, which are exposed as the acrosomal membrane disrupts on contact with the corona radiata, enable the sperm to tunnel a path through these protective barriers.
- The first sperm to reach the ovum itself fuses with the plasma membrane of the ovum, triggering a chemical change in the ovum's surrounding membrane that makes the outer layer impenetrable to the entry of any more sperm.
- The head of the fused sperm is gradually pulled into the ovum's cytoplasm by a growing cone that engulfs it.
- Penetration of the sperm into the cytoplasm triggers the final meiotic division of the secondary oocyte.

*The blastocyst implants in the endometrium through the action of its trophoblastic enzymes.*

- The zygote is rapidly undergoing a number of mitotic cell divisions to form a solid ball of cells called the morula.
- About three or four days after ovulation, progesterone is being produced in sufficient quantities to relax oviduct constriction, thus permitting the morula to be rapidly propelled into the uterus by oviductal peristaltic contractions and ciliary activity.
- During the first six to seven days after ovulation, while the developing embryo is in transit in the oviduct and floating in the uterine lumen, the uterine lining is simultaneously being prepared for implantation under the influence of luteal-phase progesterone.
- By the time the endometrium is suitable for implantation, the morula has descended to the uterus and continued to proliferate and differentiate into a blastocyst capable of implantation.
- Implantation begins when the trophoblastic cells overlying the inner cell mass release protein-digesting enzymes upon contact with the endometrium.
- The trophoblast performs the dual functions of: (1) accomplishing implantation as it carves out a hole in the endometrium for the blastocyst, and (2) making metabolic fuel and raw materials available for the developing embryo as the advancing trophoblastic projections break down the nutrient rich endometrial tissue.
- After the blastocyst burrows into the decidua by means of trophoblastic activity, a layer of endometrial cells covers over the surface of the hole, completely burying the blastocyst within the uterine lining.

*The placenta is the organ of exchange between maternal and fetal blood.*

- To sustain the growing embryo/fetus for the duration of its intrauterine life, the placenta is rapidly developed.
- By day 12, the trophoblastic layer is two cell-layers thick and is called the chorion.
- Fingerlike projections of chorionic tissue extend into the pools of maternal blood.
- The developing embryo sends out capillaries into these chorionic projections to form placental villi.
- Each placental villus contains embryonic capillaries surrounded by a thin layer of chorionic tissue, which separates the embryonic/fetal blood from the maternal blood in the intervillus spaces.
- The placenta is well established and operational by five weeks after implantation.
- By this time, the heart of the developing embryo is pumping blood into the placental villi as well as to the embryonic tissues.
- During intrauterine life, the placenta performs the functions of the digestive system, the respiratory system, and the kidneys for the "parasitic" fetus.

*Hormones secreted by the placenta play a critical role in the maintenance of pregnancy.*

- The placenta has the remarkable capacity to secrete a number of peptide and steroid hormones essential for the maintenance of pregnancy.
- The most important are human chorionic gonadotropin, estrogen, and progesterone.
- One of the first events following implantation is secretion by the developing chorion of human chorionic gonadotropin (hCG), a peptide hormone that prolongs the life span of the corpus luteum.

- The maintenance of a normal pregnancy depends on high concentrations of progesterone and estrogen.
- By the tenth week of pregnancy, hCG output declines to a low rate of secretion that is maintained for the duration of gestation.
- The fall in hCG occurs at a time when the corpus luteum is no longer needed for its steroidal hormone output because the placenta has begun to secrete substantial quantities of estrogen and progesterone.
- The placenta is unable to produce sufficient quantities of either estrogen or progesterone in the first trimester of pregnancy.
- The placenta is able to convert the androgen hormone produced by the fetal adrenal cortex, DHEA, into estrogen.
- The placenta is unable to produce estrogen until the fetus has developed to the point that its adrenal cortex is secreting DHEA into the blood.
- The placenta extracts DHEA from the fetal blood and converts it into estrogen, which it then secretes into the maternal blood.
- The placenta is simply too small in the first 10 weeks of pregnancy to produce sufficient quantities of progesterone to maintain the endometrial tissue.

*Maternal body systems respond to the increased demands of gestation.*

- During gestation, a number of physical changes take place within the mother to accommodate the demands of pregnancy.
- The most obvious change is uterine enlargement.
- The breasts enlarge and develop the capability of producing milk.
- The volume of blood increases and the cardiovascular system responds to the increasing demands of the growing placental mass.
- Respiratory activity is increased.
- Urinary output increases, and the kidneys excrete the additional wastes from the fetus.

*Changes during late gestation prepare for parturition.*

- Parturition requires: (1) dilation of the cervical canal to accommodate passage of the fetus from the uterus through the vagina and to the outside, and (2) contractions of the uterine myometrium that are sufficiently strong to expel the fetus.
- Cervical softening is believed to be caused by relaxin, a peptide hormone produced by the corpus luteum of pregnancy and by the placenta.
- Relaxin also "relaxes" the pelvic bones.
- Rhythmic, coordinated contractions begin at the onset of true labor.
- The contractions occur with increasing frequency and intensity.
- After having dilated the cervix sufficiently for passage of the fetus, these contractions force the fetus out through the birth canal.
- A dramatic progressive increase in the concentration of myometrial receptors for oxytocin, probably induced by the increasing levels of estrogen during pregnancy, is ultimately responsible for initiating labor.

*Parturition is accomplished by a positive-feedback cycle.*

- When the hypothalamus reaches a critical level of maturation it signals the anterior pituitary to increase ACTH secretion.
- ACTH causes increased cortisol secretion by the adrenal cortex.
- Fetal cortisol promotes the conversion of progesterone into estrogen, which is secreted into the maternal blood.
- Since progesterone inhibits uterine contractility and estrogen enhances it, this fetal induced shift in circulating maternal hormones brings about uterine contractions.
- This and the oxytocin receptor mechanism may both contribute to the onset of parturition.
- An alternative proposal is the idea of a "placental clock" that ticks out a predetermined length of time until parturition.
- The placental clock is believed to be measured by placental secretion of corticotropin releasing hormone (CRH).
- CRH secreted by the hypothalamus regulates anterior pituitary secretion of ACTH, but CRH is also secreted by the placental chorion.
- CRH levels rise in maternal blood as pregnancy progresses.
- Parturition can be predicted by measuring maternal plasma CRH levels.
- Myometrial contractions incessantly increase as labor progresses because of a positive feedback cycle involving oxytocin and prostaglandins.
- Labor is divided into three stages: (1) cervical dilation, (2) delivery of the baby, and (3) delivery of the placenta.
- After delivery, the uterus shrinks to its pregestational size, a process known as involution.
- Involution occurs largely because of the precipitous fall in circulating estrogen and progesterone when the placental source of these steroids is lost at delivery.
- Postpartum depression may be due to low levels of CRH.

- During pregnancy, levels of CRH of placental origin in the maternal blood rise to as much as three times the nonpregnancy CRH level.
- Cortisol secreted in response to the elevated CRH helps the mother cope with the stress of pregnancy and parturition.
- CRH helps combat depression.

*Lactation requires multiple hormonal inputs.*
- During gestation, the mammary glands, or breasts, are prepared for lactation.
- The high concentration of estrogen promotes extensive duct development, whereas the high level of progesterone stimulates abundant alveolar-lobular formation.
- Elevated concentrations of prolactin and human chorionic somatomammotropin also contribute to mammary gland development by inducing the synthesis of enzymes needed for milk production.
- The high estrogen and progesterone concentrations during the last half of pregnancy prevent lactation by blocking prolactin's stimulatory action on milk secretion.
- Prolactin is the primary stimulant of milk secretion.
- Two hormones are critical for maintaining lactation: (1) prolactin, and (2) oxytocin.
- Release of both these hormones is stimulated by a neuroendocrine reflex triggered by suckling.
- Suckling of the breast by the infant stimulates sensory nerve endings in the nipple, resulting in initiation of action potentials that travel up the spinal cord to the hypothalamus.
- The hypothalamus triggers a burst of oxytocin release from the posterior pituitary.
- Oxytocin stimulates contractions of the myoepithelial cells in the breasts to bring about milk ejection.
- In addition to prolactin, at least four other hormones are essential for their permissive role in milk production: cortisol, insulin, parathyroid hormone, and growth hormone.

*Breast-feeding is advantageous to both the infant and the mother.*
- Milk is composed of water, triglyceride fat, the carbohydrate lactose, a number of proteins, vitamins, and the minerals calcium and phosphate.
- Colostrum, the milk produced for the first five days postpartum, contains lower concentrations of fat and lactose but higher concentrations of proteins, most notably lactoferrin and immunoglobulins.
- All human babies acquire some passive immunity during gestation through the passage of antibodies across the placenta.
- Breast-fed babies gain additional protection during this vulnerable period through a variety of mechanisms: (1) Breast milk contains an abundance of immune cells that produce antibodies and destroy the pathogens outright. (2) Secretory IgA, a special type of antibody, is present in great amounts in breast milk. (3) Some components in mother's milk, such as mucus, adhere to potentially harmful microorganisms. (4) Lactoferrin is a breast-milk constituent that thwarts growth of harmful bacteria by decreasing the availability of iron. (5) Bifidus factor in breast milk promotes the multiplication of the nonpathogenic microorganism *Lactobacillus bifidus* in the infant's digestive tract. (6) Other components in breast milk promote the maturation of the body's digestive system so that it is less vulnerable to diarrhea-causing bacteria and viruses. (7) Still other factors in breast milk hasten the development of the infant's own immune capabilities.
- Breast-feeding is also advantageous to the mother.
- Oxytocin release triggered by nursing hastens uterine involution.
- Suckling suppresses the menstrual cycle by inhibiting LH and FSH secretion.
- Lactation prevents ovulation and serves as a means of preventing pregnancy, thus permitting all the mother's resources to be directed toward the newborn instead of being shared with a new embryo.

*The end is a new beginning.*
- The single cell resulting from the union of male and female gametes divides mitotically and differentiates into a multicellular individual made up of a number of different organ systems that interact cooperatively to maintain homeostasis.
- All of the life-supporting homeostatic processes introduced throughout this book begin all over again at the start of a new life.

## Key Terms

acrosome
adolescence
afterbirth
ampulla
androgens
aromatase
atresia
autosomal chromosomes
blastocyst
blood-testes barrier
chorion
climateric
colostrum
corpus albicans
corpus luteum
corticotropin-releasing hormone
cryptorchidism
decidua
dihydrotestosterone (DHT)
diploid
dysmenorrhea
ectopic (abdominal) pregnancy
embryo
emission
endocrine disrupters
erectile tissue
estrogen receptor modulators (ERM)
eunuch
expulsion
fertilization
fetoplacental unit
follicular phase
gametes (reproductive, germ cells)
gestation
haploid
homologous chromosomes
h-y antigen
inhibin
involution
leydig cells
LH surge
luteal phase
luteinization
meiosis
menarche
menstrual (uterine) cycle
morning sickness
morula
oogonia
orgasm
ovulation
parturition
placenta
postpartum depression
primary follicle
primary oocyte
primary spermatocytes
progestational phase
proliferative phase
prostaglandins
reproduction
secondary oocyte
secondary sexual characteristics
seminiferous tubules
spermatids
spermatogenesis
spermiogenesis
tenting effect
trophoblast
tubal pregnancy
vasectomy

## Review Exercises

*Answers are in the appendix.*

**True/False**

_____ 1. Formation of the gonads in the embryo is known as gametogenesis.

_____ 2. The clitoris is a small erotic structure composed of tissue identical to that of the penis.

_____ 3. Fallopian tubes are the same as oviducts.

____ 4. Somatic cells are haploid.

_____ 5. Gametes are haploid.

_____ 6. During meiosis, a diploid germ cell undergoes one chromosome replication followed by two nuclear divisions.

_____ 7. Gonadal specificity appears during the seventh week of intrauterine life.

_____ 8. SRY antigen directs differentiation of the embryological gonad into testes.

_____ 9. Testosterone induces the male pattern of hair growth.

_____ 10. Sertoli cells secrete testosterone.

_____ 11. Each primary spermatocyte has 23 doubled chromosomes.

_____ 12. The acrosome is an enzyme-filled vesicle at the tip of the head of the sperm.

_____ 13. The tight junctions between adjacent Leydig cells form a blood-testes barrier.

_____ 14. Sertoli cells phagocytize cytoplasm extruded from spermatids.

_____ 15. Seminiferous tubule fluid is secreted by Sertoli cells.

_____ 16. FSH acts on the Leydig cells to regulate testosterone secretion.

_____ 17. FSH is required for spermatid remodeling.

_____ 18. The bulbourethral glands secrete an alkaline fluid that neutralizes the acidic vaginal secretions.

_____ 19. Prostaglandins have been medically administered in treating gastric ulcers.

_____ 20. Increased parasympathetic activity to the penile arterioles produces an erection.

_____ 21. Progesterone in the female is responsible for maturation and maintenance of the entire female reproductive system.

_____ 22. The primary oocyte within a primary follicle is a haploid cell.

_____ 23. Sperm entry into the secondary oocyte is needed to trigger the second meiotic division.

_____ 24. The corpus luteum operates in the first half of the cycle to produce a mature egg ready for ovulation at midcycle.

_____ 25. Thecal and granulosa cells function as a unit to secrete estrogen.

_____ 26. Estrogen secretion in the follicular phase and progesterone secretion in the luteal phase are essential for preparing the uterus for implantation.

_____ 27. Antrum formation is induced by LH.

_____ 28. The LH surge causes differentiation of follicular cells into luteal cells.

_____ 29. No progesterone is secreted during the follicular phase.

_____ 30. Under the influence of estrogen, the mucus secreted by the cervix is abundant, clear, and thin.

**Fill in the Blank**

31. In females, reproductive potential ceases during middle age at a time known as ________________.

32. Before birth, each primary oocyte is surrounded by a single layer of ______________ to form a(n) ______________________.

33. A follicle is destined for one of two fates once it starts to develop; it will reach maturity and ovulate, or it will degenerate to form scar tissue, a process known as ______________.

34. The thecal and granulosa cells, collectively known as ______________________, function as a unit to secrete estrogen.

35. _____________ is the principle ovarian estrogen.

36. The uterus consists of two layers: the ____________________, which is the outer smooth-muscle layer, and the ____________________, which is the inner lining that contains numerous blood vessels and glands.

37. Menstrual cramps are also known as ____________________.

38. This entire period of transition from sexual maturity to cessation of reproductive capability is known as the __________________.

39. Fertilization normally occurs in the ______________.

40. The ______________________ (1) supply fructose, which serves as the primary energy source for ejaculated sperm and (2) secrete prostaglandins.

41. The ______________ phase returns the genitalia and body systems to their prearousal state.

42. The penis consists almost entirely of ______________________ made up of three columns of spongelike vascular spaces.

43. Failure to achieve an erection in spite of appropriate stimulation is referred to as ______________.

44. The orgasmic contractions occur most intensely in the engorged lower third of the vaginal canal called the ____________________________.

45. Reproduction depends on the union of male and female ______________.

46. _______________ contain 46 chromosomes.

47. Twenty-two of the chromosome pairs are ___________________________.

48. In males, the reproductive tract develops from the ______________________. In females, the ____________________ differentiate into the reproductive tract.

49. When a testis remains undescended into adulthood, this condition is known as _________________.

50. The endocrine cells that produce testosterone in males are ____________________ or _________________________.

51. Production of extremely specialized, mobile spermatozoa from spermatids requires extensive remodeling, or packaging, of cellular elements, a process known as _________________________.

52. A spermatozoon has four parts: ____________, _____________, ________________, and _________.

53. The tight junctions between adjacent Sertoli cells form a(n) ______________________.

54. The testes are controlled by the two gonadotropic hormones secreted by the anterior pituitary, _________________________, and ___________________________.

55. A(n) ________________ is a single-layered sphere of cells encircling a fluid-filled cavity.

56. The endometrial tissue so modified at the implantation site is called the ___________.

57. When the trophoblastic layer is two cell-layers thick it is then called _____________.

58. ________________________________________________ is a peptide hormone that prolongs the life span of the corpus luteum.

59. The period of gestation is about _____ weeks from conception.

60. ______________ is a peptide hormone that is produced by the hypothalamus and is a key role in the progression of labor.

61. After delivery, the uterus shrinks to its pregestational size, a process known as _______________.

62. ________________ is the milk produced for the first five days postpartum.

**Matching**

*Match the function to the placental hormone using the code on the right. Some functions may require multiple hormones*

a. estrogen
b. progesterone
c. hCG
d. relaxin
e. human chorionic somatomammotropin

_____ 63. helps prepare mammary glands for lactation
_____ 64. loosens connective tissue between pelvic bones in preparation for parturition
_____ 65. promotes formation of cervical mucus plug
_____ 66. maintains corpus luteum
_____ 67. softens cervix in preparation for cervical dilation at parturition
_____ 68. suppresses uterine contractions to provide quiet environment for fetus
_____ 69. stimulates secretion of testosterone by developing testes in XY embryos
_____ 70. stimulates growth of myometrium, increasing uterine strength for parturition

**Multiple Choice**

71. Which of the following contraceptive practices, products, or methods is no longer popular due to reported complications including pelvic inflammatory disease and permanent infertility?
a. birth-control pills
b. morning after pills
c. IUDs
d. chemical spermides
e. contraceptive sponge

72. Which of the following contraceptive practices, products, or methods is used as a method of fertility?
a. barrier method
b. coitus interruptus
c. oral contraceptives
d. rhythm method
e. none of the above

73. Which of the following devices used as barriers is the easiest to use?
a. condom
b. diaphragm
c. cervical cap
d. condomlike contraceptive for women
e. vaginal cap

74. Which of the following contraceptive practices, products, or methods increase the risk of intravascular clotting, especially in women who also smoke tobacco?
    a. IUDs
    b. morning after pill
    c. foams
    d. oral contraceptives
    e. contraceptive sponge

75. Which of the following contraceptive practices, products, or methods should be used only once due to increased risk of cardiovascular disease?
    a. sterilization
    b. birth-control pills
    c. IUDs
    d. morning after pill
    e. subcutaneous implantation of synthetic progesterone

76. Which is the longest acting method of contraception?
    a. diaphragms
    b. IUDs
    c. morning after pills
    d. birth-control pills
    e. subcutaneous implantation of synthetic progesterone

77. Which of the following contraceptive techniques has the highest failure rate?
    a. implanted contraceptives
    b. barrier methods
    c. rhythm method
    d. intrauterine device
    e. oral contraceptives

78. Which of the following hormones is the primary stimulant of milk secretion?
    a. human chorionic somatomammotropin
    b. prolactin
    c. estrogen
    d. progesterone
    e. prostaglandins

79. Which of the following hormones promotes fat deposition?
    a. prolactin
    b. human chorionic somatomammotropin
    c. estrogen
    d. progesterone
    e. none of the above

80. Which of the following hormones secreted by the corpus luteum softens the cervix?
    a. estrogen
    b. progesterone
    c. oxytocin
    d. relaxin
    e. none of the above

81. Which of the following mechanisms or factors in breast milk promotes multiplication of nonpathogenic bacteria in the infant's digestive tract?
    a. secretory IgA
    b. lactogerrin
    c. colostrum
    d. bifidus factor
    e. BST

82. Which of the following hormones promotes the conversion of progesterone into estrogen?
    a. ACTH
    b. fetal cortisol
    c. FSH
    d. CRH
    e. LH

83. Approximately how many menopausal and postmenopausal women in the U.S. take supplemental estrogen to counter the effects of their natural decline in estrogen?
    a. 10 percent
    b. 20 percent
    c. 30 percent
    d. 40 percent
    e. 50 percent

84. Which of the following hormones gives women a 15- to 20-year advantage over males in evading atherosclerotic coronary artery disease?
    a. PIH
    b. estrogen
    c. CRH
    d. progesterone
    e. oxytocin

85. Which of the following hormones is measured in the "placental clock"?
    a. CRH
    b. ACTH
    c. LH
    d. FSH
    e. hCG

86. Which of the following hormones enhances uterine contractility?
    a. progesterone
    b. estrogen
    c. CRH
    d. fetal cortisol
    e. BST

87. Which of the following hormones shields premenopausal women from osteoporosis?
    a. estrogen
    b. progesterone
    c. CRH
    d. ACTH
    e. relaxin

88. Which of the following hormone secretions is suppressed by suckling?
    a. ACTH
    b. CRH
    c. FSH
    d. hCG
    e. IgA

89. Which of the following hormones prevents "hot flashes"?
    a. progesterone
    b. cortisol
    c. LH
    d. FSH
    e. estrogen

90. Which of the following is known as the "abortion pill"?
    a. PRF
    b. mifepristone
    c. progesteriol
    d. biferrin
    e. DHEA

**Modified Multiple Choice**

*Write the appropriate letter in the blank by using the following answer code to indicate the relative magnitude of the two items in question.*

A = A is greater than B
B = B is greater than A
C = A and B are approximately equal

91._____ A. The length of the luteal phase of the average menstrual cycle
B. The length of the follicular phase of the average menstrual cycle

92. _____ A. The length of the luteal phase of the ovarian cycle
B. The length of the secretory phase of the menstrual cycle

93. _____ A. The number of chromosomes in the ovulated secondary oocyte
B. The number of chromosomes in a primary oocyte

94. _____ A. The number of primary follicles present in the ovary at any time during the menstrual cycle
B. The number of mature follicles just before ovulation

95. _____ A. Plasma concentration of estrogen during the folllicular phase of the ovarian cycle
B. Plasma concentration of estrogen during the luteal phase of the ovarian cycle

96. _____ A. Plasma concentration of progesterone during the follicular phase of the ovarian cycle
B. Plasma concentration of progesterone during the luteal phase of the ovarian cycle

*Indicate which item in the following answer code best matches the description in question by writing the appropriate letter in the blank. There is only one correct answer per question. Each answer may be used more than once. Some answers may not be used at all.*

A = erection
B = acrosome
C = Leydig cells
D = seminal vesicles
E = oviduct
F = uterus
G = inner cell mass
H = seminiferous tubules
I = midpiece of sperm
J = head of sperm
K = vagina
L = prostate gland
M = trophoblast
N = corpus luteum
O = epididymis & ductus deferens
P = ovarian follicle
Q = ejaculation
R = tail of sperm
S = Sertoli cells
T = endometrium
U = Placenta

_____ 97. Formation is stimulated by LH after ovulation.
_____ 98. Site of spermatogenesis.
_____ 99. Store sperm prior to ejaculation.
_____ 100. Occurs as a result of vascular engorgement.
_____ 101. Site of fertilization.
_____ 102. Destined to become the fetus.
_____ 103. Fills with glycogen and becomes richly vascularized under the influence of progesterone.
_____ 104. Tissue that secretes testosterone.
_____ 105. Contains enzymes utilized for penetration of the ovum.
_____ 106. Secretes HCG.
_____ 107. Secretes prostaglandins into semen.
_____ 108. Responsible for final maneuvering for penetration of ova.
_____ 109. Supplies fructose utilized by sperm for energy.
_____ 110. Destined to become the fetal portion of the placenta.
_____ 111. Semen expelled from penis.
_____ 112. Contains primarily the nucleus that contains the sperm's genetic information.
_____ 113. Serves as a route by which nutrients reach developing sperm cells.
_____ 114. Organ of exchange between mother and fetus.
_____ 115. Secretes alkaline fluid to neutralize acidic vaginal secretions.
_____ 116. Portion of blastocyst responsible for accomplishing implantation.
_____ 117. Structure within the ovum develops prior to ovulation.
_____ 118. Site of initial maturation of sperm for motility and fertility in the male.
_____ 119. Contains mitochondria for energy production within sperm.
_____ 120. Secretes estrogen but not progesterone.
_____ 121. Secretes estrogen and progesterone during the first 10 weeks of gestation.
_____ 122. Secretes estrogen and progesterone during the last seven months of gestation.

## Points to Ponder

1. It is very commonplace to discuss abnormalities in hormone secretion of all endocrine glands and tissues except the gonads. Is it because such abnormalities concerning the gonads do not exist or is it due to a social attitude?

2. After carefully studying this chapter, do you think gay people could have abnormal hormone levels?

3. Why do you think most oral contraceptives are for women? Is there a physiological reason?

4. Is it possible for a woman to become pregnant if the male does not ejaculate during sexual intercourse?

5. Which contraceptive methods are the most effective? Explain why. The least effective? Explain why.

6. Explain the vascular mechanism by which penile erection occurs. How does Viagra help this process?
7. What are some of the first signs of pregnancy?
8. When can the sex of a fetus first be detected?
9. How is hCG related to the various pregnancy tests?
10. Why are some babies born prematurely?

## Clinical Perspectives

1. Explain why birth control pills are prescribed to women who have complications at the onset of menopause.
2. Explain how an infant girl's breasts can begin to enlarge.
3. During a routine physical examination of a male, why does a physician assistant insert a finger in the inguinal rings of the scrotum and ask the male to cough?
4. Why do nursing babies sometimes refuse milk after the mother has exercised?
5. Do women with small breasts produce less milk than large-breasted women? Explain your answer.
6. After missing a menstrual period, your roommate purchases a home pregnancy test. The test is positive and she makes an appointment with the school health physician. The physician tells her to bring a urine sample in with the appointment. After giving her a pelvic exam, the physician told her she is pregnant and should begin taking vitamins and minerals and to return in one month. Why is a missed menstrual cycle one of the first signs of pregnancy? What is in the urine that indicates a positive test for pregnancy? Why was your roommate told to take vitamins and minerals?
7. Based on what you know about female physiology, what is the principle behind the surgical removal of the ovaries as a treatment for breast cancer?
8. What serum enzyme is likely to be elevated in advanced prostate cancer?
9. Your sister asks you to explain the proposed mechanisms whereby the act of a mother nursing her baby results in lactation. How would you answer this question? By what mechanisms might the sound of a baby crying elicit the milk-ejection reflex?
10. According to your college roommate, there is a female birth control pill and not a male birth control pill because the medical establishment is run by men. Do you agree with her theory? Explain your answer.

## PhysioEdge Activities

**Related to Text:**
Media Exercise 20.1: Male Reproductive Anatomy.
Media Exercise 20.2: Female Reproductive Anatomy.
Media Exercise 20.3: Male Reproductive Physiology.
Media Exercise 20.4: Female Reproductive Physiology.

**Related to Figures:**
*Figure 20.1.* For an interaction related to this figure, see Media Exercise 20.1: Male Reproductive Anatomy.
*Figure 20.2.* For an interaction related to this figure, see Media Exercise 20.2: Female Reproductive Anatomy.
*Figure 20.7.* For an interaction related to this figure, see Media Exercise 20.3: Male Reproductive Function.

*Figure 20.16*. For an interaction related to this figure, see Media Exercise 20.4: Female Reproductive Physiology.

## *Media Resources*

**PhysioEdge CD-ROM**
For a visual review of concepts in this chapter, check out:

Tutorial: Neural and Hormonal Communication – Hormones tab
Media Exercise 20.1: Male Reproductive Anatomy
Media Exercise 20.2: Female Reproductive Anatomy
Media Exercise: 20.3: Male Reproductive Physiology
Media Exercise: 20.4: Female Reproductive Physiology

**Book Companion Website**
The website for this book contains a wealth of helpful study aids, as well as many ideas for further reading and research. Log on to:

**http://www.brookscole.com/sherwoodhp6**

Select Chapter 20 from the drop-down menu, or click on one of the many resources areas, including **Case Histories**, which introduce clinical aspects of human physiology. For this chapter check out Case History 23: Pregnancy Tests.

For Suggested Readings, consult **InfoTrac® College Edition**, your online research library, at:

**http://infotrac.thomsonlearning.com**

# *Appendix: Answers*

## *Chapter 1: Homeostasis: The Foundation of Physiology*

### True/False

1. True
2. False—It's reversed.
3. False—Some organs are in several systems.
4. True
5. False—Extrinsic control.
6. True
7. True
8. False—A teleological explanation of why a person sweats is to cool off. A mechanistic approach is that when temperature-sensitive nerve cells signal the hypothalamus that there has been a rise in body temperature, the hypothalamus stimulates the sweat glands to produce sweat, evaporation of which cools the body.
9. True
10. True
11. False—The internal environment must be maintained in a dynamic steady state instead of a fixed state. Fluctuations around an optimal level are minimized by compensatory physiological responses so that the homeostatically maintained factors are kept within the narrow limits that are compatible with life.
12. True
13. True
14. False—Exocrine glands.
15. False—Intrinsic and extrinsic.
16. True
17. False—Teleological approach.
18. True
19. True
20. False—Four.
21. True
22. True
23. False—Three.
24. True
25. True

### Fill in the Blank

26. Exocrine
27. Lumen
28. Mechanistic
29. Homeostasis
30. Interstitial
31. Digestive
32. Negative feedback
33. Cell
34. Internal environment
35. Extracellular fluid, plasma, interstitial fluid
36. Negative feedback
37. Pathophysiology
38. Cell
39. EGF (epidermal growth factor)
40. Glands
41. Body systems
42. Circulatory, digestive, respiratory, urinary, skeletal, muscular, integumentary, immune, nervous, endocrine, and reproductive.
43. Physiology
44. Extracellular
45. Positive
46. Stress
47. Connective
48. Lymph
49. Cytosol
50. Lymphatic

### Matching

51. e
52. g
53. k
54. h
55. j
56. a
57. d
58. f
59. b
60. i
61. c
62. a
63. e
64. b
65. d
66. c

### Multiple Choice

67. c
68. d
69. c
70. b

71. c
72. b
73. a
74. b
75. b
76. a
77. c
78. a
79. e
80. a
81. d
82. b
83. a
84. b
85. b
86. e
87. a
88. c
89. e
90. e
91. e
92. e
93. c
94. a
95. a
96. b
97. c
98. e
99. a
100. e

## *Chapter 2: Cell Physiology*

**True/False**

1. False—Accumulation of gangliosides.
2. False—Occurs in inner mitochondrial membrane.
3. True
4. False—Glycolysis is anaerobic.
5. False—Another name for electron transport chain.
6. True
7. True
8. True
9. False—Plant cells, not human or other animal cells, are surrounded by a cell wall.
10. True
11. True
12. True
13. False—Single cell.
14. False—DNA.
15. True
16. False—Ribosomes.
17. True
18. False—Ribosomes found only on the rough ER.
19. True
20. True
21. False—Golgi sacs.
22. True
23. True
24. False—Double membrane.
25. False—ATP.
26. False—Two molecules of ATP.
27. True
28. True
29. True
30. False—Coated vesicles capture specific cargo, not a representative mixture of proteins, before budding off from the Golgi sac.
31. True
32. True
33. True
34. False—The protective, waterproof outer layer of skin is formed by the persistence of tough keratin. Keratin is a protein component of special intermediate filaments found in skin cells.
35. True
36. False—Many ribosomes can attach simultaneously to a single mRNA molecule, with each ribosome, one behind the other, synthesizing the same type of protein as specified by the mRNA code.

**Fill in the Blank**

37. Ribophorins
38. Transport vesicles
39. Microtrabecular lattice
40. Cytosol
41. Kinesin
42. Basil body
43. Peroxisomes
44. Plasma membrane, nucleus, cytoplasm
45. Cytoskeleton
46. Proteins, lipids
47. Leader sequence, ribophorin
48. Smooth
49. Exocytosis
50. Hydrolytic
51. Endocytosis
52. Catalase
53. DNA
54. Apoptosis
55. Vaults

**Matching**

56. g
57. k
58. j
59. a
60. i
61. h
62. b
63. c
64. f
65. e
66. d

**Multiple Choice**

67. b
68. e
69. a
70. b
71. a
72. d
73. b
74. a
75. e
76. a
77. b
78. b
79. d
80. b
81. b
82. a
83. b
84. b
85. c
86. b
87. b
88. d
89. a
90. d
91. d
92. c
93. b
94. a
95. b
96. a
97. a
98. b
99. b

## *Chapter 3: The Plasma Membrane and Membrane Potential*

**True/False**

1. True
2. True
3. False—ICF.
4. False—Phagocytosis.
5. False—Membrane components are arranged asymmetrically.
6. False—A tremendous amplification of the initial signal occurs as a result of a cascade effect.
7. False—There is no net movement; at equilibrium an equal number of molecules are moving in opposite directions across the membrane.
8. True
9. False—A potential of –70mV is of greater magnitude than a potential of +30mV. The sign refers to the charge on the inside of the cell.
10. True
11. True
12. True
13. False—Outer surface.
14. True
15. False—It is possible.
16. False—Channel regulation mechanism.
17. True
18. True
19. True
20. False—Collagen, elastin, fibronectin.
21. True
22. False—Fibroblast instead of fibronectin.
23. False—Tight junction.
24. True
25. True
26. False—Heart muscle.
27. False—Both charged and uncharged molecules are soluble in water. The size of the particle is the second property that influences permeability.
28. False—Eight..
29. False—The faster the rate of net diffusion.
30. True
31. True
32. False—Lighter molecules bounce farther.
33. False—The slower the rate of diffusion.
34. True
35. True
36. True
37. True
38. False—Opposite charges not like charges.
39. True
40. False—Greater the ions equilibrium potential.

### Fill in the Blank

41. Cystic fibrosis
42. Second messenger
43. First messenger
44. G protein
45. Phosphorylation
46. Cyclic guanosine monophosphate (cyclic GMP)
47. Extracellular matrix
48. Cell adhesion molecules in the cells plasma membranes, the extracellular matrix, and specialized cell junctions
49. Spot desmosomes
50. Collagen, elastin, fibronectin
51. Fibroblasts
52. Desmosomes, tight junctions, gap junctions
53. Luminal border, basolateral border
54. Tight junctions
55. Collagen
56. Plasma membrane
57. Membrane potential, negative
58. Phospholipids
59. Hydrophilic, hydrophobic
60. Cholesterol
61. Lipid bilayer, proteins
62. Carrier molecules, receptor sites
63. Elastin
64. Gap junctions, connexons
65. Fibronectin
66. Heart, smooth
67. Net diffusion
68. State of equilibrium, steady state
69. Impermeable
70. Carrier-mediated transport
71. Cysteinuria
72. Transport maximum
73. Primary active transport
74. Secondary active transport
75. Endocytosis
76. Phagocytosis
77. Membrane potential
78. Cations, negatively, anions, positively
79. Electrical gradient
80. Electrochemical gradient
81. Concentration
82. Osmosis
83. Osmotic pressure

### Matching

84. d
85. e
86. a
87. d
88. d
89. a
90. b
91. d
92. c
93. a

### Multiple Choice

94. c
95. b
96. b
97. c
98. d
99. a
100. a
101. d
102. b
103. e
104. b
105. d
106. c
107. c
108. b
109. c
110. b
111. d
112. b
113. c
114. c
115. b
116. c

## *Chapter 4: Principles of Neural and Hormonal Communication*

### True/False

1. False—Some neurotransmitters function through intracellular second messengers.
2. False—Synapses operate in one direction only.
3. False—A given synapse is either always excitatory or always inhibitory.
4. True
5. True
6. False—Larger the graded potential.
7. False—Repolarization.
8. True
9. False—50mV to –55mV is the threshold potential.
10. True
11. True
12. True
13. True

14. False—50 times.
15. False—Urgent information is carried by myelinated fibers, less urgent by unmyelinated.
16. False—Never regenerates.
17. True
18. True
19. True
20. False—Lowest
21. True
22. False—Deficiency.
23. True

**Fill in the Blank**

24. Resting membrane potential
25. Dendrites, axon, nerve fibers
26. Nodes of Ranvier
27. Gastin, glucagon, gonadotropin-releasing hormone (GnRH)
28. Cytosol, endoplasmic reticulum/Golgi complex
29. Polarization
30. Saltatory conduction
31. Synaptic cleft
32. 0.5 to 1 msec
33. Temporal summation, spatial summation
34. Voltage-gated, chemical messenger-gated, mechanically-gated
35. Axon hillock
36. Oligodendrocytes, Schwann cells
37. Schwann cells, regeneration tube
38. Glycine, glutamate, gamma-aminobutyric acid (GABA)
39. Spike
40. Cell body, dendrites
41. Lipids
42. All-or-none law
43. Subsynaptic membrane
44. Cholecystokinin (CCK)
45. Divergence

**Matching**

46. b
47. a
48. a
49. b
50. b
51. b
52. a
53. b
54. b
55. a
56. a
57. a
58. b

**Multiple Choice**

59. b
60. a
61. d
62. c
63. c
64. b
65. d
66. c
67. b
68. e
69. a
70. d
71. e
72. e
73. a
74. d
75. c
76. d
77. e
78. b
79. e
80. e
81. e
82. b
83. c
84. c
85. a
86. b
87. a
88. b
89. e
90. c

**Modified Multiple Choice**

91. a
92. a
93. b
94. b
95. b
96. b
97. a
98. a
99. b
100. a

**True/False**

101. True
102. True
103. True
104. True
105. True
106. False—Inside the cell.
107. True
108. True
109. True
110. False—The nervous system coordinates rapid responses; the endocrine regulates activities that require duration, rather than speed.

## Chapter 5: The Central Nervous System

**True/False**

1. False—Smallest.
2. False—Pyramidal cells.
3. True
4. False—Opposite.
5. True
6. False—Left.
7. True
8. True
9. False—Afferent.
10. True
11. False—Afferent neuron.
12. False—Efferent neuron.
13. True
14. True
15. True
16. True
17. False—Cerebrocerebellum.
18. False—Cerebellum.
19. False—Glial cells or neuroglia.
20. True
21. False—Oligodendrocytes.
22. True
23. True
24. False—Choroid plexuses.
25. True
26. False—Cannot.
27. True
28. False—18 inches.
29. False—Five lumbar nerves.
30. False—Descending tracts.
31. False—Afferent.
32. True
33. True
34. True
35. True
36. True
37. False—Decreasing.
38. False—Activity is not reduced.
39. False—20 percent.
40. False—Paradoxical.
41. True
42. False—Basal nuclei.
43. True
44. True
45. False—Long-term memory.
46. True
47. True
48. True
49. True
50. False—Rapid.
51. True
52. False—Nervous system.
53. True
54. False—Synaptic cleft.

**Fill in the Blank**

55. Spinocerebellum
56. Vestibulocerebellum
57. Cerebrocerebellum
58. Simple (basic reflexes) and acquired (conditioned reflexes)
59. Cervical, thoracic, lumbar, sacral, coccygeal
60. Cauda equina
61. Afferent, efferent
62. Ganglion, center or a nucleus
63. Spinal nerve
64. Nerve
65. Hypothalamus, thalamus
66. Cerebellum
67. Cerebral cortex
68. Grey, white
69. Parietal lobes, somesthetic sensations
70. Broca's, frontal lobe, Wernick's, cortex
71. Glial cells, neuroglia
72. Astrocytes
73. Oligodendrocytes
74. Microglia
75. Pia mater
76. Choroid plexuses
77. Medulla, pons, midbrain
78. Reticular formation
79. Consciousness
80. Coma
81. Paradoxical
82. Slow-wave
83. REM (paradoxical)
84. Basal nuclei, thalamus, hypothalamus
85. Thalamus
86. Hypothalamus

87. Motivation
88. Learning
89. Consolidation
90. Neuroendocrinology
91. Brain, spinal cord, peripheral nervous system
92. Afferent division
93. Afferent neurons, efferent neurons, interneurons
94. CNS

**Matching**

95. b
96. d
97. a
98. a
99. a
100. a
101. c
102. a
103. b
104. a
105. a

**Multiple Choice**

106. c
107. c
108. b
109. d
110. b
111. c
112. a
113. e
114. d
115. c

**Modified Multiple Choice**

116. b
117. e
118. d
119. d
120. b
121. c
122. d
123. b
124. a
125. b
126. a
127. e
128. c

## *Chapter 6: The Peripheral Nervous System: Afferent Division; Special Senses*

**True/False**

1. False—None of the nociceptors.
2. True
3. False—Unmyelinated fibers.
4. True
5. False—Substance P.
6. False—Afferent fibers.
7. True
8. True
9. False—Sour.
10. True
11. True
12. True
13. True
14. True
15. False—Open.
16. True
17. False—Phasic receptors.
18. False—It can.
19. True
20. True
21. False—Vitreous humor.
22. True
23. True
24. True
25. False—Rod shaped.
26. False—Open.
27. False—Other media, such as water.
28. True
29. False—Pinna.
30. True
31. False—Three.
32. False—Endolymph.
33. False—Basilar membrane.
34. True
35. True
36. False—Cones.

**Fill in the Blank**

37. Sclera
38. Radial
39. Accommodations
40. Cataract
41. Optic disc
42. Opsin
43. Optic tracts
44. Taste pore
45. Cortical gustatory area
46. Olfactory nerve

47. Taste buds
48. Subcortical route, thalamic-cortical route
49. Olfactory receptors, supporting cells, basal cells
50. Sensory input
51. Phasic
52. Pacinian corpuscle
53. Visceral afferent
54. Sensory information
55. First-order sensory neuron
56. Phantom pain
57. Vestibular apparatus
58. Cerumen
59. Endolymph
60. Vestibular
61. Auditory nerve
62. Basilar
63. Otolith organs
64. Mechanical nociceptors, thermal nociceptors, polymodal nociceptors
65. A-delta fibers
66. Substance P
67. Prostaglandins
68. Somatosensory cortex, thalamus, reticular formation
69. Opiate receptors
70. Placebo

**Matching**

71. g
72. a
73. j
74. b
75. c
76. i
77. d
78. h
79. f
80. e

**Multiple Choice**

81. c
82. e
83. a
84. d
85. d
86. a
87. e
88. b
89. b
90. d
91. b
92. e
93. d

**Modified Multiple Choice**

94. d
95. a
96. g
97. b
98. f
99. c
100. h
101. e

## *Chapter 7: The Peripheral Nervous System: Efferent Division*

**True/False**

1. True
2. False—Cells do not touch neuromuscular joint.
3. True
4. False—Opening not closing.
5. True
6. False—Just reversed.
7. False—Turned off.
8. True
9. False—Ventral horn.
10. True
11. False—Somatic nervous system.
12. True
13. False—Long.
14. False—Parasympathetic.
15. True
16. True
17. True
18. False—Sympathetic.

**Fill in the Blank**

19. Motor neurons
20. Common pathway
21. Neuromuscular junction
22. Muscle fiber
23. Terminal
24. Motor end-plate
25. ACh
26. Botulinum toxin
27. Organophosphates
28. Myasthenia gravis
29. Postganglionic fiber, preganglionic fiber
30. Adrenergic fibers
31. Varicosities

32. Epinephrine (adrenaline)
33. Nicotinic receptors
34 Muscarinic receptors

**Matching**

35. a
36. b
37. b
38. a
39. a
40. b
41. b
42. b
43. a
44. b

**Multiple Choice**

45. a
46. c
47. d
48. a
49. a
50. b
51. d
52. e
53. a
54. a
55. d
56. b
57. c
58. c
59. d
60. d
61. e
62. b
63. d
64. c
65. e
66. d
67. b
68. e
69. b
70. a

**Modified Multiple Choice**

71. c
72. d
73. b
74. a
75. a
76. b
77. b
78. a
79. b
80. a
81. c
82. d
83. c
84. d
85. b
86. a
87. d
88. c
89. e
90. d
91. d
92. c
93. b
94. d
95. e
96. a
97. e
98. b
99. e
100. a

## *Chapter 8: Muscle Physiology*

**True/False**

1. True
2. False—Intrafusal.
3. True
4. False—Negative feedback.
5. False—Externally applied tension.
6. True
7. True
8. False—Single nucleus.
9. True
10. True
11. True
12. True
13. False—Calcium concentration.
14. False—Myosin ATPase.
15. True
16. False—Myoglobin.
17. False—Pyruvic acid.
18. True
19. False—Oxidative fibers.
20. False—Muscle fiber.
21. True
22. True
23. False—Actin spiral.
24. True

25. True
26. False—Calcium binds.
27. True
28. True
29. False—Connective tissue.
30. False—Produces sustained contractions.
31. True
32. False—Motor-unit recruitment.
33. True
34. False—Origin.
35. False—Lengthens.
36. True
37. False—Sarcoplasmic reticulum.
38. False—Lateral sacs.
39. True
40. False—Latent period.
41. True
42. False—Calcium.

**Fill in the Blank**

43. Excitation-contraction coupling
44. Transverse tubule
45. Sarcoplasmic reticulum
46. Lateral sacs
47. Actin binding, ATPase
48. Rigor mortis
49. Latent period
50. Calmodulin
51. Multiunit smooth muscle
52. Neurogenic
53. Functional syncytium
54. Slow-wave potentials
55. Latch-phenomenon
56. Muscle fiber
57. Myofibrils
58. A band
59. Sarcomere
60. Myosin, actin
61. Tropomyosin
62. Tropomyosin, troponin
63. Hemiplegia
64. Quadriplegia
65. Paraplegia
66. Gamma motor neuron, alpha motor neurons
67. Golgi tendon organs
68. Motor unit
69. Tension
70. Series-elastic
71. Origin, insertion
72. Isotonic
73. Concentric, eccentric
74. Sphincter
75. Work
76. Lever, fulcrum
77. Creatine phosphate
78. Oxidative phosphorylation
79. Myoglobin
80. Disuse
81. Muscular dystrophy

**Matching**

82. b
83. a
84. c
85. a
86. c
87. a
88. c
89. a
90. b
91. b
92. a
93. c

**Multiple Choice**

94. c
95. a
96. a
97. e
98. d
99. b
100. b
101. e
102. b
103. c
104. c
105. a
106. e

**Modified Multiple Choice**

107. d
108. f
109. a
110. f
111. a
112. f
113. a
114. f
115. d
116. c
117. b
118. d
119. b
120. f

121. a
122. d
123. g
124. d
125. b
126. d
127. b
128. a
129. b
130. c
131. a
132. b,c
133. a

## *Chapter 9: Cardiac Physiology*

**True/False**

1. False—Contractile cells.
2. False—SA node.
3. True
4. True
5. True
6. False—Long.
7. True
8. False—P wave.
9. True
10. False—Decrease.
11. False—Sympathetic.
12. True
13. True
14. False—LDL.
15. True
16. True
17. True
18. True
19. False—Late.
20. True
21. True
22. True
23. False—Produces no sound.
24. True
25. True
26. True
27. False—Contain an abundance of myoglobin.
28. False—None are produced.
29. True
30. False—Systemic circulation.
31. False—Function as two separate pumps.
32. False—Right side.
33. True
34. False—Pulmonary circulation.
35. True
36. True

**Fill in the Blank**

37. Vascular spasm
38. Atherosclerosis
39. thrombus, embolus
40. Collateral circulation
41. HDL, LDL, VLDL
42. Fibrillation
43. Interatrial
44. T
45. Tachycardia
46. Bradycardia
47. Atrial
48. Apex
49. Cardiopulmonary resuscitation
50. Ventricles
51. Veins
52. Aorta
53. Tricuspid valve
54. Epicardium
55. Latent pacemakers
56. Cardiac output
57. Cardiac reserve
58. Intrinsic control
59. Frank-Starling Law of the Heart
60. Heart failure
61. End-diastolic volume
62. Isovolumetric ventricular contraction
63. End-systolic volume (ESV)
64. Murmurs
65. Systolic murmur

**Matching**

66. b
67. a
68. b
69. b
70. a
71. a
72. a
73. a
74. b
75. a

**Multiple Choice**

76. e
77. d
78. b
79. c
80. b
81. c
82. b

83. c
84. d
85. a
86. c

**Modified Multiple Choice**

87. 1
88. 8
89. 11
90. 3
91. 12
92. 7
93. 2
94. 4
95. 9
96. 10
97. 6
98. 5
99. b
100. a
101. b
102. a
103. a
104. b
105. c
106. a
107. c
108. a
109. Increased
110. Increased
111. Greater than before
112. More than
113. Increases
114. a
115. d
116. b
117. b
118. a
119. c

## *Chapter 10: The Blood Vessels and Blood Pressure*

**True/False**

1. True
2. True
3. False—Vary from organ to organ.
4. True
5. False—Exchanges are not made directly.
6. True
7. False—Is a function of the lymphatic system.
8. False—Right.
9. True
10. False—The greater the rate of flow.
11. False—Does not fluctuate.
12. True
13. True
14. True
15. False—Transient hypotension.
16. False—Septic shock.
17. True
18. False—Short term.
19. True
20. False—Unchanged.
21. False—Does not respond and adapts or reset to operate at a higher level.
22. True
23. True
24. False—Can either be chemical or physical.
25. True
26. True
27. False—Except the brain.
28. True
29. True
30. False—Decreasing.
31. True
32. False—Can only be driven forward.
33. True
34. False—5 mm Hg.

**Fill in the Blank**

35. Hypertension
36. Hypotension
37. Cardiovascular, renal, endocrine, neurogenic
38. Orthostatic hypertension
39. Cardiogenic
40. Hypovolemic
41. Irreversible shock
42. Capacitance vessels
43. Venous capacity
44. Venous return
45. Varicose veins
46. Baroreceptors
47. Carotid-sinus, aortic-arch baroreceptors
48. Arteries
49. Microcirculation
50. Flow rate
51. 120 mm Hg
52. Pulse pressure
53. Mean arterial pressure
54. Arterioles
55. Diffusion
56. Metarteriole
57. Bulk flow

58. Capillary blood pressure
59. Lymphatic system
60. Edema
61. Vasoconstriction
62. Vascular tone
63. Endothelial-derived relaxing factor (EDRF)
64. Endothelial cells
65. Cardiovascular control center

**Matching**

66. a
67. a
68. a
69. d
70. c
71. a
72. b
73. a
74. b
75. a
76. a
77. a
78. b
79. b

**Multiple Choice**

80. c
81. d
82. a
83. e
84. c
85. b
86. c
87. e
88. b
89. e

**Modified Multiple Choice**

90. c
91. a
92. d
93. b
94. c
95. d
96. a
97. b
98. e
99. a
100. d
101. b
102. e
103. a
104. c
105. b
106. a
107. a
108. a
109. a
110. c
111. a
112. a

## *Chapter 11: The Blood*

**True/False**

1. True
2. True
3. False—Trillion.
4. True
5. False—Testosterone.
6. False—Pernicious anemia.
7. False—Primary.
8. False—10 days.
9. False—Thrombin.
10. True
11. False—Seven steps.
12. True
13. True
14. True
15. False—Vitamin K.
16. False—90 percent.
17. True
18. False—Sodium and chloride.
19. True
20. True
21. True
22. True
23. False—Partially responsible for changes.
24. False—Albumins.
25. True
26. False—Red dye.
27. False—An oval or kidney-shaped nucleus.
28. True
29. True
30. True
31. False—Histamine.
32. False—B lymphocytes.

**Fill in the Blank**

33. Hemostasis
34. Blood coagulation, clotting

35. Tissue thromboplastin
36. Serum
37. Plasmin
38. Emboli
39. Hemophilia
40. Glycolytic
41. Spleen
42. Erythropoiesis
43. Aplastic
44. Anemia
45. Hemolytic
46. Sickle-cell anemia
47. Nutrients, waste products, dissolved gases, hormones
48. Plasma proteins
49. Albumins, globulins, fibrinogen
50. Albumins
51. Globulins
52. Fibrinogen
53. Leukocytes
54. Polymorphonuclear granulocytes
55. Lymphocytes
56. Pathogens
57. Neutrophils
58. Basophils
59. Cell-mediated immune response
60. Monocytes

**Matching**

61. h
62. a
63. j
64. b
65. c
66. i
67. d
68. f
69. g
70. e

**Multiple Choice**

71. b
72. d
73. b
74. c
75. e
76. e
77. e
78. a
79. d
80. b
81. d
82. c
83. a
84. b
85. e
86. a
87. c
88. d
89. b
90. d
91. c
92. d
93. e
94. a
95. d
96. c
97. e
98. a
99. c
100. d

## *Chapter 12: Body Defenses*

**True/False**

1. False—Lymphocytes.
2. True
3. False—Two: Antibody-mediated immunity and cell-mediated immunity.
4. True
5. True
6. True
7. False—Five.
8. False—B cells.
9. True
10. True
11. True
12. False—No naturally occurring antibodies.
13. True
14. True
15. False—Autoimmune responses.
16. False—20 minutes.
17. False—IgE antibodies.
18. True
19. True
20. True
21. False—Helper T cells.
22. True
23. False—T cells.
24. False—Do not result in malignancy.
25. True
26. False—Has no direct blood supply.
27. True
28. False—Are able to convert.
29. False—Does have control.

30. True
31. False—Vitamin D.
32. False—B lymphocytes.
33. False—Interferon.
34. True
35. True
36. False—Acute-phase proteins.
37. True

**Fill in the Blank**

38. Gamma globulins, immunoglobulins
39. Agglutination
40. Complement system, phagocytes, killer cells
41. B cells, antigen
42. Active immunity
43. Interleukin 1
44. Cell-mediated immunity
45. Cytotoxic T cells
46. Perforin
47. Interleukin 2
48. Helper t cells
49. Myasthenia gravis
50. AIDS
51. Allergy, hypersensitivity
52. Anaphylactic shock
53. Immunity
54. Lymphoid tissue
55. Histamine
56. Fibrinogen, fibrin
57. Lactoferrin
58. Granuloma
59. Salicylates, glucocorticoids
60. Antibody-mediated, cell-mediated
61. Thymus, thymosin
62. Haptens
63. Immediate hypersensitivity
64. Spot desmosomes
65. Sebaceous glands, sebum
66. Sweat glands, the sweat pores
67. Melanocytes, melanin
68. Keratinocytes
69. Langerhans cells
70. Granstein cells
71. Epidermis
72. Alveolar macrophages

**Matching**

73. b
74. c
75. b
76. a
77. d
78. a
79. b
80. c
81. d
82. d
83. b

**Multiple Choice**

84. e
85. c
86. d
87. b
88. e
89. b
90. a
91. b
92. e
93. a

**Modified Multiple Choice**

94. a
95. b
96. a
97. b
98. a
99. a
100. c
101. b
102. b
103. d
104. b
105. a
106. b
107. b
108. b
109. c
110. c
111. a
112. b

## *Chapter 13: The Respiratory System*

**True/False**

1. True
2. False—When compliance is decreased.
3. False—5 percent.
4. False—5.7 liters.
5. True
6. True
7. True

8. False—100 mmHg.
9. True
10. True
11. True
12. False—Between the alveoli and the blood.
13. True
14. False—DRG.
15. True
16. True
17. True
18. False—Increase.
19. True
20. True
21. False—1.5 percent.
22. True
23. False—Decreased.
24. False—97.5 % saturated.
25. True
26. True
27. False—Circulatory hypoxia.
28. True
29. False—Higher pressure to lower pressure.
30. False—Lower.
31. True
32. False—Decreases.
33. False—Increases.
34. True
35. True

**Fill in the Blank**

36. Histotoxic hypoxia
37. Hypercapnia
38. Oxygen-hemoglobin dissociation curve
39. Bohr effect
40. Carboxyhemoglobin
41. Haldane effect
42. Hypoxic hypoxia
43. Pneumonia
44. Partial pressure
45. Partial pressure gradient
46. 60 mm Hg, 6 mm Hg
47. Functional residual capacity
48. Obstructive, restrictive
49. Anatomic dead space
50. Atmospheric (barometric) pressure
51. Bronchodilation
52. Emphysema
53. Compliance
54. Interdependence
55. LaPlace's law
56. Tidal volume
57. Medullary respiratory center
58. Apneustic center, pneumotaxic center
59. Dorsal respiratory group
60. Apneusis
61. Carotid bodies, aortic bodies
62. Apnea
63. Dyspnea
64. Respiration
65. Esophagus
66. Vocal folds
67. Type I alveolar cells
68. Type II alveolar cells
69. Pores of Kohn

**Matching**

70. a
71. f
72. e
73. a
74. g
75. d
76. b
77. h
78. c
79. d
80. g
81. c

**Multiple Choice**

82. c
83. e
84. a
85. b
86. e
87. b
88. c
89. e
90. b
91. a
92. d
93. b
94. a
95. b
96. b
97. b
98. a
99. b
100. b

**Modified Multiple Choice**

101. h
102. f
103. d

104. e
105. j
106. I
107. A
108. C
109. L
110. A
111. F
112. m
113. k
114. I
115. G
116. H
117. B
118. C
119. D
120. A
121. D
122. A
123. b

## *Chapter 14: The Urinary System*

### True/False

1. True
2. False—Arterial blood.
3. True
4. False—Cortex.
5. False—Glomerular membrane.
6. False—And the filtration coefficient.
7. False—Juxtaglomerular apparatus.
8. True
9. False—Cardiac atria.
10. False—Renal threshold.
11. True
12. False—Proximal tubule.
13. True
14. False—Actively.
15. False—Hydrogen and potassium.
16. True
17. True
18. False—1.5 liters.
19. False—Less.
20. False—Greater.
21. True
22. False—Hypothalamus.

### Fill in the Blank

23. Urea
24. Sodium
25. Phosphate and calcium
26. Expands
27. Renin-angiotensin-aldosterone
28. Angiotensin I
29. Angiotensin II
30. Aldosterone
31. Transepithelial
32. Mesangial cells
33. Filtration coefficient, net filtration pressure
34. Proximal
35. Osmotic
36. Impermeable
37. A. Infectious organisms
    B. Toxic agents
    C. Inappropriate immune response
    D. Obstruction of urine flow
    E. Insufficient renal blood supply
38. Peristaltic
39. Internal urethral sphincter
40. Countercurrent exchange
41. 300
42. Sodium-potassium pump
43. Potassium secretion

### Matching

44. a
45. b, c
46. a, b, c
47. a
48. b
49. a
50. a
51. a
52. b
53. a
54. a

### Multiple Choice

55. e
56. c
57. b
58. d
59. a
60. b
61. d
62. b
63. a
64. e
65. c
66. d
67. c
68. e
69. c

70. b
71. c
72. d
73. a
74. d
75. a
76. a
77. e
78. c
79. d
80. b
81. a
82. a
83. d
84. c
85. d
86. a
87. c
88. e
89. e
90. b
91. c
92. b
93. c
94. c
95. d
96. e
97. e
98. c
99. c
100. e

**Modified Multiple Choice**

101. a
102. b
103. a
104. a
105. b
106. a
107. a
108. c
109. b
110. e
111. e
112. a
113. c
114. d

## Chapter 15: Fluid and Acid-Base Balance

**True/False**

1. False—Positive balance.
2. False—Two-thirds.
3. False—Extracellular fluid.
4. True
5. True
6. True
7. True
8. False—Acids.
9. True
10. True
11. False—Chemical buffer system.
12. False—Reversed; ammonia combines with hydrogen to form ammonium ions.
13. False—Diabetes mellitus.
14. True

**Fill in the Blank**

15. Stable balance
16. Plasma, interstitial fluid, interstitial fluid
17. Capillary blood pressure, colloid osmotic pressure
18. Obligatory loss of salt in sweat, feces, urine
19. Renin-angiotensin-aldosterone
20. Insufficient water intake, excessive water loss, diabetes insipidus
21. Hypothalamic osmoreceptors
22. pH
23. Phosphate buffer system
24. Hypoventilation
25. Metabolic acidosis
26. Metabolic alkalosis
27. Hydrogen
28. Hydrogen excretion, bicarbonate excretion, ammonia secretion

**Matching**

29. c
30. d
31. a
32. a
33. c
34. c
35. b
36. c

**Multiple Choice**

37. e
38. b
39. c
40. e
41. c
42. e
43. d
44. c
45. b
46. b

**Modified Multiple Choice**

47. b
48. a
49. a
50. b
51. a
52. d
53. b
54. b
55. a
56. c
57. b
58. b
59. a
60. a
61. a
62. b
63. a
64. b
65. d
66. b
67. b
68. c
69. c
70. a
71. c
72. a
73. d
74. c
75. b
76. c
77. c
78. c
79. c
80. c
81. b
82. b
83. b
84. b
85. c
86. b
87. b
88. b
89. c
90. c
91. b
92. b
93. a
94. d
95. b
96. c

## *Chapter 16: The Digestive System*

**True/False**

1. False—Monosaccharides not disaccharides.
2. False—Polysaccharides not disaccharides.
3. False—Monoglycerides not triglycerides.
4. True
5. False—Mucosa, not submucosa.
6. False—Between the longitudinal and smooth-muscle layers not submucosa.
7. False—Auerbach's, not Haustra's.
8. True
9. False—Skeletal, not smooth.
10. False—Found in the mucosa of the cheek's.
11. False—Occlusion is how teeth fit together.
12. True
13. True
14. True
15. True
16. False—Eructation, not borborygmia.
17. False—Fundus, not antrum.
18. True
19. True
20. True
21. False—Vomiting or emesis, not malocclusion.
22. False—Pepsinogen, not mucus.
23. False—Parietal cells, not mucous neck cells.
24. True
25. False—Pepsinogen, not trypsinogen.
26. True
27. True
28. False—Water is not absorbed from the stomach contents.
29. False—Acini, not isles of Langerhans.
30. False—Enterokinase, not HCI.
31. False—Alkaline, not acidic.
32. True
33. True
34. True
35. False—Liver, not gallbladder.
36. True
37. True

38. False—Microvilli are known as brush border.
39. True
40. True
41. True
42. True

**Fill in the Blank**

43. Succus entericus
44. Enterokinase
45. Aminopeptidases
46. Villi
47. Malabsorption
48. Crypts of Lieberkuhn
49. Transferrin
50. Mastication
51. 99.5%, 0.5%
52. Xerostomia
53. Acquired or conditioned, salivary reflex
54. Palate
55. Nitroglycerin
56. Bolus
57. Oropharyngeal stage
58. Gastroesophageal sphincter
59. Peristalsis
60. Secondary peristaltic waves
61. Achalasia
62. Appendix
63. Haustral contractions
64. Colon, cecum, appendix, rectum
65. Borborygmi
66. GIP (gastric inhibitory peptide)
67. Fundus
68. Antrum
69. Gastric pits
70. Gastrin
71. Zymogen granules
72. Gastric
73. Peptide ulcer
74. Steatorrhea
75. Liver
76. Sphincter of Oddi
77. Bile salts
78. Choleretic
79. CCK
80. Cirrhosis
81. Hydrolysis

**Matching**

82. j
83. a
84. d
85. g
86. l
87. n
88. b
89. m
90. e
91. h
92. k
93. i
94. f
95. c

**Multiple Choice**

96. c
97. b
98. e
99. a
100. e
101. c
102. e
103. b
104. d
105. d

**Modified Multiple Choice**

106. d
107. c
108. d
109. d
110. a
111. d
112. a
113. c
114. b
115. c
116. a
117. a
118. c
119. b
120. b
121. a
122. b
123. e
124. d
125. a
126. e
127. c
128. d
129. c
130. e
131. b
132. a
133. d

134. e
135. c
136. a

## Chapter 17: Energy Balance and Temperature Regulation

### True/False

1. False—External work.
2. False—Energy can be converted to another form.
3. True
4. False—Hypothalamus.
5. True
6. True
7. True
8. True
9. False—Normally rectal temperature is a degree higher.
10. True
11. False—Conduction uses direct contact.
12. True
13. True
14. True

### Fill in the Blank

15. Increased
16. Epinephrine, thyroid hormone
17. Endogenous pyrogen
18. Anterior, decreasing, vasodilation
19. Prostaglandins
20. Sustained exercise
21. Sympathetic
22. Convection
23. Lack of appetite
24. Lipostatic
25. External work, internal work
26. Positive
27. Renal failure, cancer, tuberculosis
28. Radiation

### Matching

29. d
30. a
31. f
32. b
33. c
34. e
35. g

### Multiple Choice

36. c
37. e
38. b
39. d
40. c
41. c
42. b
43. d
44. c
45. d
46. b
47. b
48. d

### Modified Multiple Choice

49. c
50. b
51. d
52. a
53. b
54. e
55. f
56. c
57. a
58. d
59. b
60. e
61. g
62. f
63. c
64. d
65. a

## Chapter 18: Principles of Endocrinology: the Central Endocrine Glands

### True/False

1. False—First two years.
2. True
3. True
4. True
5. True
6. False—Insulin.
7. False—Estrogens.
8. True
9. False—Neurohypophysis.
10. False—Vasopressin.
11. True
12. True

13. False—Paracrines.
14. False—Non-tropic hormone.
15. False—Several, not just one.
16. False—Lipophilic.
17. True
18. False—Lipophilic.
19. True
20. True

**Fill in the Blank**

21. Pituicytes
22. Gonadotropins
23. Pituitary gland
24. Infundibulum
25. Short-loop negative feedback
26. Androgens
27. Osteoblasts
28. Cartilage
29. Diaphysis, epiphysis
30. Periosteum
31. Dwarfism
32. Gigantism
33. Hormones
34. Growth-hormone-releasing-hormone
35. Melanocyte-stimulating hormone
36. Hyperplasia, hypertrophy
37. Endocrine system
38. HRE (hormone response element)
39. Paracrines
40. Tropic
41. Steroids
42. Ribosomes, rough ER
43. Zeitgebers
44. Diurnal, circadian rhythm
45. Permissiveness
46. Synergism
47. Antagonism
48. Hypophysiotropic hormones
49. Hypothalamus, anterior pituitary
50. Amines
51. Chondrocytes
52. Acromegaly

**Matching**

53. s
54. b
55. k
56. r, j
57. d
58. c
59. r
60. r
61. l
62. n
63. e
64. n
65. q
66. j
67. o
68. i
69. n
70. e
71. f
72. s
73. j, r
74. p
75. c
76. m
77. n
78. e
79. a
80. n
81. g
82. c
83. s
84. n
85. b
86. e
87. a
88. h
89. g
90. e
91. e
92. h

**Multiple Choice**

93. c
94. b
95. e
96. d
97. e
98. d
99. a
100. e
101. d
102. d

**Modified Multiple Choice**

103. b
104. a
105. c
106. c
107. a
108. a

109. a
110. b
111. a
112. c
113. a
114. d
115. b
116. a
117. a
118. b
119. b

## *Chapter 19: The Peripheral Endocrine Glands*

### True/False

1. True
2. False—$T_3$.
3. False—Thyroxine-binding globulin, albumin, thyroxin-binding prealbumin.
4. True
5. False—Increase.
6. False—Cretinism.
7. True
8. True
9. False—80 percent.
10. True
11. False—Especially muscle.
12. False—Hypernatremia.
13. True
14. False—Primary, not secondary.
15. True
16. True
17. True
18. True
19. False—Four hours.
20. False—Beta.
21. True
22. False—Site of somatostatin synthesis.
23. True
24. True
25. False—Type I.
26. True
27. True
28. False—Increase.
29. True
30. True
31. False—Decreases.
32. True
33. True
34. False—Increases.
35. True

### Fill in the Blank

36. Metabolism
37. Anabolism
38. Catabolism
39. Islets of Langerhans
40. Ketone bodies
41. Glycogenesis
42. Low
43. Lamellae
44. Osteolytic-osteoblastic bone membrane
45. Calcitonin
46. Cholecalciferol
47. Rickets, osteomalacia
48. Osteoporosis
49. Thyroglobulin
50. Euthyroidism
51. Grave's disease
52. Goiter
53. Adrenal gland
54. C-cells
55. Monoiodotyrosine
56. Thyroxine
57. Zona glomerulosa, zona fasciculata, zona reticularis
58. Cortisol
59. Cushings
60. Hirsutism
61. Addison's
62. Chromaffin granules
63. Pheochromocytoma

### Matching

64. b
65. d
66. e
67. b
68. a
69. c
70. c
71. e
72. b
73. a
74. d

### Multiple Choice

75. c
76. b
77. e
78. a
79. e
80. c

81. a
82. c
83. e
84. e

**Modified Multiple Choice**

85. a
86. a
87. b
88. b
89. c
90. a
91. a
92. a
93. a
94. a
95. a
96. b
97. b
98. a
99. c
100. c
101. b
102. b
103. b
104. a
105. a
106. a
107. c
108. c
109. a
110. a
111. c
112. c
113. a
114. a
115. a
116. b
117. a
118. a
119. b
120. a

## *Chapter 20: The Reproductive System*

**True/False**

1. False—Gametogenesis is the formation of sperm and eggs.
2. True
3. True
4. False—Somatic cells are diploid.
5. True
6. True
7. True
8. False—H-Y antigen directs the differentiation.
9. True
10. False—Leydig cells secrete testosterone.
11. False—Forty-six doubled chromosomes.
12. True
13. False—Sertoli cells, not Leydig cells.
14. True
15. True
16. False—LH, not FSH.
17. True
18. False—The prostate, not the bulbourethral glands.
19. True
20. True
21. False—Estrogen, not progesterone.
22. False—diploid, not haploid.
23. True
24. False—Follicle, not corpus luteum.
25. True
26. True
27. False—FSH, not LH.
28. True
29. True
30. True

**Fill in the Blank**

31. Menopause
32. Granulosa cells, primary follicle
33. Atresia
34. Follicular cells
35. Estradiol
36. Myometrium, endometrium
37. Dysmenorrhea
38. Climacteric
39. Ampulla
40. Seminal vesicles
41. Resolution
42. Erectile tissue
43. Impotence
44. Orgasmic platform
45. Gametes
46. Somatic cells
47. Autosomal chromosomes
48. Wolffian ducts, Mullerian ducts
49. Cryptorchidism
50. Leydig cells, interstitial cells
51. Spermiogenesis
52. A head, an acrosome, a midpiece, a tail
53. Blood-testes barrier
54. Luteinizing hormone (LH), follicle-stimulating hormone (FSH)

55. Blastocyst
56. Decidua
57. Chorion
58. HCG (human chorionic gonadotropin)
59. 38
60. Oxytocin
61. Involution
62. Colostrum

**Matching**

63. a, b, e
64. d
65. b
66. c
67. d
68. b
69. c
70. a

**Multiple Choice**

71. c
72. d
73. a
74. d
75. d
76. e
77. c
78. b
79. c
80. d
81. d
82. b
83. b
84. b
85. a
86. b
87. a
88. c
89. e
90. b

**Modified Multiple Choice**

91. c
92. c
93. b
94. a
95. a
96. b
97. n
98. h
99. o
100. a
101. e
102. g
103. T
104. c
105. b
106. u
107. d
108. r
109. d
110. m
111. q
112. j
113. s
114. u
115. l
116. m
117. p
118. o
119. I
120. P
121. N
122. u